FOUNDATIONS OF MATHEMATICAL SYSTEM DYNAMICS

The Fundamental Theory of Causal
Recursion and its Application to
Social Science and Economics

IFSR International Series on
Systems Science and Engineering — **Volume 2**

International Federation for Systems Research

International Series on Systems Science and Engineering

Editor-in-Chief: George J. Klir, *State University of New York at Binghamton, USA*

Published and Forthcoming Titles in the Series

Volume 1 ROSEN
Anticipatory Systems

Volume 2 AULIN
Foundations of Mathematical System Dynamics

Volume 3 HALL
A Unified Synthesis for Systems Methodology

FRIEDMAN
Exhuming Induction

KAMPIS
Component Systems: Constraints, Information, Complexity

Pergamon Titles of Related Interest

ANAND
Introduction to Control Systems, 2nd edition

GEERING & MANSOUR
Large Scale Systems: Theory and Applications 1986

ISERMANN
Automatic Control 1987, 10-volume set (also available separately)

JAMSHIDI
Linear Control Systems

SANCHEZ
Approximate Reasoning in Intelligent Systems, Decisions and Control

SINGH
Systems and Control Encyclopedia, 8-volume set

WALTER
Identifiability of Parametric Models

Pergamon Related Journals

Automatica

Computers and Industrial Engineering

Computers and Operations Research

Journal of the Operational Research Society
Problems of Control and Information Theory

Robotics and Computer-integrated Manufacturing

Systems Research

(free specimen copy gladly sent on request)

FOUNDATIONS OF MATHEMATICAL SYSTEM DYNAMICS

The Fundamental Theory of Causal
Recursion and its Application to
Social Science and Economics

ARVID AULIN

Chair of Mathematics and Methodology
University of Tampere, Finland

PERGAMON PRESS

OXFORD · NEW YORK · BEIJING · FRANKFURT
SÃO PAULO · SYDNEY · TOKYO · TORONTO

U.K.	Pergamon Press plc, Headington Hill Hall, Oxford OX3 0BW, England
U.S.A.	Pergamon Press, Inc., Maxwell House, Fairview Park, Elmsford, New York 10523, U.S.A.
PEOPLE'S REPUBLIC OF CHINA	Pergamon Press, Room 4037, Qianmen Hotel, Beijing, People's Republic of China
FEDERAL REPUBLIC OF GERMANY	Pergamon Press GmbH, Hammerweg 6, D-6242 Kronberg, Federal Republic of Germany
BRAZIL	Pergamon Editora Ltda, Rua Eca de Queiros, 346, CEP 04011, Paraiso, São Paulo, Brazil
AUSTRALIA	Pergamon Press (Australia) Pty Ltd, PO Box 544, Potts Point, NSW 2011, Australia
JAPAN	Pergamon Press, 5th Floor, Matsuoka Central Building, 1-7-1 Nishishinjuku, Shinjuku-ku, Tokyo 160, Japan
CANADA	Pergamon Press Canada Ltd, Suite No 271, 253 College Street, Toronto, Ontario, Canada M5T 1R5

First edition 1989

Library of Congress Cataloging in Publication Data
Aulin, Arvid.
Foundations of mathematical system dynamics: the fundamental theory of causal recursion and its application to social science and economics/Arvid Aulin.—
1st ed.
p. cm. — (IFSR international series on systems science and engineering; v. 2)
Bibliography: p.
Includes index.
1. Economics, Mathematical. 2. Social sciences—
Statistical methods. I. Title. II. Series.
HB.135.A865   1989
301—dc19   88-36471

British Library Cataloguing in Publication Data
Aulin, Arvid
Foundations of mathematical system dynamics: the fundamental theory of causal recursion and its application to social science and economics—(IFSR international series on systems science and engineering; vol 2)
1. Dynamical systems theory
I. Title II. Series
003

ISBN 0-08-036964-2

Printed in Great Britain by BPCC Wheatons Ltd., Exeter

Contents

Part 2

SIMPLE APPLICATIONS: CAUSAL RECURSION IN POPULATION DYNAMICS AND ECONOMIC GROWTH THEORY (Measurable Systems)

Part 3

COMPLEX APPLICATIONS: THE SELF-STEERING AND SELF-REGULATION OF HUMAN SOCIETIES AS WHOLES (Non-measurable Systems)

Part 4

HISTORICAL ILLUSTRATIONS OF SELF-STEERING IN DIFFERENT TYPES OF HUMAN SOCIETIES

Part 5

THE PROBLEM OF THE ORIGIN: THE SELF-STEERING AND STEERED-FROM-OUTSIDE LAYERS OF CONSCIOUSNESS?

Preface

Dynamics is a field emerging somewhere between mathematics and the sciences. In our view, it is the most exciting event on the concept horizon for many years (Ralph Abraham and Christopher Shaw, in their *Dynamics: The Geometry of Behavior*, Santa Cruz, 1983)

Mathematical dynamics, as developed mainly during the past two decades, is establishing itself as the foundational theoretical science of behavioural research, irrespective of the kind of behaviour we are speaking of – whether in physics, chemistry, biology, or in the social sciences. It provides the, until now, missing link between mathematics and what used to be called 'cybernetics and general system theory'. Being a formalization of the causal processes themselves, on a strictly mathematical basis, it offers essential theoretical tools for a fundamental theory of dynamical systems.

Mathematical dynamics, when extended from continuous systems to discontinuous systems, and from full causal recursions to nilpotent ones, in the way shown in Part 1, involves an *exhaustive systematics of dynamical systems having causal recursion*. This result is emphasized here as it has not been introduced elsewhere in current literature. Parts 2 and 3 serve to illustrate the perspectives opened up by mathematical dynamics in different fields of behavioural study.

From the systematization of causal processes presented in Part 1 *self-steering*, as distinguished from self-regulation, emerges as the central concept related to human action. A qualitative mathematical theory of self-steering is developed in Part 3. (The concept of self-steering comes from Oskar Lange (1965), whose mathematical formulation, unfortunately, was incorrect.) The two last parts of the book are devoted to illustrations in non-mathematical terms of the role that the concept of self-steering may play in historical processes (Part 4) and in the processes of human consciousness (Part 5).

As the theory of self-steering is another principal innovation of this book, the concept of self-steering has, of course, been given a rather comprehensive discussion that extends through Parts 3, 4, and 5. But the time-honoured theory of *self-regulation*, which in dynamics is usually treated in the form of 'stability theory', has recently been revitalized by the study of chaotic processes and strange attractors. Recent results in the field lead to a sub-classification of self-regulating systems, and of the systems steerable from outside, studied in Chapter 3.

Of the three seminal 'self' notions of system dynamics, *self-organization* is, at present, worst off: no general mathematical theory of self-organization exists as yet. Of the existing fragments, Feigenbaum and Hopf bifurcations are considered in this book.

The book has been written with the idea in mind of reaching the widest possible audience of researchers of behaviour in different fields. The level of mathematical sophistication in the representation of mathematical dynamics has been adjusted accordingly: the formalism of algebraic topology is not generally used, only its major findings are given in a language and terminology that are familiar from the tradition of 'cybernetics and general system theory'.

Helsinki, February 1988   ARVID AULIN

Part 1

Fundamental Dynamics: The General Theory of Causal Recursion

CHAPTER 1

Causal Recursion in Theoretical Physics

1.1 Causal Relation, Causal Law, and Causal Recursion

There are three primary types of causality, only two of which play an essential role in exact natural science.

Causal relation is the weakest of the three. We have a causal relation between two states of affairs, x and y, as soon as there is a time-ordered material implication between them, so that $x(\tau) \Rightarrow y(t)$, where $\tau < t$. This is the common usage of the term 'causality' in everyday life as well as in the study of history. But it has little use in the mathematical sciences, because of the difficulties that are mostly associated with its verification.

Causal law implies a set of causal relations holding for the values of two measurable variables, x and y (that are mathematically representable as vectors of the two respective Euclidean spaces). We now have

$$y(t) = T^{t\tau}(x(\tau)), \quad \text{where } x \in X, \, y \in Y, \text{ and } \tau < t. \tag{1.1}$$

Here X and Y are subsets of some Euclidean spaces.

If, in a causal law, the dependence of the causal function T on time is explicitly indicated, the law always implies that the effect $y(t)$ follows the cause $x(\tau)$ after the same interval of time, $t - \tau$. We then have what Ashby (1972) called a 'regular system'. But sometimes the time-ordering between cause and effect is left unspecified, and only implied. Examples of this kind of causal law are Ohm's law, Coulomb's law, Biot-Savart's law, and the laws (Boyle, Gay-Lussac, etc.) that characterize the thermodynamic equilibrium. Examples of time-specified causal laws are, of course, plenty. Among them are the law of the freely falling body and other laws of mechanics, as well as the laws of electrodynamics. Common to all the laws of physics mentioned above is that they are 'phenomenological laws', i.e. more or less conceived of as direct inductive generalizations from experimental results (or, if they are not, they can still be considered as such generalizations).

Causal recursion is the type of causality required at the fundamental level of physical theory, and thus at that level of natural science generally. It implies a complete state-description of the dynamical system concerned, given by a total state x, as a function of which any property x of the system at any moment t can be expressed: $z(t) = z(x(t))$. Causal recursion is defined for the total state x if there is a transitive recursion of $x(t)$ to any past state $x(\tau)$, i.e. if

$$x(t) = \varphi^{t\tau}(x(\tau)), \quad \varphi^{tt'} \cdot \varphi^{t'\tau} = \varphi^{t\tau} \text{ for } t > t' > \tau. \tag{1.2}$$

Thus, a system having causal recursion is what Ashby (1972) called a 'state-determined system'.

Note that it does not matter whether causal recursion is 'strict' or probabilistic: in the latter case the total state is simply a probability distribution, or equivalent to such a distribution (as in quantum theory). Causal recursion may even appear on a less than fundamental level of science, for instance in a model. The Markov process would be an example of probabilistic causal recursion used in a model. The possibility of using causal recursion even in the model-theoretical sense is not, of course, in contradiction with its most important use in fundamental physical theory, where it is a *sine qua non*.

In contemporary physics there are three different complete state-descriptions, and thus causal recursions, referring to three different levels of theoretical concepts.

1.2. Causal Recursion in Classical Physics: the Hamiltonian Formalism

All the fundamental laws of physics that involve an explicit time dependence are causal recursions, and accordingly define a state-determined system. A case in point is the Hamiltonian formalism and the conservative 'Hamiltonian systems' it defines that have played a remarkable part, historically, in classical physics.

The total state in a Hamiltonian system is defined by a set of canonical co-ordinates, $x = (q, p)$, and the causal recursion is determined by differential equations of the form

$$dq_i/dt = \partial H(p, q)/\partial p_i, \quad dp_i/dt = -\partial H(q, p)/\partial q_i, \quad i = 1, 2, \ldots, r. \tag{1.3}$$

The co-ordinates q_i define 'positions' and the p_i 'momenta' of the parts of the system concerned. H is the Hamiltonian function; usually in physical applications it is the total energy of the system.

The Hamiltonian equations (1.3) can be generalized by defining q and p as field variables, i.e. $q = q(\bar{r}, t)$ and $p = p(\bar{r}, t)$, where $\bar{r}$ is the position vector in three-dimensional physical space. In this generalization Maxwell's equations of the electromagnetic field also can be given a formulation in terms of Eqns (1.3).

In an implicit form, the solution of (1.3) can be written, with an initial state $x(0)$, as

$$q_i(t) = (\exp tD)(q_i(0)), \quad p_i(t) = (\exp tD)(p_i(0)), \quad i = 1, 2, \ldots, r. \tag{1.4}$$

Here the differential operator is

$$D = \sum_{i=1}^{r} \left(\frac{\partial H}{\partial p_i} \frac{\partial}{\partial q_i} - \frac{\partial H}{\partial q_i} \frac{\partial}{\partial p_i} \right). \tag{1.5}$$

Thus, for any variable z in a Hamiltonian system we have

$$dz/dt = Dz = \{z, H\}, \tag{1.6}$$

which is called the Poisson bracket of the variable z. It follows that H is an invariant in every Hamiltonian system.

Thus, the Hamiltonian formalism is capable of incorporating the universal law of conservation of energy, which is why Sir William Hamilton in 1834–35 introduced it in the first place.

Example 1 (Galilei's Law). The implicit solution (1.4) of the Hamiltonian equations can be turned explicit only in simple cases. Mostly more sophisticated tools are needed in the solution of (1.3). For the state of the art see, for example, Aubin, Bensoussan, and Ekeland (1983). But for polynomial Hamiltonians H the exponential series is a finite one, and we get the explicit solution from it. A case in point is that of a freely falling body, the historical 'Galilei's law', in which case

$$H = p^2/2m + mgq. \tag{1.7}$$

Here m is the mass of the body, q its height from the earth, p its momentum (so that the velocity is $v = p/m$), and g is the constant acceleration due to gravity.

Hence we get successively:

$$D = v\partial/\partial q - g\partial/\partial v, \qquad Dq = v, \qquad D^2q = -g, \qquad D^3q = 0,$$

$$Dv = -g, \qquad D^2v = 0, \tag{1.8}$$

so that the solution of (1.3) now reads

$$q(t) = q(0) + v(0)t - gt^2/2, \qquad v(t) = v(0) - gt. \tag{1.9}$$

Here we can recognize the old Galilei's law, which used to be taught at school, in those ancient times when school teaching was still good.

Example 2 (Kepler's Laws of Planetary Motion). The Hamiltonian formalism, of course, is nothing but a reformulation of 'Newtonian physics' in terms of causal recursion. Thus, the derivation of Kepler's laws from it is just another way of pointing out that these laws are, counter to the statements of some philosophers, neither 'independent' of (as claimed by Stove, 1982, p. 96) nor 'contradictory' with (counter to Popper, 1972, p. 198) Newtonian physics. The Hamiltonian function now takes the form

$$H = \tfrac{1}{2}(p_x^2 + p_y^2) - \mu m M(x^2 + y^2)^{-\frac{1}{2}}. \tag{1.10}$$

Here m is the mass of the planet, M that of the sun, x and y are the orthogonal co-ordinates of the planet on the plane of its orbit, the sun being at the origin, and p_x and p_y are the respective components of the momentum of the planet (so that its velocity components are $v_x = p_x/m$ and $v_y = p_y/m$). The positive constant μ depends on the choice of the units of measurement.

The Hamiltonian (1.10) gives the differential generator

$$D = v_x \partial/\partial x + v_y \partial/\partial y - \mu M(x^2 + y^2)^{-3/2}(x\partial/\partial v_x + y\partial/\partial v_y), \qquad (1.11)$$

which gives

$$Dh = 0 \quad \text{for} \quad h = xv_y - yv_x. \qquad (1.12)$$

Thus, we have found another invariant h, together with H, of planetary motion. While H stands for the total energy, the first term in (1.10) being the kinetic energy and the latter part the potential energy of the planet, the invariant h expresses Kepler's second law: the radius of the planet covers equal areas in equal intervals of time. This is obvious, as h takes, in polar co-ordinates (r = radius, α = angle), the form

$$h = r^2 d\alpha/dt, \quad \text{so that} \quad dA/dt = \tfrac{1}{2}h = \text{const.}, \qquad (1.13)$$

where $A(t)$ is the area covered by the radius in time t.

Kepler's first law is obtained from the Hamiltonian equations (1.3) with the substitution of (1.10) into them. This gives

$$d^2x/dt^2 = -\mu Mx(x^2 + y^2)^{-3/2},$$
$$d^2y/dt^2 = -\mu My(x^2 + y^2)^{-3/2}. \qquad (1.14)$$

together with $dx/dt = p_x/m$ and $dy/dt = p_y/m$, which are simply the definitions of velocity components in terms of those of momentum. By trivial, but tedious, calculation Eqs (1.14) can be easily solved, with the result that the orbit of the planet is found to be an ellipse, parabola, or hyperbola, depending on whether the total energy H is negative, zero, or positive. In view of the expression (1.10) of H, 'Newtonian physics' thus predicts Kepler's first law for

$$v_0 < (2\mu M/r_0)^{\frac{1}{2}}, \qquad (1.15)$$

which is the only case of planetary motion it implies. Here $v_0 = (v_x^2 + v_y^2)^{\frac{1}{2}}$ and $r_0 = (x^2 + y^2)^{\frac{1}{2}}$ are simultaneous values of the velocity and the distance from the sun of the planet.

The same calculation gives the axes of the ellipse the following lengths:

$$a = \mu mM/2|H|, \quad b = h(2|H|/m)^{-\frac{1}{2}}. \qquad (1.16)$$

The total area of the ellipse is πab, which on the other hand, in view of Eq. (1.13), i.e. Kepler's second law, is equal to $\tfrac{1}{2}hT$, where T is the period of the planetary motion. Thus (1.16) gives

$$T = 2^{-\frac{1}{2}}\pi\mu Mm^{3/2}|H|^{-3/2} = (2\pi/\mu^{\frac{1}{2}}M^{\frac{1}{2}})a^{3/2}. \qquad (1.17)$$

Hence, we immediately get Kepler's third law:

$$(T/T')^2 = (a/a')^3. \qquad (1.18)$$

1.3 Causal Recursion in Quantum Physics: the Unitary Time-translations in a Hilbert Space of States

In quantum theory the idea of causal recursion was incorporated into the state-description of any dynamical system from the very outset. In all forms of quantum theory, time-translation is implemented by a unitary operator U_t that transforms the state $|x_\tau\rangle$ of the system at the moment τ to the state $|x_t\rangle$ at the moment t:

$$|x_t\rangle = U_{t\tau}|x_\tau\rangle. \qquad (1.19)$$

The states are unit vectors in the respective Hilbert space of states of the system concerned, so that $\langle x|x\rangle = 1$ for any such state x. Observables are represented by Hermitean operators, and the eigenvalues of these operators are the possible observable values. If Λ is an observable, and λ one of its eigenvalues, the probability of obtaining the result λ in a measurement performed when the system is in the state $|x_t\rangle$ is given by

$$|\langle x_t|\lambda\rangle|^2 = \text{prob}(\lambda), \quad \text{with } |\lambda\rangle = \text{eigenvector.} \qquad (1.20)$$

This is the 'probabilistic interpretation' of quantum theory. It is valid in all the forms of that theory.

Example 1 (Causal Recursion in Quantum Mechanics). In quantum mechanics, i.e. in the simplest form of quantum theory coping with such stable structures as atoms, molecules, and nuclei, there is a theoretical analogy to classical physical theory. Thus, a 'Hamiltonian operator' H can be introduced into the causal recursion (1.19) with the substitution

$$U_{t\tau} = \exp ihH(t - \tau). \qquad (1.21)$$

By derivation Eq. (1.19) then gives

$$\frac{\mathrm{d}|x_t\rangle}{\mathrm{d}t} = ihH|x_t\rangle, \quad \text{(Schrödinger's equation)} \qquad (1.22)$$

which is the abstract form of the basic equation of quantum mechanics. Here h is the Planck constant divided by 2π.

Example 2 (Scattering of Elementary Particles). In more advanced quantum theory the analogy to classical physics, required for the construction of the Hamiltonian operator, does not work. Thus, for instance, in the physics of elementary particles, causal recursion is applied in the form of the 'scattering operator'

$$S = \lim_{\substack{\tau \to -\infty \\ t \to +\infty}} U_{t\tau}. \tag{1.23}$$

In a typical scattering experiment of elementary particles, their mutual interactions are investigated by directing rays of various particles against each other, and measuring the densities of scattered showers of particles. The process is theoretically described in terms of the 'S-matrix elements' $\langle x_{out}|S|x_{in}\rangle$, where $|x_{in}\rangle$ is the initial state (for $t \to -\infty$) of the system of particles, and $|x_{out}\rangle$ is the final state (for $t \to +\infty$). In view of the probabilistic interpretation of quantum theory, eq. (1.20), we have

$$|\langle x_{out}|S|x_{in}\rangle|^2 = \text{prob}(|x_{in}\rangle \to |x_{out}\rangle). \tag{1.24}$$

This gives the theoretical predictions to be compared with the experimental scattering results.

Note (The Three-valued Logic of Quantum Theory).   From the probabilistic interpretation of quantum theory, Eq. (1.20), we can also see that the result of measurement is strictly determined only, if:

(a) the system happens to be in the eigenstate $|\lambda\rangle$ corresponding to the eigenvalue λ, in which case we can say that the value λ of the observable is true; or if

(b) the actual state of the system, $|x_t\rangle$, is 'orthogonal' to the eigenvector $|\lambda\rangle$, i.e. $\langle x_t|\lambda\rangle = 0$, in which case we know that the observable certainly does not have the value λ.

But in the remaining cases, when

(c) $0 < |\langle x_t|\lambda\rangle| < 1,$ \hfill (1.25)

the probability of obtaining the value λ in the measurement of the observable Λ is also between 0 and 1. This is the case of 'indeterminacy', where the statement 'observable Λ has the value λ' is neither true nor untrue but indeterminate. Thus, other than in classical physics, the two-valued logic, where every state of affairs is at any moment either true or false, is not valid in quantum theory. Instead there is a three-valued logic involving the state of indeterminacy as the third 'truth value'.

1.4   The Molecular State-description: a Third Level of Causal Recursion

In addition to the macrophysical (Hamiltonian) and quantum-theoretical complete state-descriptions there is a third one used in fundamental physical theory, viz. the molecular state-description defined in statistical thermodynamics. It is midway between the other two, in the sense that while every macrostate can be thought of as divided into a set of molecular states (see Fig. 1), each molecular state can in turn be conceived of as being divisible into a set of quantum states. It is even midway chronologically: Sir William

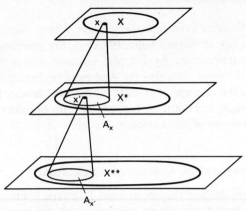

FIG. 1.   The mutual relations between the different complete state-descriptions ('paradigms') of modern physics: macrophysical (X), molecular (X^*), and quantum-theoretical (X^{**}). Formally we could write $X^* = \cup\{A_x;\ x \in X\}$, $X^{**} = \cup\{A_{x_*};\ x' \in X^*\}$. (Kuhn in his theory of scientific knowledge failed to notice the accumulation of theoretical knowledge from X to X^* to X^{**}.)

Hamilton introduced his formalism ten years before the birth of Ludwig Boltzmann (1844–1906), who formulated statistical dynamics towards the end of the nineteenth century. Quantum theory first began to take a systematic shape in the quantum mechanics of Louis de Broglie, Erwin Schrödinger, Werner Heisenberg, and Paul Dirac in the 1920s and 1930s.

The three complete state-descriptions of contemporary physics and the respective three causal recursions stand for three different conceptual levels of theoretical analysis. In the molecular state-description we penetrate 'deeper' into the structure of matter than we do with macrophysical analysis, and a further step into the fine structure of matter is taken with quantum theory.

Sometimes the mutual relations between the different levels of physical theory can be given a concrete form. One such case is the so-called ideal gas, where the total state in the macrophysical state-description is defined by any two of the variables measuring temperature, pressure, and volume. But the same temperature and pressure within the same volume of gas is, according to statistical thermodynamics, produced by a number of alternative molecular configurations, each of which defines a molecular state of the ideal gas. The set of the latter states thus corresponds to a single macrostate.

Usually the hierarchy of complete state-descriptions illustrated in Fig. 1 is much harder to work out explicitly. Still, it is important to notice that all the information that was contained in the 'old' state-description is included in the new, more comprehensive one: thus, theoretical knowledge is accumulated in the changes of complete state-descriptions (changes of 'paradigms') of physics. In fact, the accumulation of theoretical knowledge is a unique property of all the mathematical sciences. It is due to the method of progress in these sciences being based on generalizing proofs that leave the old mathematical theories valid in a special case. I have discussed this point in more detail in my criticism (Aulin, 1987c) of the philosophies of extreme empiricism, as represented by Popper, Lakatos, Kuhn, Feyerabend, von Bertalanffy, and others. (See also the Appendix of this book.)

To return to the concept of causality: on all the levels of complete state-description causal recursion is the concept of causality underlying the fundamental theories of exact science.

1.5   Conservative and Dissipative Systems

A conceptual distinction that originated in physical theory, but has later come to play a certain role in mathematical dynamics more generally (cf. Schuster, 1984), is that between *conservative* and *dissipative* systems. A system having causal recursion is called conservative, if the volume element $\Delta V(x)$ of state-space keeps constant in the causal process that the system is undergoing:

$$\mathrm{d}\Delta V(x)/\mathrm{d}t = 0 \quad \text{(conservative system)}. \tag{1.26}$$

A dynamical system with causal recursion is dissipative, if the volume element shrinks in the causal process:

$$\mathrm{d}\Delta V(x)/\mathrm{d}t < 0 \quad \text{(dissipative system)}. \tag{1.27}$$

For a Hamiltonian system we get:

$$\begin{aligned}
\mathrm{d}\Delta V(x)/\mathrm{d}t &= \Delta V(x)\mathrm{div}\frac{\mathrm{d}x}{\mathrm{d}t} \\
&= \Delta V(x)\left[\sum_{i=1}^{r}\left(\frac{\partial}{\partial q_i}\frac{\mathrm{d}q_i}{\mathrm{d}t} + \frac{\partial}{\partial p_i}\frac{\mathrm{d}p_i}{\mathrm{d}t}\right)\right] \\
&= \Delta V(x)\left[\sum_{i=1}^{r}\left(\frac{\partial^2 H}{\partial q_i\partial p_i} - \frac{\partial^2 H}{\partial p_i\partial q_i}\right)\right] \\
&= 0.
\end{aligned} \tag{1.28}$$

Thus, every Hamiltonian system is conservative.

In a Hamiltonian system the function $H(x)$ is invariant. If we drop this assumption, and require only that the system has causal recursion and continuous state-trajectories in a state-space embedded in an Euclidean space, we have the equations of motion of the total state

$$x = (x_1, x_2, \ldots, x_n) \in X \subset R^n \tag{1.29}$$

in the more general form

$$\mathrm{d}x/\mathrm{d}t = f(x). \tag{1.30}$$

Hence we get:

$$d\Delta V(x)/dt = \Delta V(x)\,\mathrm{div}\,\frac{dx}{dt} = \Delta V(x)\cdot\mathrm{div}\,f(x). \qquad (1.31)$$

Now the conditions of conservation and dissipation read:

$$\mathrm{div}\,f(x) = \sum_{i=1}^{n}\frac{\partial f_i}{\partial x_i} = 0 \quad \text{(conservative system)}, \qquad (1.32)$$

$$\mathrm{div}\,f(x) = \sum_{i=1}^{n}\frac{\partial f_i}{\partial x_i} < 0 \quad \text{(dissipative system)} \qquad (1.33)$$

The motion of a pendulum without friction is an example of a Hamiltonian and thus conservative physical system. Its motion is periodical. But if we add the effect of friction, we have a system in which the motion of the pendulum slows down towards a rest position, and the system is dissipative: it gives up its kinetic energy transforming it into the warmth generated by friction. Originally, in their physical interpretations, conservative and dissipative systems meant systems that conserve and dissipate energy (e.g. Nicolis and Prigogine, 1977).

If causal recursion is expressed in terms of a 't to $t + 1$ map' φ, as it mostly will be in the following, we have instead of the differential equation (1.30) a still more general (as it appears) formulation by means of the difference equation

$$x(t + 1) = \varphi(x(t)). \qquad (1.34)$$

Hence we get:

$$\Delta V(x(t + 1)) = \Delta V(x(t))\,|\det \partial\varphi/\partial x| = \Delta V(x(t))\rho_1\rho_2\ldots\rho_n. \qquad (1.35)$$

Here the ρ_i, for $i = 1, 2,\ldots,n$, are the absolute values of the eigenvalues of the $n \times n$ matrix $\partial\varphi/\partial x$ at the point $x(t)$ of state-space.

The conditions of conservative and dissipative systems now read, obviously,

$$|\det \partial\varphi/\partial x| = \rho_1\rho_2\ldots\rho_n = 1 \quad \text{(conservative system)} \qquad (1.36)$$

and

$$|\det \partial\varphi/\partial x| = \rho_1\rho_2\ldots\rho_n < 1 \quad \text{(dissipative system)}. \qquad (1.37)$$

If a system is dissipative all along the trajectory started by an initial state $x(0)$, it follows from (1.35) and (1.37) that

$$\Delta V(x(t)) \to 0 \quad \text{with } t \to \infty, \qquad (1.38)$$

i.e. the volume element of the n-dimensional total state-space shrinks to zero. While this may evoke an image of a trajectory that would attract all the neighbouring trajectories, close enough to it, and make them approach it asymptotically, such an image is wrong: Eq. (1.38) only tells us that the total state-space misses one or several of its dimensions in a neighbourhood of that trajectory.

Thus, by the theorem of orbital convergence (Theorem 4.6.1, Section 4.6), the dissipative systems are not necessarily *goal-seeking* in the strict sense that will be defined in Section 2.3, neither are goal-seeking systems necessarily dissipative. Examples of this will be given in Section 8.2. There is, however, also a general relation between dissipative and goal-seeking systems, which will be discussed at the end of Part 3 (see 'Notes on Possible Extensions of the Ideas Here Introduced').

CHAPTER 2

Causal Recursion in Mathematical Dynamics

2.1. Causal Recursion in Dynamical Systems Generally

The paramount importance of the concept of causal recursion in exact natural science was emphasized in the previous chapter. Now we move from physical theory to that more general mathematical theory of dynamical systems that has been named *dynamics* (for a short history of this modern field of science, see Abraham and Shaw, 1983). As a matter of fact, mathematical dynamics can be conceived of as a generalization of macrophysical theory, from which all physical details have been eliminated, save for the fundamental fact of causal recursion. Thus, any causal behaviour in whatever field is the object of study of mathematical dynamics.

(a) Continuous dynamical systems

A continuous dynamical system or, equivalently, *continuous flow*, can be defined, with Bhatia and Szegö (1967), as a continuous function F from $E \times R \to E$, such that $F(F(x,t), s) = F(x, s + t)$ and $F(x, 0) = x$. Here E is a Euclidean space, and R is the set of real numbers (here standing for points of time). Obviously this is nothing but the causal recursion (1.2) in a new shape.

We introduce (Bhatia and Szegö, 1967) the following notational convention that will be applied throughout the rest of the book:

Simplified notation: we shall write xt for $F(x, t)$, MS for $\{xt; x \in M, t \in S\}$, and xS for $\{x\}S$, and Mt for $M\{t\}$, where $\{x\}$ and $\{t\}$ are singletons.

In this notation the transitivity of the 'motion' F reads simply: $(xt)s = x(t + s)$ and $x0 = x$. This notation can be used whenever there is no need to refer explicitly to the function F.

If x is a point of E, the set xR is the trajectory of the 'motion' induced by the flow F through x, while xR^+ and xR^- are the positive and negative half-trajectories of x, respectively. Here R^+ is the set of non-negative numbers, R^- that of non-positive numbers. The sets xR, xR^+, and xR^- are subsets of E. Hence we can also write $xR = \gamma(x)$, $xR^+ = \gamma^+(x)$, and $xR^- = \gamma^-(x)$, where γ, γ^+, and γ^- are mappings from E to 2^E, which is the set of all subsets of E.

For any subset A of E the closure, boundary, and complement are $\bar{A}$, ∂A, and $\mathscr{C}(A)$, respectively. Thus, the set-theoretical difference $\bar{A} \setminus \partial A$ is the interior of A. If y is a point of E, and a sequence x_n of points of E converge to y, we write $x_n \to y$. If the positive half-trajectory xR^+ contains a sequence xt_n that converges to y with $t \to +\infty$, we call y a positive limit point of x. The set of all positive limit points of x is its positive limit set $\Lambda^+(x)$. The negative limit set is similarly defined, with xR^- and $t_n \to -\infty$ in this case.

Thus, every point y of the perimeter C of the unit circle shown in Fig. 2a is a limit point of every point x of the spiral that approaches C asymptotically. Hence $\Lambda^+(x) = C$. But in a Riemann mapping of this circle to an infinite strip, the image C' of C consists of two parallel straight lines that still make up the positive limit set of all points of the transformed spiral, while C' is not approached asymptotically by this curve (Fig. 2b).

The latter picture also shows an example of an unbounded trajectory, whose points have non-empty positive and negative (the origin!) limit sets. Only a bounded non-empty limit set can be asymptotically approached by the respective half-trajectory. Also an unbounded trajectory xR^+ that recedes to infinity with $t \to +\infty$ can be approached asymptotically by another trajectory yR^+, but both $\Lambda^+(x)$ and $\Lambda^+(y)$ are empty for these type of trajectories (Fig. 2c).

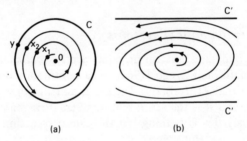

(a)   (b)

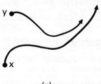

(c)

FIG. 2.   (a) The perimeter C of the unit circle as the positive limit set of a spiral that approaches it asymptotically. (b) The Riemann transforms of the curves of Picture (a): C' is the positive limit set of the transformed spiral, but is not approached asymptotically by it. (c) Two trajectories that recede to infinity and approach each other asymptotically, but have an empty positive limit set.

For any continuous dynamical system we have two mutually equivalent *normal forms* of representation of causal recursion, viz. the one given by the difference equation

$$x(t + 1) = \varphi(x, t), \quad \text{where } \varphi(x) = F(x, 1) \quad \text{for all } x \in E, \tag{2.1}$$

and the one given by the differential equation

$$dx/dt = f(x), \quad \text{where } f(x) = (\partial F(x, t)/\partial t)_{t=0} \quad \text{for all } x \in E. \tag{2.2}$$

The connection between the two representations is given by

$$\varphi(x) = x + \int_0^1 f(xt)\, dt, \quad \text{where } f(xt) = \partial F(x, t)/\partial t. \tag{2.3}$$

Note. Sometimes causal recursion is given in an implicit form, containing higher-order equations of the type

$$G(x(t + n), x(t + n - 1), \ldots, x(t + 1), xt) = 0$$

or

$$H(d^n x/dt^n, d^{n-1} x/dt^{n-1}, \ldots, dx/dt, x) = 0.$$

It can always be reduced to the respective normal form by the substitutions

$$xt = y_1 t, \, x(t + 1) = y_2 t, \, \ldots, x(t + n - 1) = y_n t,$$

or

$$x = y_1, \, dx/dt = y_2, \, \ldots, d^{n-1} x/dt^{n-1} = y_n,$$

provided that the equation

$$G(y_n(t + 1), y_{n-1} t, \ldots, y_2 t, y_1 t) = 0$$

or

$$H(dy_n/dt, y_{n-1}, \ldots, y_2, y_1) = 0$$

can be solved for $y_n(t + 1)$, or dy_n/dt. And even if it cannot be solved, the equation $G = 0$ or $H = 0$ often gives useful information about the dynamical behaviour of the system.

(b) Discontinuous dynamical systems

Causal recursion in discontinuous systems can be represented only in terms of difference equations, i.e. in the normal form (2.1), often called a 't to $t + 1$ map' of the system. Being applicable to both continuous and discontinuous dynamical systems (2.1) can be called the *standard representation* of causal recursion. Starting with an initial state x, a successive application of the causal recursion φ obviously gives a set, say xT^+, of points located on the positive half-trajectory xR^+. Let us call it the standard discrete representation of xR^+.

Let us consider, by way of an example, the dynamical system generated by a linear homogeneous causal recursion,

$$x(t + 1) = \varphi(xt), \quad \text{where } \varphi(x) = Ax, \tag{2.4}$$

A being a non-singular n-square matrix. If all the eigenvalues $\rho_1, \rho_2, \ldots, \rho_n$ are single and real, their eigenvectors $v_1, v_2, \ldots, v_n$ of unit length span the state-space of the system. We can then express any initial state $x0 = x$ as

$$x = \sum_{j=1}^{n} c_j v_j, \tag{2.5}$$

where the c_j are real numbers. Hence the eigenvalue equations $Av_j = \rho_j v_j$ (for $j = 1, 2, \ldots, n$) give

$$xt = A^t x = \sum_{j=1}^{n} c_j \rho_j^t v_j, \quad t = 0, 1, 2, \ldots \tag{2.6}$$

If all the eigenvalues are positive, this gives for $t = 0, 1, 2, \ldots$ the standard discrete representation xT^+ of the continuous half-trajectory xR^+. The continuous dynamical system in question is defined by

$$F(x, t) = \sum_{j=1}^{n} c_j \rho_j^t v_j, \quad F(x, 0) = x = \sum_{j=1}^{n} c_j v_j, \quad t \in R. \tag{2.7}$$

This, of course gives the causal recursion, as expected, as

$$\varphi(x) = F(x, 1) = \sum_{j=1}^{n} c_j \rho_j v_j = Ax. \tag{2.8}$$

But as soon as at least one of the eigenvalues is negative, Eq. (2.6) gives a discontinuous trajectory *rapidly vibrating* with period 2, from one of the continuous trajectories to another, and back again, there being (see Fig. 3)

$$xT^+ = xT^+_{\text{even}} \cup yT^+_{\text{odd}}. \tag{2.9}$$

Here T^+_{even} is the set of even non-negative integers, and T^+_{odd} that of odd non-negative integers, and

$$y = \sum_{j=1}^{n} (-1)^{b_j} c_j v_j. \tag{2.10}$$

Here $b_j = 0$ for $\rho_j > 0$ and $b_j = 1$ for $\rho_j < 0$.

We shall call a dynamical system *reducibly discontinuous*, if its trajectories are rapidly vibrating, with period 2, between two continuous trajectories. Otherwise a discontinuous dynamical system generated by a causal recursion is *irreducibly discontinuous*. Obvious cases of irreducibly discontinuous dynamical systems are finite systems, i.e. systems with only a finite number

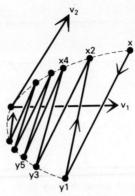

FIG. 3. A 2-dimensional example of a rapidly vibrating trajectory in a reducibly discontinuous system.

of states: they will be discussed in Section 3.9. But we shall soon see, in Section 2.2, that there are also irreducibly discontinuous dynamical systems that are defined on the whole space E.

From what has been said above we can conclude that the class of all dynamical systems having causal recursion is much wider than the class of continuous dynamical systems. As a general expression of causal recursion we can use only the difference form (2.1), not the differential form (2.2).

(c) The internal coupling structure of causal recursion

Except for being a more general concept than that of a continuous dynamical system, causal recursion has another notable property. It can always be associated with a coupling structure as follows.

Let N *causal elements* $T_1, T_2, \ldots, T_N$ be defined as functions

$$T_i: X_i \times R \to Y_i, \quad y_i(t + 1) = T_i(x_i t), \quad i = 1, 2, \ldots, N. \qquad (2.11)$$

Here the X_i and Y_i are Euclidean spaces or their subsets. Let the elements be coupled with one another by means of a *coupling function*

$$C: Y_1 \times Y_2 \times \ldots \times Y_N \to X_1 \times X_2 \times \ldots \times X_N, \qquad (2.12)$$

which together with the *causal function* T gives the causal recursion of the total coupling system (see Fig. 4) as follows:

$$\varphi(x) = (C \circ T)(x), \quad T(x) = (T_1(x_1)T_2(x_2), \ldots, T_N(x_N)). \qquad (2.13)$$

This defines the *causal structure* (C, T) associated with the causal recursion φ.

Any causal recursion either has an internal causal structure composed of the mutual coupling of at least two causal elements, or else we can conceive

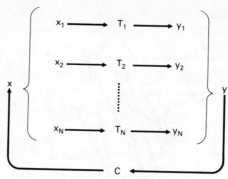

FIG. 4. The coupling of N causal elements $T_1, T_2, \ldots,$ and T_N to a feedback system.

the causal recursion itself as defining a causal feedback element $(C = I, T = \varphi)$ as follows:

$$\varphi(xt) = y(t + 1), \quad x(t + 1) = I(y(t + 1)) = y(t + 1). \tag{2.14}$$

Thus, causal recursion is not only a concept that unifies continuous and discontinuous dynamical systems, but it also involves a causal coupling structure that sometimes may be useful, as will be seen in the following section.

2.2 Dynamical Systems With Nilpotent Causal Recursion

Let φ be a causal recursion for which the formulae

$$\varphi'(x) = x_0 \quad \text{for all } x \in E \quad \text{and} \quad \varphi(x_0) = x_0 \tag{2.15}$$

hold true. Here r is a positive integer, φ^r an r-fold successive application of φ, and x_0 a fixed point of the Euclidean space E. If E is n-dimensional, any n-square nilpotent matrix A of index $r \leqslant n$ defines, by $\varphi(x) = Ax$, a causal recursion of the type (2.15), the point x_0 being the origin. Hence we shall call the function φ defined by (2.15) *nilpotent causal recursion* (Aulin, 1985, 1986b).

Theorem 2.2.1. A nilpotent causal recursion generates an irreducibly discontinuous dynamical system.

Proof. It follows from Eq. (2.15) that whatever be the initial state $x \neq x_0$ the state xt reaches the equilibrium point x_0 at the moment $t = r$. Thus, the trajectory xT^+ consists of the two parts (cf. Fig. 5) $B = \{xt; t = 0, 1, 2, \ldots, r - 1\}$ and $C = \{xt = x_0; t = r, r + 1, \ldots\}$. In order that $xT^+ = B + C$ be either the standard representation of a trajectory induced by a continuous

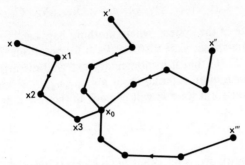

FIG. 5. Four trajectories of a nilpotent dynamical system.

flow or a trajectory that vibrates rapidly, with period 2, between two trajectories of a continuous flow F, it is necessary that $F(x_0, t) = x_0$ for $t \geq 0$. On the other hand, the transitivity of F gives $F(F(x_0, t) - t) = F(x_0, 0) = x_0$. Thus, $F(x_0, t) = x_0$ would imply that $F(x_0, -t) = x_0$ as well, so that any part $B \neq \{x_0\}$ would be excluded. The trajectory $xT^+ = B + C$ accordingly cannot be a representation of a continuous flow, nor reducible to rapid vibrations between two trajectories of such a flow. (This is equivalent to the statement that in a continuous system an equilibrium state x_0 cannot be reached from an initial state $x \neq x_0$ in a finite time.) □

The theorem indicates that the theory of continuous flows can offer no help for the study of nilpotent causal recursions. Instead, we must start from their coupling structure. Let us study a representative example.

Example 1 (Mechanistic System). Let a dynamical system be called mechanistic, as distinct from a cybernetic one, if there is no feedback between its causal elements. Then these elements, $T_1, T_2, \ldots, T_N$, can be supposed to have been ordered so that the coupling function C takes the form (see Fig. 6)

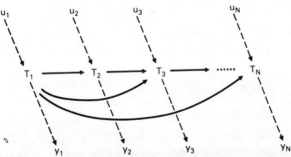

FIG. 6. The coupling of causal elements in a mechanistic dynamical system. The external inputs u_i to each element T_i and the outputs y_i are also indicated in the picture.

$$x_i = C_i(y_1, y_2, \ldots, y_{i-1}); \quad i = 2, 3, \ldots, N; \quad C_1 = 0. \tag{2.16}$$

Let the rest state of the system, where 'nothing happens', be represented by the origin of state-space: $x_0 = (0, 0, \ldots, 0)$. It follows from the peculiar form of the coupling (2.16) that if a momentary external disturbance affects the system, in the form of an external input $u = (u_1, u_2, \ldots, u_N)$ it receives at the moment $t = 0$, the total state xt will be back in the rest state at the moment $t = N$:

$$xN = \varphi^N(u) = x_0. \tag{2.17}$$

The system, accordingly, is nilpotent of (at highest) index N. The state-space may be *any subset* $X \subset E$ as well as E itself.

Since $xt = \varphi^t(u)$, the total output yt is given by

$$yt = T(x(t-1)) = (T \circ \varphi^{t-1})(u). \tag{2.18}$$

Thus, the 'response' yt of the system to the 'stimulus' u starts with $T(u)$ at the moment $t = 1$, and is finished with $(T \circ \varphi^{N-1})(u)$ at the moment $t = N$, lasting N units of time. If the same stimulus is offered again, the system gives the same finite total response. Thus, it is a memoryless system that does not learn from experience.

A simple case in point is a juke-box, or indeed any of those coin-operated automata giving us food, or music, or cigarettes in return for coins. But there is ample need for mechanistic systems even in science, where such a system with its strictly invariant repeatable total response to the same stimulus offers an ideal experimental arrangement for the verification of causal laws. Indeed, any time-involving causal law can be represented in a mechanistic system, as we can make the number N as large, and the unit of time as small, as is needed in each case for a faithful reproduction of data.

General Interpretation (Nilpotent Causal Recursions as Performers of Finite Tasks).   There is a more general interpretation that can be carried from mechanistic systems to any systems with nilpotent causal recursion. Quite generally, we can establish the convention that $x_0 = (0, 0, \ldots, 0)$ and $T_i(0) = 0$ for all $i = 1, 2, \ldots, N$. In the general interpretation of nilpotent causal recursions (Aulin, 1986b) the initial change of state stands for the *formulation of a task* to be performed, and the total response is the *performance of the task*:

$$x_0 \to x0 = u \quad \text{(Formulation of the Task)}, \tag{2.19}$$

$$\{yt = (T \circ \varphi^{t-1})(u); \quad 1 \leqslant t \leqslant r - 1\}$$
$$\text{(Performance of the Task)}. \tag{2.20}$$

The latter process, of course, is determined by the finite half-trajectory uT^+.

The whole operation of receiving and performing a finite task can be now conceived of as a process, where first an external disturbance occurring at the moment $t = 0$ throws the system out of its rest state x_0 to a perturbed state $x0 = u$, after which the causal recursion conducts the total state xt along the half-trajectory uT^+ until, at the moment $t = r$, the system is back in the rest state x_0. It is the finiteness and repeatability of such a stimulus–response process that makes the words 'task' and 'performance of a task' appropriate verbal descriptions in actional terms of what happens in a system having a nilpotent causal recursion.

Although mechanistic systems no doubt form the major category of systems that have a nilpotent causal recursion, they are not the only examples of such dynamical systems. Just to prove the existence of *cybernetic* (i.e. feedback) systems having nilpotent causal recursion we can construct the following simple (and rather artificial, I admit) mathematical example (Aulin, 1985).

Example 2 (A Cybernetic System with Nilpotent Causal Recursion).   Let the causal recursion φ be associated with the internal causal structure

$$T_i(x_i) = l_i x_i \quad \text{for } i = 1, 2, \ldots, N; \tag{2.21}$$

$$C_i(y_1, y_2, \ldots, y_N) = k_i \cdot \sum_{j=1}^{N} y_j, \tag{2.22}$$

so that

$$\varphi_i(x) = k_i \cdot \sum_{j=1}^{N} l_j x_j, \quad i = 1, 2, \ldots, N. \tag{2.23}$$

Here the components x_i and y_i, as well as the parameters k_i and l_i, are real numbers.

This defines a linear homogeneous causal recursion,

$$\varphi(x) = Ax, \quad \text{with } A = \|k_i l_j\|. \tag{2.24}$$

The matrix A is of rank one and, accordingly, the determinant as well as all the principal minors of A are zero, except for the 1×1 principal minors, i.e. the elements in the main diagonal. Hence, the characteristic equation is

$$\mu^N - (Tr A)\mu^{N-1} = 0. \tag{2.25}$$

It suffices to put

$$Tr A = \sum_{i=1}^{N} k_i l_i = 0, \tag{2.26}$$

and we have a matrix A all the eigenvalues of which are zero. This matrix is nilpotent, indeed of index two, and we get:

$$\varphi^2(x) = A^2 x = \| \sum_{s=1}^{N} k_i l_s k_s l_j \| x = \|k_i l_j\| \cdot \sum_{s=1}^{N} l_s k_s \| x = 0. \tag{2.27}$$

The system is therefore nilpotent of index 2.

For the response of the system to a given stimulus $x0 = u$ we get, successively:

$$y1 = T(x0) = T(u) = (l_1u_1, l_2u_2, \ldots, l_Nu_N), \tag{2.28}$$

$$y2 = T(x1) = T(C(y1)) = T(k_1 \cdot \sum_{j=2}^{N} y_j1, k_2 \cdot \sum_{j=2}^{N} y_j1,$$

$$\ldots, k_N \cdot \sum_{j=2}^{N} y_j1)$$

$$= \left(\sum_{j=1}^{N} l_ju_j\right)(k_1l_1, k_2l_2, \ldots, k_Nl_N). \tag{2.29}$$

Note. More-concrete examples of nilpotent cybernetic systems are computers that have been programmed to solve a finite problem, i.e. a problem that can be solved in a finite number of steps of computation in the machine. But computers can also be programmed to simulate systems that have a full causal recursion. Thus, nilpotence is not an inherent property of a computer, but is achieved only by a certain kind of programming. (In the case of a computer the nilpotent system is, of course, not defined on the Euclidean space but on a finite set X.)

2.3. Dynamical Systems with Full Causal Recursion

The mathematical definition of 'goal' is based on an infinite process, and thus on a full causal recursion. The systems with full (i.e. non-nilpotent) causal recursion can be classified on the basis of a classification derived from continuous systems on a Euclidean space. Therefore, here we can apply the topological method, and formulate the definitions of the different types of systems accordingly, their generalizations being trivial.

To define exactly the difference between a goal and a task, let us again assume that an external disturbance throws the system at the moment $t = 0$ from an unperturbed state x (in this case not necessarily a rest state!) to a perturbed state p. Corresponding to the alternative cases, related to the behaviour of the Euclidean distance $\rho(pt, xR^+)$ of the point pt from the half-trajectory xR^+ and to the boundedness or unboundedness of xR^+, we have the following four types of systems with full causal recursion (Aulin, 1986b):

Definitions. (1) If, for a small enough δ-neighbourhood $S(x, \delta)$ of x, the Euclidean distance $\rho(pt, xR^+) \to 0$ with $t \to +\infty$ for all $p \in S(x, \delta)$, and if the positive half-trajectory xR^+ is unbounded, the system is called *self-steering in the state x.*

(2) If the convergence of $\rho(pt, xR^+)$ is as above, but the half-trajectory xR^+ is bounded, the system is called *self-regulating in the state x.*

(3) If, for a small enough δ-neighbourhood $S(x, \delta)$ of x, the Euclidean

distance $\rho(pt, xR^+)$ remains finite for all $p \in S(x, \delta)$, but does not for all $p \in S(x, \delta)$ converge to zero with $t \to +\infty$, the system is called *steerable from outside in the state x*.

(4) If, in any δ-neighbourhood $S(x, \delta)$ of x, there is a point p for which $\rho(pt, xR^+) \to \infty$ with $t \to +\infty$, the system is called *disintegrating in the state x*.

Here $S(x, \delta)$ is the open sphere with the centre x and radius δ. The four definitions obviously exclude one another, and together exhaust the class of all dynamical systems having a full (i.e. non-nilpotent) causal recursion.

Note: A dynamical system may be self-steering in some domains of state-space, self-regulating in other domains, steerable from outside in still other domains, and disintegrating in the rest of it.

Examples of such domains will be given in Chapter 3. A full causal recursion may even reduce to a nilpotent one for some values of parameters. An example is given in Chapter 5.

The mutually excluding cases 1-4 – excluding as far as the same point x of state-space is in question – are illustrated in Fig. 7. In cases 1 and 2 all the perturbed half-trajectories pR^+ for a small enough initial displacement $p - x$ approach asymptotically the unperturbed half-trajectory xR^+. Of course, they approach each other as well. Note that the asymptotic approach means only approach in the end: they do not necessarily approach each other or the unperturbed trajectory all the time. Note also that the convergence involved here is orbital: it does not imply the convergence of states, i.e. the time function $pt - xt$ needn't converge to zero.

In case 3 all the perturbed half-trajectories pR^+ for a small enough $p - x$ remain within a finite distance from the unperturbed half-trajectory xR^+, so that the half-trajectory $(p - x)R^+$ is 'positively Lagrange stable' (Bhatia and Szegö, 1967), but not all of them approach xR^+ asymptotically. This kind of external disturbance accordingly may have a permanent but finite effect on the state-trajectory, by transforming it into a new one that, however, keeps within a finite distance from the old. Hence, such a system is steerable from outside by means of those external disturbances.

In cases 1–3 the system will be called *goal-directed* (or *purposive*) in the state x, and in cases 1 and 2 (actively) *goal-seeking* in that state.

In case 4, however, a small external disturbance may set the system on a perturbed half-trajectory pR^+ such that $(p - x)t$ recedes to infinity with $t \to \infty$. For energetic reasons any such material system would disintegrate in the end, which explains the name chosen for it.

The disintegrating systems, together with the systems that are steerable from outside, form the class of *dynamically unstable systems* with a full causal recursion. The self-steering and self-regulating systems are *dynamically stable*. The classes of goal-directed, disintegrating, and nilpotent systems together

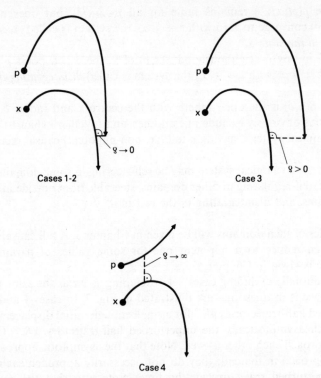

FIG. 7.   The behaviour of perturbed (p) and unperturbed (x) trajectories in a
dynamical system that is goal-seeking (Cases 1–2), or steerable from outside (Case
3), or disintegrating (Case 4).

exhaust the class of all dynamical systems having causal recursion. The so-
obtained *exhaustive systematics of causal systems* is illustrated in Fig. 8.

In a philosophical terminology, our cybernetic systematics may be classified
to the category of *terminal causality*, where causal processes are classified on
the basis of what happens in the system when $t \to \infty$.

2.4.   Self-organization vs Rigid-structured Dynamical Systems

The function that defines the causal recursion of a dynamical system is
usually given in the form that contains some constants or coefficients. Thus,
the causal recursion actually is a function

$$\varphi: X \times \Lambda \to X. \tag{2.30}$$

Here X is the state-space and Λ the parameter space associated with the

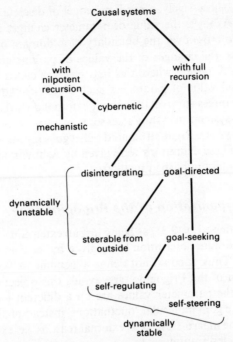

FIG. 8. An exhaustive classification of the dynamical systems that have causal recursion. Each of the indicated types of system, i.e. each of the *causal types*, may appear both as rigid-structured and as self-organizing versions. All the causal types shown in this figure, save the mechanistic ones, are cybernetic systems, i.e. systems involving circuits of feedback in their coupling structure.

causal recursion φ. Both of them are Euclidean spaces or parts of Euclidean spaces. The vector $\lambda \in \Lambda$ is the parameter vector of the causal recursion φ.

Definition. A dynamical system is *self-organizing*, if it is able itself to change the values of its parameters, otherwise it is *rigid-structured*.

There have been two different understandings of the content of being 'able itself to change the values of its parameters'. Accordingly, we could distinguish between two kinds of self-organization.

(1) Self-organization of the Ashby type

The parameter may be coupled with the rest of the system as one of the causal elements. In this case we can, without any loss of generality, rewrite the state-transition equation (2.1) as follows:

$$\lambda(t+1) = \varphi_0(xt, \lambda t), \quad x(t+1) = \varphi(xt, \lambda t). \tag{2.31}$$

An example of this kind of dynamical system is Ashby's homeostat (Ashby, 1972), where a partial system is programmed to change the parameter so as to take on, successively, a random sequel of discrete values. In this original scheme of Ashby's, the value of parameter changes every time the state xt threatens to cross over the boundary of a 'domain of survival' in the state-space. Thus, the changes of the values of parameter secure the survival of the system. A system where this happens was called by Ashby 'ultrastable'. For the final value of parameter, such a system finds an equilibrium state, which guarantees survival (until next external disturbance at least). Thus, an ultrastable system, in Ashby's sense, is in our terms a self-organizing, self-regulating system. Rigid-structured self-regulating systems were called 'stable' by Ashby. These definitions were given by Ashby in his books (Ashby, 1970, 1972), originally published in 1950s.

(2) Self-organization of the Prigogine type

One can criticize Eq. (2.31) as being just an extension of the state-description, the complete state-description being given by the combination (x, λ), and not by x alone. Thus, it could not define a 'genuine' self-organization. The self-organization of the Prigogine type avoids this objection. Now the cause of changes of the parameter values are on a different level of complete state-description, e.g. in molecular 'fluctuations' instead of at the macro-level, and thus affect the macro-state xt as external reasons, i.e. external to the complete macro-state-description x.

An example of self-organization of the Prigogine type is the 'trimolecular model' of chemical reactions (Nicolis and Prigogine, 1977), defined by the differential equations

$$dx/dt = a - (b + 1)x + x^2y + d_1\nabla^2x,$$

and

$$dy/dt = bx - x^2y + d_2\nabla^2y. \tag{2.32}$$

Here (x, y) defines the total state, there being $x = x(\bar{r}, t)$ and $y = y(\bar{r}, t)$, where $\bar{r}$ is the position vector in the three-dimensional physical space. The differential operator ∇^2 operates on $\bar{r}$. The parameter vector is $\lambda = (a, b, d_1, d_2)$. When the value of b exceeds a critical value b_c, the equilibrium state (x_0, y_0) bifurcates into two states. In the case of a chemical reaction this may happen as a consequence of strengthened molecular fluctuations which, at the same time, make one of the bifurcated states a new equilibrium state (x'_0, y'_0). While the system for $b < b_c$ was self-regulating in a neighbourhood of the old steady state (x_0, y_0), for $b > b_c$ it may be self-regulating in a neighbourhood of the new steady state (x'_0, y'_0).

Another example of self-organization of the Prigogine type would be the mutations affecting the animal populations of an ecosystem, as a consequence of which the nature of the ecosystem may be radically changed.

As a general rule, we can say that self-organization changes the trajectories as well as the operation and mutual coupling of causal elements of the dynamical system. In cybernetic system dynamics it has been rather generally assumed, ever since Ashby, that

> *every living organism is a self-regulating system that is capable of self-organization,*

beginning with the smallest microbes, and even with some pre-cellular life processes (Nicolis and Prigogine, 1977). On the other hand, the stable structures of inorganic matter like atoms, molecules, and crystals are rigid-structured, not capable of self-organization.

(3) Extreme self-organization in self-steering actors

Later we shall meet an extreme form of self-organization in self-steering actors, to be introduced in Chapter 9. The remarkable properties of this ultra-self-organization will be considered there in detail. Let it be mentioned here that it implies not only an ability to change the values of parameters of causal recursion but the capability of self-creation of causal recursion by the dynamical system concerned (or, if you like, the change of all the thinkable parameters of causal recursion). The mathematical concept of the self-steering actor presupposes a combination of the theory of dynamical systems with mathematical actor theory, as will be shown in Chapter 9.

2.5 Causality, Determinism, and Indeterminism

Causal recursion, the strictest form of causality, may exist in all four of the following cases:

Strong determinism: the predictability, with certainty, of single future events in the given dynamical system. Examples: macrophysics, classical physical theory (see Section 1.2 above).

Probabilistic determinism: the predictability of the probability distributions of future events. Examples: microphysics, quantum theory (see Section 1.3 above); Markovian systems (see Example in Section 4.6 below).

Weak determinism: the predictability of the possibility distributions of future events. Examples: the dynamical systems where a generalization of the concept of entropy from probability to possibility distributions (Higashi and Klir, 1982) is valid. (This case is not discussed in the present book.)

Indeterminism: the unpredictability of all future events in the dynamical system concerned. Examples: self-steering actors (see Chapter 9 below).

Thus, the concept of causality cannot be identified simply with 'determinism', but allows, according to modern mathematical dynamics, three different degrees of determinism and, in addition to them, a case of complete indeterminism.

How are we to understand the compatibility of causality and indeterminism? Mathematically, the compatibility of these two concepts follows from Theorem 9.3.1 of Dual Causality, and from the property of expanding state-development that is characteristic of self-steering systems (see Chapter 9). But a preliminary introduction to the understanding, or rather anticipation, of the coexistence of causality with complete indeterminism can be given in terms of the concept of self-organization as follows.

In a process of self-organization, where the dynamical system itself changes some values of its parameters, the field of trajectories in the state-space may be radically altered. Even the *causal type* may change. Examples of different causal types are self-steering systems, self-regulating equilibrium systems, periodically pulsating self-regulating systems, almost periodically pulsating self-regulating systems, irregularly ('chaotically') pulsating self-regulating systems (for these types of self-regulating systems see Sections 3.2–3.5 below), the different types of systems steerable from outside (see Section 3.8 below), disintegrating systems, and the system with a nilpotent causal recursion.

Let us now make the following thought experiment. Let a dynamical system be equipped with an extreme capability of self-organization, so that the system is actually able to create its own causal recursion, i.e. to generate the expansion of its causal recursion from the set of past states to present and future states. Provided that the past states will never recur in future, we cannot use our knowledge of causal recursion, as defined for the past states, to predict any future events in such a system.

That is exactly the situation with the so-called self-steering actors, to be constructed in Chapter 9. In such a dynamical system, therefore, complete indeterminism goes with the existence of causal recursion, i.e. with the existence of causality in its strictest form. The behaviour of such a system can be explained causally, in principle at least, but only afterwards, by making use of the causal recursion that has been already generated by the system: obviously, here is an interesting message for the science of history, provided that human beings and their communities are interpreted as self-steering actors.

CHAPTER 3

The Systematics of Goal-directed Systems

Discussions in this chapter again lean on the topological method and thus only apply to continuous systems. However, the generalizations to discontinuous systems are trivial and are left for the reader, except for the analysis of finite systems (and of the Turing machine) that are considered in Section 3.9.

3.1. General Properties of the Goals of Self-regulating Systems

Definition. The *goal* G_x of a self-regulating system in the state x is the positive limit set of this state:

$$G_x = \Lambda^+(x). \tag{3.1}$$

From the definition of a self-regulating system (Definition 2, Section 2.3) it follows immediately that the discrete-time half-trajectory xT^+, the corresponding continuous half-trajectory xR^+, its closure and the positive limit set $\Lambda^+(x)$ are all bounded. As to the mutual relation between xR^+ and $\Lambda^+(x)$ there are two possibilities: either $xR^+ \cap \Lambda^+(x) = \varnothing$ (an empty set), or $xR^+ \subset \Lambda^+(x)$.

In the former case the trajectory xR^+ tends asymptotically to the positive limit set $\Lambda^+(x)$, and is called a *positively asymptotic trajectory* (Bhatia and Szegö, 1967). In this case we have

$$\Lambda^+(x) = \partial xR^+ \quad \text{(valid for } x \in \mathscr{C}\Lambda^+(x)\text{)}. \tag{3.2}$$

In the latter case the goal has already been reached, as now $x \in \Lambda^+(x)$. The trajectory xR^+ and the corresponding motion of state are called *positively Poisson-stable* (Bhatia and Szegö, 1967). Since the points of $\Lambda^+(x)$ are the limit points of the points of xR^+, we must now have:

$$\Lambda^+(x) = \overline{xR^+} \quad \text{(valid for } x \in \Lambda^+(x)\text{)}. \tag{3.3}$$

Thus, we have proved the following statement:

Theorem 3.1.1. The goal of a self-regulating dynamical system is always represented in the state-space by the closure of a positively Poisson-stable trajectory.

It also follows from the above argument that every bounded positive half-trajectory xR^+ in a Euclidean space is either positively asymptotic or positively Poisson-stable (cf. Bathia and Szegö, 1967, Theorem 1.10.11).

Let a dynamical system be self-regulating in the state x of state-space. In the spirit of Definition 2, Section 2.3, we define the *domain of self-regulation* (*or stability*) *of the system in the state* x or the *domain of validity of the goal* G_x by

$$D_x = \{y; \rho(yt, xR^+) \to 0 \text{ with } t \to +\infty\}. \tag{3.4}$$

Note that the system is self-regulating only at the interior points of D. Obviously $x \in D_x \backslash \partial D_x$.

For a point $y \in D_x$, $y \neq x$, either $\rho(y, xR^+) > 0$ or $\rho(y, xR^+) = 0$. In the former case there are again two alternatives: either $yR^+ \cap xR^+ = \varnothing$, in which case yR^+ tends to xR^+ and thus to $\Lambda^+(x)$ asymptotically, there being

$$\Lambda^+(y) = \Lambda^+(x) \quad \text{and thus } G_y = G_x; \tag{3.5}$$

or then $yt_0 = x$ for some finite $t_0 > 0$. In this case, if xR^+ is positively asymptotic, so is yR^+, and we have again $G_y = G_x = \Lambda^+(x) = \Lambda^+(y)$. But if $x \in \Lambda^+(x)$ while $yt_0 = x$ for a finite $t_0 > 0$, we can infer only that

$$xR^+ \subset yR^+, \quad \text{thus } \Lambda^+(x) \subset \Lambda^+(y) \text{ and } G_x \subset G_y. \tag{3.6}$$

Similarly, if $\rho(y, xR^+) = 0$, we have $xt_0 = y$ for some finite $t_0 > 0$, and if $x \in \Lambda^+(x)$, we can infer only that

$$yR^+ \subset xR^+, \quad \text{thus } \Lambda^+(y) \subset \Lambda^+(x) \text{ and } G_y \subset G_x. \tag{3.7}$$

On the other hand, if $\rho(y, xR^+) = 0$ and xR^+ is positively asymptotic, so is yR^+, and we have again $G_y = G_y = \Lambda^+(x) = \Lambda^+(y)$.

3.2. Self-regulating Equilibrium Systems

It follows from the discussions of the preceding section that the sub-classes of self-regulating systems, characterized by the different types of goals, correspond to the sub-classes of positively Poisson-stable continuous trajectories. The simplest case of such a trajectory is a single point x_0 of state-space or, expressed as a set, a singleton $\{x_0\}$. This corresponds to a system that is self-regulating in a state x, and has the following goal and the associated domain of validity:

$$G_x = \{x_0\} \quad (\text{an } \textit{equilibrium goal}) \tag{3.8}$$

$$D_x = \{y; \rho(yt, x_0) \to 0 \text{ with } t \to +\infty\}. \tag{3.9}$$

Let such a system be called a *self-regulating equilibrium system*.

All the positive half-trajectories starting from within the domain D_x, save for the point x_0 itself, are positively asymptotic and tend asymptotically to the *equilibrium* or *rest* or *steady state* or *point* x_0. In Fig. 9a the origin is an equilibrium goal, whose domain D of validity is the interior of the unit circle. The system is self-regulating in the domain D, disintegrating on the boundary ∂D and outside it. In Fig. 9b the equilibrium goal x_0 is located on the boundary of D, so that the system is not self-regulating in the goal state x_0. In this case the domain of validity D of the goal is the closed unit circle, and the system is self-regulating in the interior $D \backslash \partial D$ but disintegrating elsewhere on the state-plane E^2.

A third situation with an equilibrium goal is illustrated in Fig. 9c, where $D = E^2 \backslash \{0\}$ is the domain of stability of the goal state x_0. The origin O is now another equilibrium state, but there the system is steerable from outside (though only to the perturbed state x_0). The system is self-regulating elsewhere on the state-plane E^2. In Fig. 9d the origin is an equilibrium goal, and its domain of validity comprises the whole plane: $D = E^2$. In this case the system is self-regulating in every state.

In an old physical terminology, the equilibrium in the state x_0 of Fig. 9b should be called 'unstable', while in the equilibrium states shown in Figs 9a, 9c, and 9d it is 'stable'. Yet the stability of equilibrium in Fig. 9a clearly differs from that in Figs 9c and 9d. To be able to express this difference we introduce the following terminology (Aulin, 1986b):

Definitions. In a steady state x_0, the system is *locally stable*, if for a small enough neighbourhood U of x_0 the relation $p \in U$ implies that

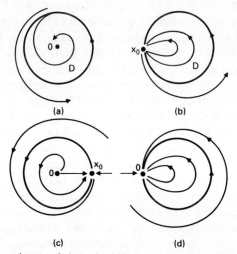

FIG. 9.   The perimeter of the unit circle: (a) as the separatrice between the domains of attraction and repulsion of a locally stable equilibrium point 0, (b) as the boundary of the domain of attraction of a locally unstable equilibrium point x_0 located on it, (c) as one of the trajectories asymptotically approaching the locally unstable equilibrium point x_0, or (d) the locally unstable equilibrium point 0.

(1) $\rho(pt, x_0) \to 0$ with $t \to +\infty$,   and

(2) $pR^+ \subset U$.

The system is *locally steerable from outside* at x_0, if only the condition (2) holds good. If neither (1) or (2) is true, the system at x_0 is *locally disintegrating*.

These definitions, which are associated with a system that is in an equilibrium state x_0, help us to articulate the differences observed between Fig. 9a, on the one hand, and Figs 9c and 9d on the other. In Fig. 9a the system is locally stable in the state O. In Fig. 9b the system is locally disintegrating as well as disintegrating in the state x_0, so that here we have nothing new to say. But in Fig. 9c the system is locally disintegrating in both equilibrium states x_0 and O, while in the 'global' terms defined in Section 3.3 it is self-regulating in x_0 and steerable from outside at the origin. In Fig. 9d the system is at the origin O locally disintegrating but globally self-regulating.

Mathematical examples of self-regulating equilibrium systems and of the associated causal recursions are plenty, as this is the case that has been most investigated in classical dynamics (cf. Abraham and Shaw, 1983). We shall see such examples in Chapters 4 and 5.

Real-world Examples.   Think, for instance, of a ball in a cup that has the form of a half-sphere. If you give the ball a not-too-strong push it will start moving, and tends back to the middle point of the cup, which marks the state of equilibrium of this macrosystem. Another example from the everyday world is a room equipped with a good thermostat, which is able to return temperature to its chosen equilibrium value after a small external disturbance.

But even stable atoms, molecules, and crystals can be considered as self-regulating equilibrium systems, provided that we confine ourselves to representing them in terms of the observable values of total energy. In such a state-description the lowest energy state E_0 is the ground state to which the atom (molecule, or crystals) returns after an external disturbance that is small enough not to throw the atom to an energy state E above the limit of ionization, E_∞ (see Fig. 10). Thus, the domain of self-regulation comprises the ground state and all the 'excited states' E_n (for $n = 1, 2, \ldots$), for which $E_n < E_\infty$. In more complicated stable microstructures even relative energy minima, and thus several equilibrium states, may appear. The atoms and molecules of stable (i.e. not radioactive) inorganic matter are rigid-structured, not able to change their parameters and causal recursions. (In a quantum-theoretical state-description atoms, molecules, and crystals are again self-regulating equilibrium systems, but as the state-space is in that case a Hilbert space, it cannot be coped with by our present formalism based on a Euclidean state-space.)

Note on Living Organisms.   It was assumed in Section 2.3 that every living organism is a self-organizing self-regulating system. We can now add that it

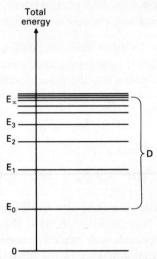

FIG. 10.   Atom as a rigid-structured self-regulating equilibrium system: The domain of stability D consists of the equilibrium ('ground') state E_0 and of the excited states E_n below the limit of ionization E_∞. The underlying dynamics, however, is quantum-mechanical, not classical.

may be simultaneously, in the different parts of state-space, an equilibrium system and any of the pulsating ones (see Sections 3.3–3.5).

3.3   Periodically Pulsating Self-regulating Systems

The other 'classical case' of a positively Poisson-stable trajectory is a closed curve, i.e. a trajectory for whose every point x the formula

$$\Lambda^+(x) = xR^+ = \overline{xR^+} \quad \text{(valid for } x \in \Lambda^+(x)) \tag{3.10}$$

holds good, and which does not reduce to a single point.

If a dynamical system is self-regulating at the point x of state-space, which either is on a closed curve C_0 or on a positively asymptotic trajectory tending to C_0, we shall call it a *self-regulating periodically pulsating system*. Such a system has the 'limit-cycle goal' C_0 and the associated domain of self-regulation obeying

$$G_x = C_0 = \{y; \Lambda^+(y) = yR^+ = C_0\}, \quad \text{(a *limit-cycle goal*)} \tag{3.11}$$

$$D_x = \{z; \rho(zt, C_0) \to 0 \text{ with } t \to +\infty\}. \tag{3.12}$$

The system in Fig. 2a is self-regulating and periodically pulsating, with the unit circle as a limit-cycle, and its interior (minus the origin) as its domain of stability. The unit circle is also the limit-cycle goal of the system shown in Fig. 11, but now the domain of validity of this goal comprises the whole plane save for the origin, in which the system (as in Fig. 2a) is locally disintegrating and globally steerable from outside.

Mathematical Example (Causal Recursion with a Limit-Cycle Goal). It is not difficult to construct a causal recursion leading to the unit circle as a limit-cycle goal. Let us introduce the polar co-ordinates (r, α) on the plane of Fig. 11. The causal recursion defined by

$$\varphi: \varphi_r(r) = r^{\frac{1}{2}}, \quad \varphi_\alpha(\alpha) = \alpha + \alpha_0 \tag{3.13}$$

generates the motion of state given by

$$rt = r^{1/2^t}, \quad \alpha t = \alpha + t\alpha_0, t = 0, 1, 2, \dots \tag{3.14}$$

Obviously, for any initial state (r, α) with $r \neq 0$ the sequence rt converges to one, and we have the trajectories of the type shown in Fig. 11. Note that while all the spirals approach each other as curves, their states as functions of time do not: the initial difference in the phase α remains. The system is continuous, with the flow

$$F: F_r(r, t) = r^{1/2^t}, \quad F_\alpha(\alpha, t) = \alpha + t\alpha_0, t \in R^+. \tag{3.15}$$

By derivation we get the differential equations that generate this flow:

$$dr/dt = -kr \log r, \text{ with } k = \log 2, \, d\alpha/dt = \alpha_0. \tag{3.16}$$

Real-world Examples (Periodically Pulsating Phenomena of Life). Living organisms, including very small ones, usually have some periodically pulsating systems that tend back to the same rhythm after being disturbed. The heart could be an example of this. A biologist can multiply the list – let us leave it to him. All such systems are periodically pulsating self-regulating systems in the sense that we have defined them here.

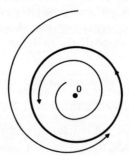

FIG. 11. The perimeter of the unit circle as a limit cycle, with the domain of stability consisting of the whole plane reduced by the origin.

3.4 Almost Periodically Pulsating Self-regulating Systems

The remaining type of positively Poisson-stable trajectory is one for which Eq. (3.3) holds good but not Eq. (3.10), i.e.

$$\Lambda^+(x) = \overline{xR^+} \neq xR^+ \quad \text{(valid for } x \in \Lambda^+(x)\text{)}. \tag{3.17}$$

Such a trajectory visits every δ-neighbourhood of its every point x an infinite number of times without being a closed curve. But obviously an appropriate name for it is *quasi-closed curve*, as on each of its infinitely many rounds it only just avoids being closed.

The simplest example of quasi-closed curve is the curve defined by the equation

$$\psi = k \cdot \chi, \quad \text{with a positive irrational number } k, \tag{3.18}$$

on a two-dimensional torus (see Fig. 12). The closed curves $\psi = $ const. are the 'meridians' and $\chi = $ const. the 'latitudes' on the torus, each of them being of unit length, by a convention of measurement. Hence the curve (3.18) would be closed for a rational number $k = m/n$, since the point $(\psi = m, \chi = n)$ is equal to the origin $(0,0)$ of the torus for any integers m and n. With an irrational k we get an infinite curve, whose closure fills up the whole torus. Thus, for every point x on the curve (3.18) Eq. (3.17) is valid.

The simplest type of the motion of state along a quasi-closed curve is called *almost periodical*, and was discovered by Bohr (Bohr, 1924, 1925, 1926). An example is the uniform motion along the trajectory (3.18) determined by

$$d\psi/dt = k, \quad d\chi/dt = 1. \tag{3.19}$$

The almost periodically pulsating self-regulating systems are the simplest ones, where the goal of self-regulation is a *non-trivial attractor limit set* $\Lambda^+(x)$ for each point x within the domain of regulation (or on its boundary). In saying so, by trivial attractor limit sets is meant an equilibrium point or a limit cycle (cf. Fig. 13). An example of a non-trivial attractor limit set is the torus of Fig. 12, provided that it is an attractor for some positively asymptotic

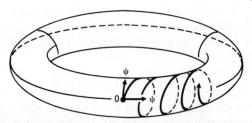

FIG. 12.   A 2-dimensional torus embedded in a 3-dimensional Euclidean space, and a 'linear' trajectory on it starting from the origin.

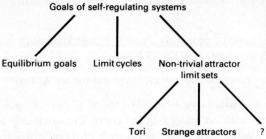

FIG. 13.   The systematics of the goals of self-regulating systems. It is not known whether the present concept of strange attractor together with that of torus together exhaust the class of non-trivial attractor limit sets.

trajectories obeying Eq. (3.2), with the torus being the limit set $\Lambda^+(x)$, and thus the boundary of some asymptotic trajectories xR^+ in a neighbourhood of the torus in a three-dimensional Euclidean state-space. The motions of state along all of these asymptotic trajectories then tend to an almost periodic motion along one of the infinite curves of the type (3.19) on the torus.

The question of the possible real-world examples will be discussed in the following section.

3.5   Irregularly ('Chaotically') Pulsating Self-regulating Systems

Again we are concerned with positively Poisson-stable motions along quasi-closed curves on a non-trivial attractor limit set, or motions of state tending to such motions from outside the attractor limit set. Again we can use, by way of introduction, the infinite curves of the type (3.18) on the torus of Fig. 12. If, instead of the uniform motion of state, Eq. (3.19), we assume a motion defined by Nemytskii and Stepanov (1972)

$$d\psi/dt = k \cdot f(\psi, \chi), \quad d\chi/dt = f(\psi, \chi), \tag{3.20}$$

where f is a continuous and positive-valued function, we have an example of what is called an *irregular or 'chaotic'* motion of state.

While in uniform, almost periodic, motions of states along two trajectories xR^+ and yR^+ of the type (3.18), the mutual distance $\rho(xt, yt)$ of states remain constant, it may become very large with time in the motions of the type (3.20) (see Fig. 14):

$$\exists t : t \gg 0, \quad \text{such that } \rho(xt, yt) \gg \rho(x, y). \tag{3.21}$$

This, indeed, is the defining property of what is called an *irregular* or *chaotic motion of states* in a dynamical system (cf. Schuster, 1984). It follows from this property that the long-term predictions become sensitively dependent

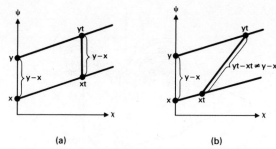

(a)   (b)

FIG. 14. An illustration of the difference between a uniform (Picture a) and non-uniform (Picture b) motion of state along pseudolinear trajectories on a 2-dimensional torus. The motions are further complicated by the periodicity of the co-ordinates, i.e. that $\psi + n = \psi$ and $\chi + n = \chi$ for each integer n, which is not indicated in the pictures.

on the initial states in such systems, the more so the more time has passed: even a tiny difference in the initial states x and y can cause an enormous difference $\rho(xt, yt)$ with a sufficiently long time t.

Generally speaking, an irregularly pulsating self-regulating system is a dynamical system, in which a non-trivial attractor limit set attracts positively asymptotic trajectories, located outside it, and the motions of states on such trajectories asymptotically approach irregular motions along a quasi-closed trajectory belonging to the attractor limit set. The tori of any dimensionality may appear as attractors, but they are not the only types of non-trivial attractor limit sets (i.e. goals) that are possible in connection with irregularly pulsating self-regulating systems. There are also *strange attractors* for which no generally accepted formal definition exists as yet, but which have been illustrated in an ever-increasing number of examples in mathematical literature, especially since the 1970s (for this literature, see Schuster, 1984).

The first appearance of a strange attractor took place in a paper by the meteorologist E. N. Lorenz (1963), based on the following simple dynamical system:

$$dx/dt = -ax + ay, \quad dy/dt = -xz + bx - y, \quad dz/dt = xy - cz. \quad (3.22)$$

Here a, b, and c are positive coefficients. A picture of the resulting attractor-limit set is shown in Fig. 15. There are much more strange-looking attractors in later literature, e.g. those shown in Fig. 16a. 'These systems of curves, these clouds of points suggest sometimes fireworks or galaxies, sometimes strange and disquieting vegetal proliferations', D. Ruelle writes (1980), adding: 'A realm lies here to explore and harmonies to discover.'

Real-world Examples.   The most periodically or nonperiodically pulsating self-regulating systems seem to be partial systems in living organisms. The frequent appearance of pulsation in the biological world may be an arrangement for saving energy. Among the pulsating self-regulating systems

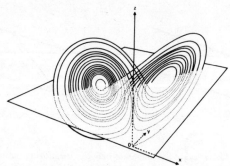

FIG. 15.   The historically first strange attractor obtained for the Lorenz system (3.22), according to a computer calculation by Lanford (1977).

in living organisms may be strange attractors too, but no details are known as yet. On the other hand, even the inorganic world contains such systems. A meteorological example is the Lorenz attractor. The ultimate reason for the pulsation of aerial masses, observed in meteorology, probably is the pulsating flow of energy from the sun, due to the rotation of the earth.

Terminological Note: What is 'Equilibrium'?   Earlier in physical theory it was customary to call even periodic states 'equilibrium states', sometimes 'stationary equilibrium states' as distinct from steady states. An extension of this terminological usage could lead us into calling any goal states in self-regulating systems, i.e. any positively Poisson-stable states, 'equilibrium states'.

However, now that strange attractors, not to speak of limit cycles, are becoming commonplace in both mathematical and physical dynamics, such a terminological usage is confusing. Hence, the term 'equilibrium state' has been limited here to the steady states, represented in state-space by a rest point.

A further terminological problem is connected with the different state-descriptions of modern physics. What is a 'nonequilibrium system' from the point of view of statistical thermodynamics (e.g. in Nicolis and Prigogine, 1977), may be an equilibrium system, with a steady-state equilibrium, or a limit-cycle system, or any self-regulating or even (in theory at least) self-steering system in terms of macrophysical state-description. However, this problem is avoided here, as all the discussions from Chapter 2 on refer to macrostates.

3.6.   Self-steering Systems

If a positive half-trajectory xR^+ on a Euclidean space is unbounded, and attracts the trajectories in its neighbourhood, all these trajectories, including

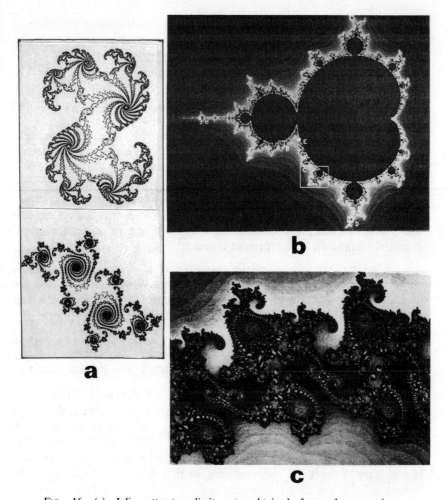

FIG. 16. (a) Julia attractor limit sets obtained from the recursion $z(t + 1) = z^2(t) + c$ on the complex z-plane for the parameter values $c = 0.32 + 0.43i$ (the upper figure) and $c = -0.194 + 0.6557i$ (the lower figure). (b) Mandelbrot's set of the parameter values c for which the Julia attractor sets are connected. (c) A detail from the 'tail of the seahorse' in Picture (b). (Source: Peitgen and Richter, 1984.)

xR^+ itself, recede to infinity with $t \rightarrow +\infty$, and all their points have empty positive limit sets. (Note that the unbounded trajectory with the non-empty positive limit set in Fig. 2b is not an attractor.) This explains the scarcity of mathematical results concerning self-steering, as most recent research in dynamics has been focused on attractor limit sets, and thus on self-regulating systems.

Where the self-steering system fundamentally differs from a self-regulating one is that a *self-steering trajectory*, i.e. a positive half-trajectory xR^+ through a state x in which the system is self-steering,

(1) has no finite end in the form of a non-empty limit set, and

(2) never visits the past states again: *the same total state never returns* on a self-steering trajectory, not even in the approximative sense as it happens with a quasi-closed trajectory in self-regulating non-periodically pulsating systems.

In other words: every total state on a self-steering trajectory is *unique*; it has never appeared before, and will never turn up again. Thus, a self-steering system is all the time creating new states and new goals.

Hence, it is more appropriate to speak of *self-steering goals* rather than of the goals of self-steering. If a definite goal has to be indicated, it cannot be other than the future trajectory of the state x in which the system is at the 'present moment' $t = 0$, admitting that this future course of the system is entirely unknown at the 'present moment':

$$G_x = xR^+ \setminus \{x\}. \tag{3.23}$$

The fact that it still is a genuine goal is confirmed by the fact that a non-empty region of attraction is associated with it, viz. the set

$$D_x = \{y; \rho(yt, xR^+) \to 0 \text{ with } t \to +\infty\}. \tag{3.24}$$

It defines the *domain of self-steering* of the system in the 'present state' x.

General Interpretation (Self-steering Intellectual Processes).   If the uniqueness of the states of mind, along with the goal-oriented nature of thought processes, is typical of human consciousness, the only thinkable causal representation of what happens in the human mind in an alert state is the self-steering process. For a reason that is explained in Section 3.7 below, it is necessary to limit the interpretation so that what is self-steering in the human mind is the *total* intellectual process. Not all partial processes need be self-steering.

Real-world Examples (Human Development, Social Interaction).   The intellectual total process in the human brain is the ultimate source of human development and the basis of all social interaction between human beings. Hence, self-steering has been proposed to play a prominent role in the conceptual foundations of social science (Aulin, 1982, 1985, 1986a, 1987b; Geyer and van der Zouwen, 1986). Mathematically this can be proved only in the cases where causal recursions can be given a mathematical representation. Such cases appear in the theory of economic growth: examples will be given in Chapter 8. But a more fundamental application of the concept of self-steering may be the *qualitative theory of social development* it suggests, as will be shown in Part 3. Real-world examples of the significance of self-steering in social development will be given in Part 4. The interesting psychological and epistemological problem of the source of self-steering in human consciousness will be investigated in Part 5.

Note on Goal Functions. The goal functions used in the earliest attempts at a mathematical definition of self-steering (Lange, 1965; Aulin, 1982, 1985) have been abandoned here, as they were in a previous paper (Aulin, 1986b). After all, only topological properties such as orbital convergence are needed, not the convergence of (the 'motions' of) states. If goal functions are used, you have to introduce a different dependence on time into the goal functions $g_x(t)$ of different states x. This makes the goal functions a redundant tool that can be dispensed with.

3.7. The Ashby–Lange Effect

It follows from the properties of a self-steering system, discussed above, that the total state x in such a system can never settle down in a state of equilibrium (neither can it approach any periodical or other positively Poisson-stable motion). What happens, if a partial process $x_1 t$ tends to a partial state of equilibrium, x_{1_0}, in such a system?

To answer the question, we have to consider the two parts S_1 and S_2 of the given system, having no common parts between them, and the corresponding decomposition of the total state x:

$$S = S_1 + S_2, \quad x = (x_1, x_2). \tag{3.25}$$

If $x_1 t \to x_{1_0}$, the similar behaviour is denied of the remaining partial state, since the total state cannot be in equilibrium: $x_2 t \nrightarrow x_{2_0}$. Hence the process $x_2 t$ goes on, and through its interaction with the process $x_1 t$ compels the latter, too, out of equilibrium. Thus, in a self-steering system only temporary approaches to partial states of equilibrium can appear, while the total process 'denies' the total stop of any of its partial processes.

This is what W. Ross Ashby (1972) observed in his 'homeostat', which was an electromagnetic apparatus constructed to model a self-regulating system. Indeed, in a self-regulating system, too, the phenomenon described above appears if a partial state $x_1 t$ approaches a premature state of equilibrium, which does not lead to an equilibrium of the total state. Ashby called it the *power of veto* (1972, p. 79), of the part S_2 upon the part S_1. Oskar Lange (1965, pp. 72–73) derived the effect from his concept of self-steering.

We can now give an interpretation to the Ashby–Lange effect in connection with the total intellectual process occurring in the human brain in an awake state. We can conceive of a partial process $x_1 t$ approaching a partial state of equilibrium x_{1_0} as a process of *problem-solving* striving for the answer x_{1_0}. As soon as the answer is clear, the total intellectual process compels the partial problem-solving to move again, in the restless pursuit of 'truth', characteristic of the total process.

One can extend the analogy between problem-solving and the power of veto to the development of science: every solved problem opens up new perspectives with new problems and thus new paths to go down.

3.8.   Systems that are Steerable from Outside

A dynamical system that is steerable from outside can be considered as a case of degeneration of either a self-steering or self-regulating system, where the domain of self-steering or self-regulation has shrunk to a single trajectory, viz. the positive half-trajectory through the state x of the system at the 'present moment' t:

$$D_x = xR^+. \tag{3.26}$$

The goal of the system at the 'present moment' $t = 0$ accordingly is either

$$G_x = \Lambda^+(x) \quad \text{or} \quad G_x = xR^+\backslash\{x\}, \tag{3.27}$$

depending on whether the trajectory xR^+ is bounded or unbounded, i.e. whether the system is a degeneracy of a self-regulating or self-steering system, respectively.

By definition 3, Section 2.3, any small enough state-displacement $x \rightarrow p \in \mathscr{C}(xR)$ in a system that is steerable from outside throws the state on to another goal-directed trajectory pR^+, which remains within a finite distance from the initial trajectory xR^+ without converging it. Therefore, the perturbed trajectory pR^+ could be called a *satellite* of the unperturbed one.

The transition to a satellite also engenders a change of goal:

$$x \rightarrow p \in \mathscr{C}(xR) \Rightarrow G_x \rightarrow G_p. \tag{3.28}$$

In the perturbed state p the system may still be steerable from outside, but it can also be self-regulating or self-steering. In any case, the initial goal G_x, according to (3.28), is steerable from outside, as it can be changed by a displacement of state, $x \rightarrow p$ due to an external disturbance.

A system that is steerable from outside can be

(1)   an equilibrium system, or
(2)   periodically pulsating, or
(3)   non-periodically pulsating, or
(4)   a system with unbounded trajectories.

Type 1 is illustrated in Fig. 17a. The trajectories are straight lines parallel to the x-axis on the plane E^2. The points of intersection with the y-axis are the equilibrium goal states. A disturbance throwing the system from an initial state $z = (x, y)$ to a perturbed state $p = (x', y')$ also induces a change of goal if $y' \neq y$:

$$G_z = \{(0, y)\} \rightarrow G_p = \{(0, y')\}. \tag{3.29}$$

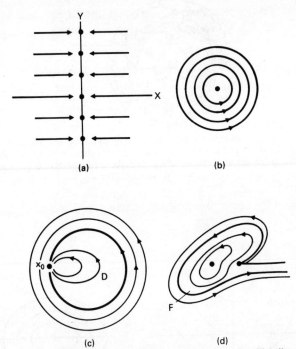

FIG. 17. (a) An equilibrium system that is steerable from outside all over the plane. (b)–(d) Perodically pulsating systems that are steerable from outside either all over the plane (in Picture b), or outside the closed circle D (in Picture c), or inside the unbounded domain F (in Picture d), respectively.

Type 2 is represented by Fig. 17b, where the system is steerable from outside all over the plane, by Fig. 17c, where the *domain of steerability* consists of the boundary and exterior of the unit circle, and by Fig. 17d, where it is the interior of the region F.

Type 3 has quasi-closed curves as trajectories in its (bounded) domain of steerability. Often in such systems the infinite trajectories with irregular ('chaotic') motions of state are densely interweaved with finite, closed trajectories along which the motion of state is periodical (Schuster, 1984). Such a situation is schematically illustrated in Fig. 18.

Type 4 is illustrated by the parabolas shown in Fig. 19a. A system of this type can have either trajectories that are receding to infinity (Fig. 19a) or trajectories that are not (Fig. 2b).

Mathematical Examples (Hamiltonian Systems).   A large class of dynamical systems that are steerable from outside are conservative Hamiltonian systems (see Section 1.1). Such a system is either disintegrating or steerable from outside in the region of state-space, where the Hamiltonian function $H(x)$ is continuous. The equation

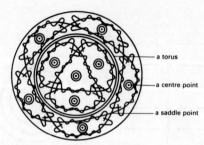

FIG. 18. Trajectories of a system that is steerable from outside, and pulsates periodically in some parts of state-space (e.g. near the centre points), and non-periodically (chaotically) elsewhere (e.g. near the saddle points), in an illustration given by Schuster (1984).

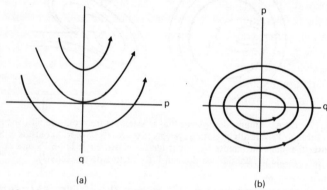

(a)   (b)

FIG. 19   (a) A system with unbounded parabola trajectories that is disintegrating, produced, for instance, by a freely falling body thrown upwards. (b) A periodically pulsating system steerable from outside, produced by a harmonc oscillator.

$$H(x) = c, \text{ with } c \in R, \tag{3.30}$$

defines an invariant curve, an invariant surface, or an invariant supersurface in the state-space, depending on whether this space has two, three, or more dimensions. Let us call them the 'shells' in the state-space. If x is the initial state of the system at the moment $t = 0$, we can always choose a perturbed state p so that it is not on the same shell: $H(p) \neq H(x)$. If $\rho(pt, xR^+) \to 0$ with $t \to +\infty$ were true, the continuity of the function H would imply that $H(pt) \to H(x)$, which is impossible, since $H(pt) = H(p) \neq H(x)$. It follows, by Definition 3, Section 2.3, that our system is either disintegrating or steerable from outside everywhere where H is continuous.

Consider, for instance, the system of a freely falling body thrown upward (Example 1, Section 1.1). Equation (3.30) then takes the form

$$p^2/2m + mgq = c. \tag{3.31}$$

For different values of the parameter c this gives the family of parabolas like that shown in Fig. 19a.

Or take a 'one-dimensional harmonic oscillator' defined by the equation

$$d^2q/dt^2 = -(k/m)q^2. \tag{3.32}$$

Here m is the mass of the oscillating body, and k is a positive constant. Equation (3.30) now has the shape

$$p^2/2m + \tfrac{1}{2}kq^2 = c, \tag{3.33}$$

which defines in the state-space E^2 an ellipse having the main axes

$$a = (2c/k)^{\tfrac{1}{2}} \quad \text{and} \quad b = (2mc)^{\tfrac{1}{2}}. \tag{3.34}$$

The pattern of trajectories is illustrated in Fig. 19b. The harmonic oscillator thus is a periodically pulsating system that is steerable from outside.

Or take the Keplerian motion of a planet round the sun, which also defines a conservative system and thus a system that is steerable from outside. The state-space is four-dimensional, and we cannot illustrate the trajectories visually. But we know that the nature of the system depends only on whether v_0^2 is smaller than, equal to, or larger than $2\mu M/r_0$, where v_0 is the velocity and r_0 the distance from the sun of the planet (cf. Eq. (1.15)). Thus, the trajectories are closed curves, and we have a planetary motion, in the shaded area of Fig. 20. This is the area, where we have a periodically pulsating system that is steerable from outside. On the boundary $v_0^2 = 2\mu M/r_0$ and above it the system is disintegrating and now its trajectories are unbounded and recede to infinity.

Real-world Examples (A Flying Ball, Frictionless Oscillator, Robot). If you throw a ball, the height and distance reached, and the path through the

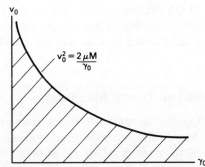

FIG. 20.   The (shaded) area that gives the Keplerian motion of a planet round the sun, and produces a periodically pulsating system that is steerable from outside (for the variables see the text).

air it makes all depend on the initial conditions, i.e. the place from where the ball has been thrown and the velocity you give it, including the initial direction of the ball. If the resistance of the air is negligible, the system of the flying ball is entirely steerable from outside, i.e. depending only on the initial condition you determine.

The frictionless oscillation for instance of a pendulum is only approximatively true in this imperfect world. But often it is true enough to justify such a pendulum being considered as a system that is steerable from outside: every push you give it sets the system on a new trajectory that satisfies the conditions of Definition 3, Section 2.3. Such a pendulum, of course, approximates a periodically pulsating system that is steerable from outside.

Even planetary motion is an example of the real-world systems that are steerable from outside – by a mighty giant who is able to push a planet to another trajectory. If he gives it a light push the perturbed motion will be an ellipse in the physical space. If the push is strong enough he can set the planet on to a perturbed path that is either a parabola or hyperbola, and cause the planet to leave our solar system. But in each case the planetary system is steerable from outside, and remains so even after the push.

A far more interesting real-world example, however, is a robot that has been programmed to do something, and is entirely governable from outside. A modern advanced industrial robot performs quite a sequence of various acts in a prescribed order, after which it repeats them as many times as the 'on-button' is pressed, or for as long a time as the energy supply works. The state-trajectory of such a robot is a closed curve, and the system is a periodically pulsating system steerable from outside.

Also, the construction of robots of other kinds is at least thinkable. In principle, one could make robots corresponding to each of the types of systems that are steerable from outside. 'Robots', of course, is also a metaphor used to describe people who seem to be doing things without thinking about their actions. The theory given here indeed gives support to the image by suggesting that even a self-steering being, such as man, may be similar to a robot, if the domain of self-steering shrinks. But we shall postpone a closer inspection of the sinister possibility of steering people from outside to Part 5, where the source of self-steering in man, and the possible obstacles to it, will be discussed.

3.9   Finite Systems and Turing Machines

The classification in Fig. 8 of dynamical systems having causal recursion is an exhaustive one. Thus, it also includes finite systems, i.e. discontinuous systems that have a finite number of states. Such a system may have a nilpotent causal recursion, in which case what was said in Section 2.2 applies, with trivial modifications. Let us discuss briefly, by way of an example, those finite systems with a full causal recursion.

Our system consists of two causal elements, of the kind discussed at the end of Section 2.1. However, we shall now assume the following finite sets of input and output sets:

$$X_1 = \{0, a, b, c, d\}, \quad Y_1 = \{0, a, b\}, \tag{3.35}$$

$$X_2 = \{0, a, b\}, \quad\quad Y_2 = \{0, c, d\}, \tag{3.36}$$

the causal functions of the two elements being defined by

$$T_1(0) = 0, \quad T_1(a) = b, \quad T_1(b) = a, \quad T_1(c) = a, \quad T_1(d) = b, \text{ and} \tag{3.37}$$

$$T_2(0) = 0, \quad T_2(a) = d, \quad T_2(b) = c, \tag{3.38}$$

respectively. The input or output 0 marks the event that 'nothing happens'.
Let the coupling function be given simply by

$$C_1(y_2) = y_2, \quad C_2(y_1) = y_1. \tag{3.39}$$

Thus, both of the elements feed their outputs immediately back to the input of the other element. This gives the following causal recursion:

$$x_1(t + 1) = C_1(y_2(t + 1)) = y_2(t + 1) = T_2(x_2 t), \tag{3.40}$$

$$x_2(t + 1) = C_2(y_1(t + 1)) = y_1(t + 1) = T_1(x_1 t). \tag{3.41}$$

It follows from (3.35)–(3.36) that our system has 15 total states $x = (x_1, x_2)$. By application of the causal recursion (3.40)–(3.41) we can see that the state-space $X = X_1 \times X_2$ is divided to three parts (here + means the set-theoretical sum):

$$X_1 = D_0 + D_1 + D_2. \tag{3.42}$$

D_0 consists of a single isolated rest point, D_1 is the domain of self-regulation (or stability) belonging to a limit cycle LC_1, and D_2 is the domain of self-regulation belonging to another limit cycle LC_2 as follows:

$$D_0 = \{(0,0)\}, \tag{3.43}$$

$$D_1 = \{LC_1, (a, 0), (b, 0)\}, \tag{3.44}$$

$$D_2 = \{LC_2, (a, a), (a, b), (b, a), (b, b)\}. \tag{3.45}$$

The limit cycles are

$$LC_1 = \{(0, a) \to (d, 0) \to (0, b) \to (c, 0) \to (0, a)\} \tag{3.46}$$

$$LC_2 = \{(c, a) \to (d, a) \to (d, b) \to (c, b) \to (c, a)\}, \tag{3.47}$$

respectively. From the other states of the domains of stability D_1 and D_2 the respective limit cycles are reached by just one step of the causal process, in the following way:

$$LC_1: (a, 0) \to (0, b), \quad (b, 0) \to (0, a), \tag{3.48}$$

$$LC_2:(a,a)\to(d,b),\quad (a,b)\to(c,b),$$

$$(b,a)\to(d,a),\quad (b,b)\to(c,a). \tag{3.49}$$

In whichever of its 15 total states the system may be, an external 'disturbance' may throw it to another state but, as we see by (3.43)–(3.49), the system will end up either being in its rest state $(0,0)$, or doing one of its two limit cycles, LC_1 or LC_2. Thus, applying, with trivial modifications, the definitions 1–4 of the categories of systems with causal recursion given in Section 2.3, we find out that our finite system can be either

(1) steerable from outside, or
(2) a self-regulating system.

In the state $(0,0)$ our system is always steerable from outside, as only two types of external disturbances are possible, viz.:

$$e_{01}:D_0\to D_1 \quad \text{and} \quad e_{02}:D_0\to D_2, \tag{3.50}$$

and both of them throw the system permanently out of its unperturbed state $(0,0)$, either in the domain D_1 or in the domain D_2, where the perturbed trajectory certainly keeps within a finite distance from the unperturbed state $(0,0)$.

But even for all the other external disturbances of the type that throw the system from one of the domains of stability into another, viz.:

$$e_{10}:D_1\to D_0,\quad e_{20}:D_2\to D_0, \tag{3.51}$$

$$e_{12}:D_1\to D_2,\quad e_{21}:D_2\to D_1, \tag{3.52}$$

we must consider the system, in view of the definitions of Section 2.3, as being steerable from outside.

The only cases where our system, in view of the definitions of Section 2.3, is self-regulating, are those where an external disturbance only changes the state of the system within the domain D_1 or within the domain D_2:

$$e_{11}:D_1\to D_1 \quad \text{or} \quad e_{22}:D_2\to D_2. \tag{3.53}$$

An obvious alternative to the application of the local definitions of Section 2.3 would be to apply here a global definition, by stating that any dynamical system whose state-space consists of some domains of stability only is *polystable*. Obviously, such a global definition can be applied to continuous or reducibly discontinuous systems as well (even though Ashby, 1972, who introduced the term, was discussing only finite systems).

A finite system cannot be self-steering, and it can be neither a disintegrating nor a non-periodically pulsating self-regulating system, because all these system categories are possible only in a state-space containing an infinite number of states.

Note on the Turing machine. As was noted at the end of Section 2.2, a computer, and of course the Turing machine (the ideal computer), can be programmed so as to have a nilpotent causal recursion. But the Turing machine can also be programmed to have a full causal recursion. In fact, the Turing machine is a discontinuous system with an infinite, but enumerable set of total states. Can it be programmed so as to produce, as necessary, any of the types of causal systems having causal recursion?

As far as the measurable systems go (see Part 2), the answer seems to be in the affirmative. This covers all the cases where causal recursion can *ex ante* be given an explicit mathematical expression.

But for the non-measurable systems (Part 3), i.e. for self-steering actors that are able to generate their causal recursion step by step, so that causal recursion can be known only *ex post*, the answer is negative: the unpredictable, indeterministic behaviour of self-steering actors can never be programmed *beforehand* in a computer, not even in the ideal computer, the Turing machine.

Note on 'closed' and 'open' systems. Every dynamical system is closed in a complete state-description. Thus the difference between open and closed systems is not an inherent property of the system but a convention concerning the state-description.

CHAPTER 4

The Modes of Asymptotic
Approach of Goals

Note: The dynamical systems spoken of in this chapter are all, if not stated otherwise, goal-directed systems, either self-regulating, self-steering, or steerable from outside, and either continuous or reducibly discontinuous.

We shall start by applying 'Lyapounov's first method', and consider the linear representation of the causal recursion $x(t + 1) = \varphi(xt)$ near an equilibrium point x_0:

$$x(t + 1) - x_0 = (\partial\varphi/\partial x)_{x_0}(xt - x_0). \tag{4.1}$$

Later, we shall point out the limits of applicability of this approximation when encountering those limits.

Throughout this chapter we shall use the notation

$$(\partial\varphi/\partial x)_{x_0} = A. \tag{4.2}$$

4.1. Causal Suction: The Ideal Type

Let us first study the causal recursion $z(t + 1) = Azt$ or, the same thing expressed another way,

$$zt = A^t z, \quad t = 0, 1, 2, \ldots, \tag{4.3}$$

on a Euclidean plane E^2 under the assumption that both eigenvalues ρ_1 and ρ_2 of A are single and obey

$$0 < \rho_1 < 1, \quad 0 < \rho_2 < 1. \tag{4.4}$$

Then the corresponding unit eigenvectors v_1 and v_2 span the state-space, and any initial state can be expressed as their linear combination:

$$z = c_1 v_1 + c_2 v_2, \tag{4.5}$$

where c_1 and c_2 are real numbers. By a successive application of the eigenvalue equations $Av_i = \rho_i v_i$ (for $i = 1, 2$) to (4.5) we get

$$A^t z = c_1 \rho_1^t v_1 + c_2 \rho_i^2 v_2, \quad t = 0, 1, 2, \ldots \tag{4.6}$$

By introducing on the state-plane E^2 the orthogonal co-ordinates (x, y) so that v_1 lies on the x-axis, we have

$$z = \begin{pmatrix} x \\ y \end{pmatrix}, \quad v_1 = \begin{pmatrix} 1 \\ 0 \end{pmatrix}, \quad v_2 = \begin{pmatrix} \cos \beta \\ \sin \beta \end{pmatrix}, \tag{4.7}$$

where β is the angle between the eigenvectors v_1 and v_2. In the orthogonal co-ordinates Eq. (4.3), with the substitution (4.6), takes the form

$$xt = c_1 \rho_1^t + c_2 \rho_2^t \cos \beta, \quad yt = c_2 \rho_2^t \sin \beta. \tag{4.8}$$

For the eigenvalues obeying (4.4) we have $xt \to 0$ and $yt \to 0$ with $t \to +\infty$, so that the point zt approaches asymptotically the origin with $t \to +\infty$. The mode of approach defined by (4.8) and illustrated in Fig. 21a is the *ideal form of causal suction*.

By derivation we first get from (4.8):

$$dx/dt = c_1 \rho_1^t \log \rho_1 + c_2 \rho_2^t \cos \beta \log \rho_2,$$
$$dy/dt = c_2 \rho_2^t \sin \beta \log \rho_2. \tag{4.9}$$

Here we can eliminate the expressions $c_1 \rho_1^t$ and $c_2 \rho_2^t$ by means of Eq. (4.8), which gives:

$$c_1 \rho_1^t = xt - yt - \cot \beta, \quad c_2 \rho_2^t = yt/\sin \beta. \tag{4.10}$$

Substituting this into Eq. (4.9) gives the differential equations

$$dx/dt = x \log \rho_1 + y \cot \beta \log \rho_2/\rho_1, \quad dy/dt = y \log \rho_2 \tag{4.11}$$

or, in a more concise form:

$$dz/dt = Bz, \quad \text{with } B = \begin{pmatrix} \log \rho_1 & \cot \beta \log \rho_2/\rho_1 \\ 0 & \log \rho_2 \end{pmatrix}. \tag{4.12}$$

The pattern of causal suction shown in Fig. 21a, accordingly, can also be generated by means of these differential equations.

Since causal suction is a continuous mode of asymptotic approach, the same equations (4.8) define both the continuous half-trajectories xR^+ (when read for $t \in R^+$) and their φ-representations xT^+ (when read for $t = 0, 1, 2, \ldots$). The form of these trajectories, illustrated in Fig. 21a, is not of course affected by orthogonal transformations of co-ordinates. Hence the pattern of causal suction, as shown by Fig. 21a, is the same for any orthogonal co-ordinates on the plane E^2, even though the form of the defining equations, (4.8) and (4.12), will change.

For $\rho_1 = \rho_2 = \rho$ Eq. (4.6) becomes

$$A^t z = \rho^t(c_1 v_1 + c_2 v_2) = \rho^t z \quad \text{for } t = 0, 1, 2, \ldots \tag{4.13}$$

This pattern of trajectories is generated by the causal recursion

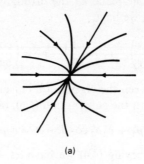

(a)

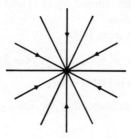

(b)

FIG. 21.   (a) The trajector field of causal suction around a tangential node. (b)
The star field around a stellar node.

$$z(t + 1) = \begin{pmatrix} \rho & 0 \\ 0 & \rho \end{pmatrix} \cdot zt \quad \text{for } t = 0, 1, 2,\ldots \tag{4.14}$$

and the corresponding continuous trajectories by the differential equation

$$dz/dt = \begin{pmatrix} \log \rho & 0 \\ 0 & \log \rho \end{pmatrix} z. \tag{4.15}$$

For $0 < \rho < 1$ the states zt approach the origin asymptotically along the rays starting from the origin (Fig. 21b). The emerging pattern of rectilinear trajectories is a special case of the ideal form of causal suction.

In the tradition of mathematical dynamics, an equilibrium point of the type shown in Fig. 21a is called a 'node', and that of Fig. 21b a 'star' or 'stellar node'.

4.2.   The Generalization of Causal Suction to *n* Dimensions and to Nonlinear Causal Recursions in an Equilibrium System

Let us now discuss causal suction in a general self-regulating equilibrium system:

Definition (Causal Suction Toward an Equilibrium Goal).   In a self-regulating equilibrium system, a family of positively asymptotic continuous half-trajectories xR^+ or their φ-representations xT^+ tending to an equilibrium goal state x_0 of the Euclidean state-space E is called *causal suction*, if the projection of each trajectory on every plane $S^2 \subset E$ through the point x_0 turns round this point by less than $180°$, or if the space E is one-dimensional.

Thus, a continuous mode of asymptotic approach of an equilibrium goal in one-dimensional state-space is always causal suction. For $n > 2$ dimensions it is advisable to discuss linear and nonlinear recursions separately.

Case 1 (Linear Causal Recursion).   For a linear causal recursion $\varphi(x) = Ax$ having an equilibrium point x_0, i.e. $Ax_0 = x_0$, we have $\varphi(z) = Az$ in terms of the variable $z = x - x_0$, the equilibrium point being so transferred to the origin:

$$\varphi(z) = z1 = x1 - x_0 = Ax - Ax_0 = Az, \quad z_0 = x_0 - x_0 = 0. \quad (4.16)$$

This gives as a generalization of Eq. (4.6),

$$zt = A^t z = c_1 \rho_1^t v_1 + c_2 \rho_2^t v_2 + \cdots + c_n \rho_n^t V_n \to 0 \text{ for } t \to +\infty, \quad (4.17)$$

when

$$0 < \rho_1 < 1, \quad 0 < \rho_2 < 1, \ldots, 0 < \rho_n < 1. \quad (4.18)$$

It follows from the form of Eq. (4.17), and from the results of Section 4.1 that, according to the above definition, the mode of asymptotic approach is again causal suction as soon the condition (4.18) is valid.

What happens if one or more of the eigenvalues ρ_j become zero? Obviously, the corresponding dimensions in the expression (4.17) get lost, while causal suction still obtains, if the remaining non-zero eigenvalues obey (4.18). When all eigenvalues are zero we have a nilpotent matrix A, and thus a nilpotent causal recursion. This case was discussed in Section 2.2.

The case of linear causal recursion with rectilinear causal suction in an n-dimensional Euclidean space is obtained from (4.17) on the assumption that

$$0 < \rho_1 = \rho_2 = \cdots = \rho_n = \rho < 1 \quad (4.19)$$

holds good. This gives

$$zt = \rho^t(c_1 v_1 + c_2 v_2 + \cdots + c_n v_n) = \rho^t z, \quad dz/dt = z \log \rho, \quad (4.20)$$

or, in terms of the A- and B-matrices,

$$z(t + 1) = Azt \quad \text{with } A = \begin{pmatrix} \rho & 0 & \cdots & 0 \\ 0 & \rho & \cdots & \\ \cdots & \cdots & \cdots & \cdots \\ 0 & 0 & \cdots & \rho \end{pmatrix}, \quad (4.21)$$

FMSD—C

$$dz/dt = Bz \qquad \text{with } B = \begin{pmatrix} \log \rho & 0 & \ldots & 0 \\ 0 & \log \rho & \ldots & 0 \\ \ldots & \ldots & \ldots & \ldots \\ 0 & 0 & \ldots & \log \rho \end{pmatrix}. \qquad (4.22)$$

The trajectories are again rays starting from the equilibrium point (i.e. the origin $z = 0$), which is a star or stellar node like that in the two-dimensional picture of Fig. 21c.

Case 2 (Nonlinear Causal Recursion).   For a nonlinear causal recursion $\varphi(x)$ we can write, assuming the analyticity at the point of equilibrium x_0,

$$\varphi(x) = \varphi(x_0) + (\partial\varphi/\partial x)_{x_0}(x - x_0) + \tfrac{1}{2}(\partial^2\varphi/\partial x^2)_{x_0}(x - x_0)(x - x_0)^{\mathsf{T}}$$

$$+ \text{ higher terms of the Taylor series.} \qquad (4.23)$$

Since $\varphi(x_0) = x_0$ this gives, using the variable $z = x - x_0$,

$$\varphi(z) = (\partial\varphi/\partial x)_{x_0}z + \tfrac{1}{2}(\partial^2\varphi/\partial x^2)_{x_0}zz^{\mathsf{T}} + \cdots \qquad (4.24)$$

Here the symbol T indicates transpose.

For the corresponding continuous flow we have the differential equation

$$dx/dt = f(x) = f(x_0) + (\partial f/\partial x)_{x_0}(x - x_0)$$

$$+ \tfrac{1}{2}(\partial^2 f/\partial x^2)_{x_0}(x - x_0)(x - x_0)^{T}$$

$$+ \text{ higher terms in the Taylor series.} \qquad (4.25)$$

Or, written for the variable $z = x - x_0$, in view of $f(x_0) = 0$;

$$dz/dt = (\partial f/\partial x)_{x_0}z + \tfrac{1}{2}(\partial^2 f/\partial x^2)_{x_0}zz^{T} + \cdots \qquad (4.26)$$

All this provided that the function $f(x)$, i.e. dx/dt as a function of x, is analytic at the point x_0.

Thus, the linear equations

$$z(t + 1) = Azt \quad \text{and} \quad dz/dt = Bz \qquad (4.27)$$

are now the first-order approximations based on the assumptions that

$$\varphi(z) = (\partial\varphi/\partial x)_{x_0}z, \quad (\partial\varphi/\partial x)_{x_0} = A \neq 0, \qquad (4.28)$$

$$f(z) = (\partial f/\partial x)_{x_0}z, \quad (\partial f/\partial x)_{x_0} = B \neq 0, \qquad (4.29)$$

respectively.

The 'Lyapounov approximation' (4.28)–(4.29) worked well in Section 4.1, and gave us the patterns of trajectories of causal suction both in their curved and rectilinear ideal forms. For a nonlinear causal recursion, however, these ideal types are valid approximations only in a small enough neighbourhood of the equilibrium point x_0. We shall soon see, in a numerical example, how the causal suction generated by a nonlinear causal recursion on a Euclidean plane E^2 gradually disappears, giving way to another kind of pattern, when we recede from the equilibrium point (see Example 1 below).

But in the nearest neighbourhood of the equilibrium goal, the Lyapounov approximation works even in the general case of an n-dimensional Euclidean space, provided that both the matrices A and B in Eq. (4.27) are different from zero. Thus, on this condition the results reported in Case 1 above are valid for a nonlinear causal recursion as well, in a small enough neighbourhood of the point x_0.

What happens with the Lyapounov approximation when either $A \to 0$ in Eq. (4.28) or $B \to 0$ in Eq. (4.29)? A case in point is obtained when the n-fold eigenvalue ρ of the n-square matrix A in Eq. (4.21) becomes either zero or one: it follows from (4.21) and (4.22) that

$$\rho \to 0 \Rightarrow (\partial\varphi/\partial x)_{x_0} = A \to 0 \quad \text{and} \quad (\partial f/\partial x)_{x_0} = B \to -\infty, \qquad (4.30)$$

$$\rho \to 1 \Rightarrow (\partial f/\partial x)_{x_0} = B \to 0 \quad \text{and} \quad (\partial\varphi/\partial x)_{x_0} = A \to I. \qquad (4.31)$$

In both cases the linear equations (4.27) based on the Lyapounov approximation (4.28)–(4.29) fail to give the pattern of asymptotic trajectories tending to x_0 even in the closest neighbourhood of x_0. Thus, we must consider *both* of these cases as *not locally linearizable*.

Even in the case of not locally linearizable causal recursion, however, the fundamental modes of asymptotic approach such as causal suction (and the other ones to be met later) apply, but now only to a certain limited sector of state-space near x_0. Such an equilibrium point has sometimes been called *multiple singular point* (e.g. Andronov, Vit, and Khaikin, 1966; Nicolis and Prigogine, 1977). An example will be shown below (Example 2).

As a matter of fact, we have met multiple singular points before, in connection with the topological analysis of equilibrium goals, in Section 3.2. Such a point is nothing but a *boundary equilibrium point*, i.e. an equilibrium point that is located on the boundary of its own region of attraction. Such a point, of course, receives positively asymptotic trajectories only from the region of attraction, i.e. from a certain limited sector of state-space. The trajectories outside this region pass by the point x_0. A topological example of a multiple singular point has already been given in Fig. 9b.

A 'simple singular point', where the Lyapounov approximation is valid, is, in a self-regulating equilibrium system, either a saddle point or an *interior equilibrium point*, situated in the interior of the domain of self-regulation. Thus, we have the following theorems. (For the concept of saddle point see p.77.)

Theorem 4.2.1. In a self-regulating equilibrium system, the linear Lyapounov approximation of causal recursion, (4.27)–(4.29), is valid in a small enough neighbourhood of an interior equilibrium point x_0, and the positively asymptotic trajectories tending to x_0 in this case fill the whole neighbourhood of x_0.

Theorem 4.2.2. In a self-regulating equilibrium system, every boundary equilibrium point x_0, which is a goal state and is not a saddle point, is a multiple singular point, at which causal recursion is not locally linearizable, and which receives a family of positively asymptotic trajectories only from a limited sector of its neighbourhood.

Example 1 (Effects of Nonlinearity: A Locally Linearizable Case). Let us consider the nonlinear causal recursion defined on a Euclidean plane E^2 by

$$z(t + 1) = \varphi(zt), \quad t = 0, 1, 2, \ldots; \quad z = (x, y) \in E^2; \tag{4.32}$$

$$\varphi_x(z) = a + (1 - b)x - xy^2, \quad \varphi_y(z) = bx + xy^2. \tag{4.33}$$

Here x and y are the orthogonal co-ordinates of the point z, and a and b are real constants. The condition $\varphi(z_0) = z_0$ gives the equilibrium state

$$x_0 = a/(a^2 + b), \quad y_0 = a. \tag{4.34}$$

We choose the numerical values

$$a = 0.1, \quad b = 0.49, \quad \text{thus } x_0 = 0.2, \quad y_0 = 0.1. \tag{4.35}$$

This gives

$$(\partial\varphi/\partial z)_{z_0} = A = \begin{pmatrix} 0.5 & -0.04 \\ 0.5 & 0.04 \end{pmatrix}. \tag{4.36}$$

Thus, our causal recursion is locally linearizable. The eigenvalues and the respective eigenvectors of A are

$$\rho_1 = 0.4514, \quad \rho_2 = 0.0886, \quad v_1 = (v_{1x}, v_{1y}), \quad v_2 = (v_{2x}, v_{2y}), \tag{4.37}$$

where

$$v_{1y}/v_{1x} = 1.215 \quad \text{and} \quad v_{2y}/v_{2x} = 10.285. \tag{4.38}$$

Figure 22a shows the ideal form of causal suction obtained from the linear approximation $\varphi(z) = Az$. The eigenvector directions are indicated, and illustrate the fact that the eigenvector belonging to the largest eigenvalue, here v_1, is the tangent of trajectories at the equilibrium point z_0.

In Fig. 22b the nonlinear equations (4.33) have been used, with the substitutions (4.35), to compute the numerical co-ordinates of successive points in five trajectories near the equilibrium point $(0.2, 0, 1)$. Their pattern still exhibits causal suction in the sense of the general definition given above. But it reduces to the ideal type only in the closest neighbourhood of x_0, say in the circle $S(z_0, 0.03)$ depicted in the diagram by a dashed line.

We can also see that the trajectories of Fig. 22b tend to oscillate farther from z_0. If we were to extend the picture to larger values of x and y, we would indeed observe their oscillations round z_0. This too is a lawful phenomenon: our point $(a, b) = (0.1, 0.49)$ is near the boundary of the region

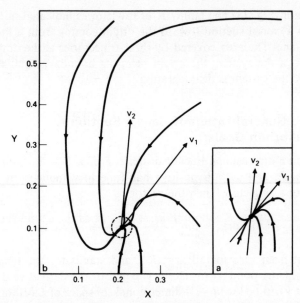

FIG. 22. (a) The ideal form of causal suction obtained close enough (roughly in the dashed-line circle of Picture b) to the equilibrium point for the nonlinear causal recursion (4.33) with the numerical values (4.35). (b) The effects of nonlinearity as observable farther from the equilibrium point in the same nonlinear system. (The directions of eigenvectors are indicated in both pictures.)

of causal suction in the parameter space (see Chapter 5), and the trajectories farther from z_0 tend to be affected by causal vortex (see Section 4.4 below), whose region begins beyond that boundary. We shall make a closer inspection of the parameter space in Chapter 5.

Example 2 (Effects of Nonlinearity: A Not Locally Linearizable Case). Historically, the first mathematical analysis of a multiple singular point concerned the trajectories generated by the differential equations

$$dx/dt = -x^6, \quad dy/dt = y^3 - x^4 y \qquad (4.39)$$

on the Euclidean plane E^2 (Frommer, 1928). The equilibrium state $z_0 = (x_0, y_0)$ is the origin $O = (0, 0)$, and for the function

$$f(z): \quad f_x = -x^6, \quad f_y = y^3 - x^4 y \qquad (4.40)$$

we get

$$(\partial f/\partial z)_{z_0} = 0, \quad (\partial^2 f/\partial z^2)_{z_0} = 0. \qquad (4.41)$$

Thus, the causal recursion of the continuous flow generated by (4.39) is not locally linearizable at the origin. The curves

$$dy/dx = (y^3 - x^4 y)/(-x^6) \qquad (4.42)$$

are shown in Fig. 23. The family K of trajectories indicated in the diagram makes up a causal suction toward the origin coming from a limited sector of state-plane. The area covered by these trajectories is the domain of self-regulation of the system, the origin being its boundary equilibrium point. Outside K the system is disintegrating.

4.3.   The Generalization of Causal Suction to Nonequilibrium Goals

Let E be an n-dimensional Euclidean space, $F : E \times R \to E$ a continuous flow defined on E, and xR^+ a positive half-trajectory induced by F and not reducing to a singleton. Then the vector

$$(\mathrm{d}F/\mathrm{d}t)_{xt} = \lim_{a \to 0} \frac{F(x, t + a) - F(x, t)}{a}, \quad a > 0 \tag{4.43}$$

is different from zero for all $xt \in xR^+$, and stands for the tangent of this trajectory at the point xt (in fact it is equal to the derivative $\mathrm{d}x/\mathrm{d}t$ at this point). Let $V_{\mathrm{d}F}(t)$ be the $(n - 1)$-dimensional subspace of E orthogonal to the vector $(\mathrm{d}F/\mathrm{d}t)_{xt}$, and intersecting the trajectory xR^+ at the point xt. Thus,

$$xt \in V_{\mathrm{d}F}(t) \quad \text{for all } t \in R^+. \tag{4.44}$$

Definition (Causal Suction Toward a Nonequilibrium Trajectory). In a Euclidean space E, a family of continuous half-trajectories yR^+, or their

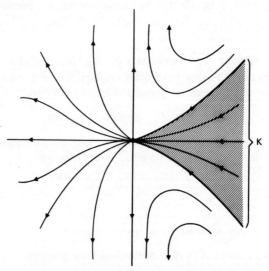

FIG. 23.   Frommer's case: the effect of nonlinearity on trajectories near a multiple singular point. Causal suction is here limited to the shaded region K.

φ-representations yT^+, that approach asymptotically the half-trajectory xR^+ (or, respectively, xT^+), which is either a closed curve or of infinite length, defines causal suction towards the trajectory xR^+ (or xT^+), if each point $z_St \in yR^+ \cap S^2(t)$ on each of the planes $S^2(t) \subset V_{dF}(t)$ through the point xt turns round the point xt, with $t \to +\infty$, by less than $180°$, or if the space E is two-dimensional.

On a plane E^2, causal suction towards a limit cycle (Fig. 24a) or a self-steering trajectory (Fig. 24b) is accordingly an obvious generalization from causal suction towards an equilibrium state. But so too are causal suctions toward non-equilibrium trajectories in a more than two-dimensional state-space. Since only the three-dimensional case is visualizable, and it does not essentially differ from the two-dimensional one, more pictorial illustrations of causal suctions to nonequilibrium trajectories are unnecessary. The target trajectory xR^+ may be either a self-steering one, a closed curve, or a quasi-closed curve.

4.4. Causal Torsion

Let us now study the linear causal recursion

$$z(t + 1) = Azt, \quad \text{i.e.} \ zt = A^tz, \quad t = 0, 1, 2, \ldots, \quad (4.45)$$

under the assumption of an n-fold real eigenvalue ρ of the n-square matrix A. Such a matrix can always, by means of an orthogonal rotation, be transformed into a triangular matrix having the eigenvalue ρ as its main diagonal elements (see e.g. Ayres, 1974). Since the form of the trajectories is invariant in such rotations, we can assume that the rotation has been already made, and confine ourselves to the study of such triangular matrices.

Case n = 2. We can start with the 2×2 matrix

$$A = \begin{pmatrix} \rho & 0 \\ \alpha & \rho \end{pmatrix}, \quad \alpha \neq 0, \quad (4.46)$$

(a)   (b)

Fig. 24.   Causal-suction approach. (a) to a limit cycle, and (b) to a self-steering trajectory.

since we have already discussed the diagonal matrix A obtained when $\alpha = 0$ (see Eq. (4.21), Section 4.2). Choosing, for convenience, the orthogonal base

$$u = \begin{pmatrix} 1 \\ 0 \end{pmatrix}, \quad v = \begin{pmatrix} 0 \\ 1 \end{pmatrix}, \tag{4.47a}$$

we can express any initial state as

$$z = c_1 u + c_2 v, \quad c_1 \in R, \quad c_2 \in R. \tag{4.47b}$$

We get, successively:

$$Az = c_1(\rho u + \alpha v) + c_2 \rho v, \quad A^2 z = c_1(\rho^2 u + 2\alpha \rho v) + c_2 \rho^2 v,$$
$$A^3 z = c_1(\rho^3 u + 3\alpha \rho^2 v) + c_2 \rho^3 v, \ldots$$
$$A^t z = c_1(\rho^t u + t\alpha \rho^{t-1} v) + c_2 \rho^t v = [c_1(u + \alpha t v/\rho) + c_2 v]\rho^t. \tag{4.48}$$

Or, if written for the orthogonal co-ordinates of the point $zt = (\xi t, \eta t)$:

$$\xi t = c_1 \rho^t, \quad \eta t = (c_1 \alpha t/\rho + c_2)\rho^t. \tag{4.49}$$

The trajectories (4.49) asymptotically approach the origin, provided that $0 < \rho < 1$, in the way illustrated in Fig. 25a. We can see that they turn round the origin by exactly 180°, and therefore stand for a new continuous mode of asymptotic approach. It can be called 'causal torsion'.

By derivation of Eq. (4.49) we find the differential equations that generate these torsion trajectories:

$$dz/dt = Bz, \quad \text{with } B = \begin{pmatrix} \log \rho & 0 \\ \alpha/\rho & \log \rho \end{pmatrix}. \tag{4.50}$$

Case $n = 3$. Now we start with the 3×3 matrix

$$A = \begin{pmatrix} \rho & 0 & 0 \\ \alpha & \rho & 0 \\ \beta & \gamma & \rho \end{pmatrix}, |\alpha| + |\beta| + |\gamma| \neq 0. \tag{4.51}$$

Introducing the base

$$s = \begin{pmatrix} 1 \\ 0 \\ 0 \end{pmatrix}, \quad u = \begin{pmatrix} 0 \\ 1 \\ 0 \end{pmatrix}, \quad v = \begin{pmatrix} 0 \\ 0 \\ 1 \end{pmatrix}, \tag{4.52}$$

we can again express the initial state as a linear combination of the base vectors,

$$z = c_1 s + c_2 u + c_3 v. \tag{4.53}$$

By repeated application of A we then get, for the components of the point $zt = (\xi t, \eta t, \zeta t)$:

$$\xi t = c_1 \rho^t, \quad \eta t = (c_1 \alpha t/\rho + c_2)\rho^t,$$

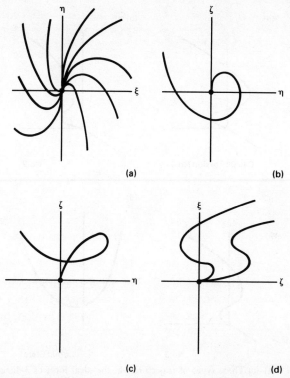

FIG. 25. The ideal form of causal torsion in a 3-dimensional state-space as illustrated by the projections of trajectories on the co-ordinate planes. (b) and (c) show alternative projections (for different values of parameters). (a) also gives the ideal form of 2-dimensional causal torsion.

$$\zeta t = \left\{ c_1 \frac{\alpha\gamma}{2\rho^2} t^2 + \left[c_1 \frac{2\beta\rho - \alpha\gamma}{2\rho^2} + c_2 \frac{\gamma}{\rho} \right] t + c_3 \right\} \rho^t \qquad (4.54)$$

Again the trajectories asymptotically approach the origin when $0 < \rho < 1$.

The projections of a trajectory (4.54) on the coordinate planes are shown in Figs 25a–25d. One of them, viz. that of Fig. 25b, turns round the origin by 360°, the others less than that. Figures 25b and 25c show two alternatives (for different values of parameters). Three types of the three-dimensional curves (4.54) are illustrated in Figs 26a–26c.

By comparing Eqs (4.49) and (4.54) with one another we find the general rule: with increasing dimensionality a new polynomial of t of a higher degree by one is added each time, so that in the nth dimension of state-space we have a polynomial of t, $P_{n-1}(t)$, of $(n-1)$th degree, the general solution of (4.45) being of the form

$$z_i t = P_{i-1}(t)\rho^t; \quad i = 1, 2, \ldots, n; \quad t = 0, 1, 2, \ldots \qquad (4.55)$$

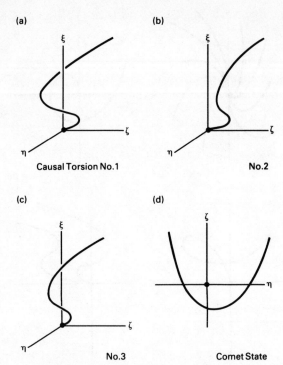

FIG. 26. (a)–(c) Three types of trajectories in the ideal form of 3-dimensional causal torsion, and (d) a comet-state trajectory obtained from the 3-dimensional torsion when the 3-fold eigenvalue becomes equal to 1.

Hence, in an n-dimensional space E the projection of a torsion trajectory on the co-ordinate plane (z_1, z_n) intersects the z_n-axis $n - 1$ times, the projection of the plane (z_1, z_{n-1}) intersects the z_{n-1}-axis $n - 2$ times, etc. This observation concerning the linear 'torsion' recursions suggests the following general definitions:

Definition (Causal Torsion Toward an Equilibrium State). A family of positively asymptotic continuous half-trajectories xR^+, or their φ-representations xT^+, tending to an equilibrium point x_0 in a Euclidean space E is called *causal torsion*, if the projection of each trajectory on every plane $S^2 \subset E$ through x_0 turns round x_0 a finite number of times, and at least by 180° on at least one of them.

An Extension (Causal Torsion Towards a Nonequilibrium Trajectory). We extend the definition of causal torsion to the mode of asymptotic approach of a nonequilibrium trajectory in the same way that causal suction was extended to that case, with the sole alteration that now the point xt referred to in the definition of causal suction (Section 4.3) has to be turned round a finite number of times, and at least by $180°$ on at least one of the planes.

The theorems 4.2.1 and 4.2.2 are, of course, valid for causal torsion as well. Thus, in a neighbourhood of a boundary equilibrium point causal torsion too is restricted to a limited sector (see Fig. 27), while for an interior rest point it fills the whole neighbourhood (Fig. 25a).

4.5. Causal Vortex

Let us again discuss a linear causal recursion $z(t + 1) = Azt$, i.e. $zt = A^t z$, on a Euclidean plane E^2, but assuming now that the eigenvalues of the matrix A are complex numbers, say μ_1 and μ_2. Their respective eigenvectors w_1 and w_2 then are complex vectors, and we have

$$Aw_1 = \mu_1 w_1 \Rightarrow Aw_1^* = \mu_1^* w_1^*, \quad \text{thus } \mu_2 = \mu_1^* \text{ and } w_2 = w_1^*. \quad (4.56)$$

It follows that we can put

$$\mu_1 = \mu = \rho e^{i\omega}, \quad \mu_2 = \mu^* = \rho e^{-i\omega}, \quad w_1 = w, \quad w_2 = w^*. \quad (4.57)$$

If v and u are the real and imaginary parts of w, respectively, we have

$$w = v + iu, \quad v = \tfrac{1}{2}(w + w^*), \quad u = -\tfrac{1}{2}i(w - w^*). \quad (4.58)$$

By expressing the initial state z with the help of the base vectors v and u we get

$$z = c_1 v + c_2 u, \quad c_1 \in R, \quad c_2 \in R. \quad (4.59)$$

The equations (4.56)–(4.59) give, after a trivial calculation,

$$zt = A^t z = [c_1(v \cos \omega t - u \sin \omega t) + c_2(u \cos \omega t + v \sin \omega t)]\rho^t. \quad (4.60)$$

We can always choose the vector v to be of unit length, but then u will not necessarily be one. However, we have two unit vectors v and u_0, where

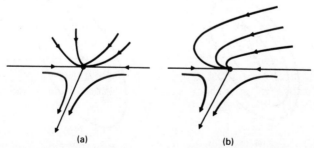

(a)   (b)

Fig. 27.   Two multiple singular points, or points of equilibrium situated on the boundary of their respective domains of stability (here: the upper half of the state-plane in both cases), with (a) causal suction and (b) causal torsion, respectively, both limited to the upper half-plane.

$$u = ku_0, \quad \text{with } k = (u_1^2 + u_2^2)^{\frac{1}{2}}. \tag{4.61}$$

Let us place the vector v along the x-axis of an orthogonal co-ordinate system (x, y) on the plane E^2 to get v and u expressed as

$$v = \begin{pmatrix} 1 \\ 0 \end{pmatrix}, \quad u = \begin{pmatrix} k \cos \beta \\ k \sin \beta \end{pmatrix}. \tag{4.62}$$

Here β is the angle between the vectors v and u.

The equations (4.60)–(4.62) now give the final result, viz. the equations of motion of the state $zt = (xt, yt)$ in a component form:

$$xt = [c_1(\cos \omega t - k \cos \beta \sin \omega t) + c_2(k \cos \beta \cos \omega t + \sin \omega t)]\rho^t,$$
$$yt = k \sin \beta(-c_1 \sin \omega t + c_2 \cos \omega t)\rho^t. \tag{4.63}$$

The trajectories (4.63) for $\rho < 1$ are spirals tending to the origin with $t \to +\infty$, of the type illustrated in Fig. 28a. They define on the plane E^2 what has been called 'causal vortex' towards an equilibrium state. The shape of the spirals is best seen by putting $\rho = 1$ in Eq. (4.63), in which case the trajectory becomes a closed curve, viz. an oval that is symmetric with respect to the origin, as shown in Fig. 28b.

To get the differential equations that generate the 'vortex' trajectories (4.63) it is best to start by derivation of Eq. (4.60), then make the substitution (4.62), and eliminate from the expressions of dx/dt and dy/dt the terms

$$\rho^t(c_1 \cos \omega t + c_2 \sin \omega t) = x - y \cot \beta, \quad \text{and}$$
$$\rho^t(c_2 \cos \omega t - c_1 \sin \omega t) = y/k \sin \beta. \tag{4.64}$$

The result is

$$dz/dt = Bz, \quad \text{with}$$

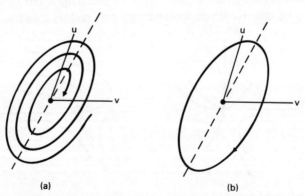

(a)   (b)

FIG. 28.   (a) The ideal form of causal vortex. (b) An oval-shaped closed trajectory obtained when the vortex system becomes steerable from outside. (The real vectors u and v connected with a complex eigenvalue of the recursion matrix are indicated in the pictures.)

$$B = \begin{pmatrix} \log \rho & 0 \\ 0 & \log \rho \end{pmatrix} + \begin{pmatrix} -\omega k \cos \beta & \dfrac{\omega}{k \sin \beta}(1 + k^2 \cos^2 \beta) \\ -\omega k \sin \beta & \omega k \cos \beta \end{pmatrix}. \qquad (4.65)$$

For $\beta = \pi/2$, $k = 1$, and $\rho = 1$ this gives a clockwise circular movement of a satellite state (cf. Fig. 29):

$$\mathrm{d}x/\mathrm{d}t = \omega y, \quad \mathrm{d}y/\mathrm{d}t = -\omega x. \qquad (4.66)$$

We can, without further examples, proceed to the general definitions:

Definition (Causal Vortex Toward, an Equilibrium State). A family of positively asymptotic continuous half-trajectories xR^+, or their φ-representations xT^+, tending to an equilibrium point x_0 in a Euclidean space E is called *causal vortex*, if the projection of each trajectory on at least one of the planes $S^2 \subset E$ through x_0 turn round the point x_0 an infinite number of times.

An Extension (Causal Vortex Towards a Nonequilibrium Trajectory). We immediately extend the definition of causal vortex to the mode of asymptotic approach of a nonequilibrium trajectory in the same way that causal suction and torsion were extended to that case, with the exception that now the point xt referred to in the definition of causal suction (Section 4.3) has to be turned round at least on one of the planes an infinite number of times.

Causal vortex in a three-dimensional state-space toward an equilibrium state is illustrated in Fig. 30a, towards a limit cycle in Fig. 30b, and towards a self-steering trajectory in Fig. 30c.

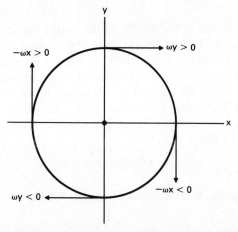

FIG. 29. The clockwise circular movement of satellite state as a special form of the degeneration of causal vortex into periodically pulsating states, when a self-regulating vortex system becomes steerable from outside.

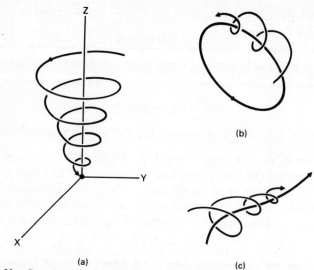

FIG. 30. Causal vortex in a 3-dimensional state-space; (a) toward an equilibrium point, (b) toward a limit cycle, and (c) toward a self-steering trajectory.

The theorems 4.2.1 and 4.2.2, of course, apply also to causal vortex, with the obvious limitation that there cannot be causal vortex toward a boundary equilibrium point on a two-dimensional state-space.

Example (Effect of Nonlinearity on Causal Vortex). For an example we can take the nonlinear causal recursion (4.32)–(4.33) that has already served us once. If we choose the numerical values

$$a = 2^{-\frac{1}{2}} = 0.707, \quad b = 1.5, \quad \text{thus } x_0 = 0.354, \quad y_0 = 0.707, \quad (4.67)$$

of parameters, we are in the middle of the vortex region in the parameter space (see Chapter 5). Here we get, with the initial state $x = y = 0.5$, the almost ideal form of causal vortex shown in Fig. 31. Moving in the parameter space closer to the boundary of the region of causal suction, we can see how the vortex trajectory, shown in Fig. 32, has also been affected by causal suction: farther from the equilibrium point ($x_0 = 0.632$, $y_0 = 0.316$) the oscillation is oblongated in the manner of a suction trajectory. This effect is obtained for the values

$$a = 0.316, \quad b = 0.4. \quad (4.68)$$

4.6. The Approach of a Goal Through Rapid Vibrations

When, in Eq. (4.8), we have

$$0 < \rho_1 < 1, \text{ but } -1 < \rho_2 < 0, \quad (4.69)$$

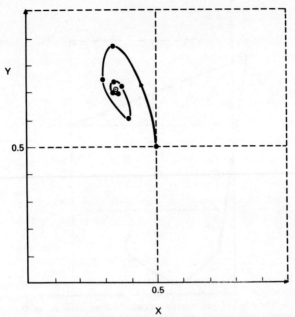

Fig. 31. The almost ideal form of causal vortex obtained for the nonlinear causal recursion (4.33) with the numerical values (4.67). These parameter values belong to the middle of the vortex region in the parameter space.

we get the trajectories

$$xt = c_1\rho_1^t + c_2(-1)^t|\rho_2|^t \cos \beta, \quad yt = c_2(-1)^t|\rho_2|^t \sin \beta. \qquad (4.70)$$

These trajectories asymptotically approach the origin through *transversal rapid vibrations* of period 2, of the type shown in Fig. 3.

If both eigenvalues are negative, i.e.

$$-1 < \rho_1 < 0 \quad \text{and} \quad -1 < \rho_2 < 0, \quad \rho_1 \neq \rho_2, \qquad (4.71)$$

we get the equations of motion

$$xt = c_1(-1)^t|\rho_1|^t + c_2(-1)^t|\rho_2|^t \cos \beta,$$
$$yt = c_2(-1)^t|\rho_2|^t \sin \beta. \qquad (4.72)$$

The trajectories now tend to the origin through *longitudinal rapid vibrations* over the origin, shown in Fig. 33.

In both the cases illustrated the vibrations occur between two trajectories of causal suction. We can have rapid vibrations between two torsion trajectories as well, but only of the longitudinal type, by the substitution of

$$\rho = -|\rho| \qquad (4.73)$$

into the equations of torsion motions, (4.49) and (4.54).

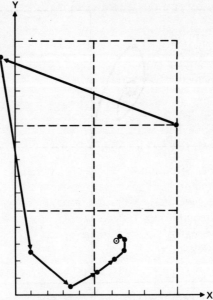

FIG. 32. The effect of nonlinearity on causal vortex obtained for the nonlinear causal recursion (4.33) with the parameter values (4.68) that define a point near the boundary of the vortex region in the parameter space. The nearby region of causal suction has affected the form of causal vortex, and made the trajectory closer in appearance to a typical suction trajectory.

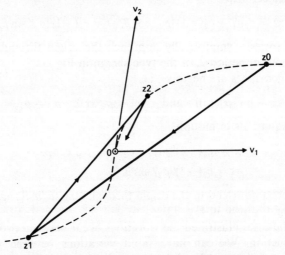

FIG. 33. Longitudinal rapid vibrations over the origin, between two suction trajectories. The vibrations are obtained when both eigenvalues of a 2×2 recursion matrix become negative.

We can straight away give the general definitions:

Definition (The Vibrational Mode of Asymptotic Approach of an Equilibrium State).   The causal recursion $z(t + 1) = \varphi(zt)$ defined in a Euclidean space generates a *vibrational mode of asymptotic approach* of an equilibrium state z_0, if all the eigenvalues of the matrix

$$(\partial\varphi/\partial z)_{z_0} = A \tag{4.74}$$

have an absolute value between one and zero, and at least one of the eigenvalues is negative.

An Extension (The Vibrational Mode of Asymptotic Approach of a Nonequilibrium Trajectory).   The causal recursion $z(t + 1) = \varphi(zt)$ defined in a Euclidean space generates a vibrational mode of asymptotic approach of a continuous nonequilibrium trajectory xR^+, if the eigenvalues of the restriction $\Omega(t)$ of the matrix $(\partial\varphi/\partial z)_{xt}$ to the space $V_{dF}(t)$ orthogonal to the trajectory xR^+ at the point xt,

$$\Omega(t) = (\partial\varphi/\partial z)_{xt} | V_{dF}(t), \tag{4.75}$$

have an absolute value between zero and a fixed positive number $\theta < 1$, and at least one of them is negative.

The extension is based on the following theorem, for the proof of which I refer to Aulin, 1986b:

Theorem 4.6.1 (Orbital Convergence).   In a continuous dynamical system F, where all the eigenvalues of the restriction $\Omega(t)$ of $(\partial\varphi/\partial z)_{xt}$ to the space $V_{dF}(t)$ orthogonal to $(dF/dt)_{xt}$, for all $t \in R^+$, have an absolute value between zero and a fixed positive number $\theta < 1$, any positive half-trajectory that starts from within a small enough neighbourhood of the nonequilibrium trajectory xR^+ approaches the latter asymptotically.

For instance, Eq. (3.13) generates spirals that asymptotically approach the perimeter of the unit circle. In this case the matrix Ω reduces to the number

$$\Omega = (d\varphi_r/dr)_{r=1} = \tfrac{1}{2} < 1. \tag{4.76}$$

Further examples will be met in Chapter 8.

Note. Theorem 4.6.1 states that when xR^+ attracts the neighbouring trajectories, the volume element $\Delta V_{dF}(t)$ shrinks to zero with $t \to \infty$. This does not imply that the total volume element $\Delta V(t)$ would shrink, neither does $\Delta V(t) \to 0$ imply $\Delta V_{dF}(t) \to 0$. Thus, an attractor system is not necessarily dissipative, nor vice versa (see also Section 8.2).

The causal recursions that generate a vibrational mode of asymptotic approach all define a reducibly discontinuous dynamical system. Hence, these causal recursions are all, for some values of their parameters, standard representations of a continuous dynamical system. Thus, a continuous system is associated with every reducibly discontinuous system. But the latter is not obtainable from the respective continuous flow or from the differential equations that generate it. The reducibly discontinuous systems are obtained only by the (discrete) causal recursions that underlie both the continuous and discontinuous dynamical systems.

The reducibly discontinuous systems, with their rapid vibrations, are not just mathematical artefacts, but they may arise unexpectedly in quite normal types of problem settings. Here is a (fictive but realistic) ecological example.

Example (Rapid Vibrations Towards an Ecological Equilibrium).   In a classical example Ashby (1970, Section 9.6) discussed a simple Markovian process, $x(t + 1) = Axt$, with the following matrix of transition probabilities:

$$A = \begin{pmatrix} 0 & 3/4 & 3/4 \\ 3/4 & 1/4 & 1/8 \\ 1/4 & 0 & 1/8 \end{pmatrix}. \tag{4.77}$$

Unfortunately, he confined himself to finding out, by the condition $Ax_0 = x_0$, the equilibrium state

$$x_0 = (0.429, 0.449, 0.122), \tag{4.78}$$

without performing a formal stability analysis or recording the mode of asymptotic approach. Had he done so, he would have observed the rapid vibrations generated by the quite 'normal' transition probabilities (4.77).

First we compute the eigenvalues of the matrix A, finding the following:

$$\rho_1 = 1, \quad \rho_2 = 1/8, \quad \rho_3 = -3/4. \tag{4.79}$$

The respective eigenvectors of unit length are

$$v_1 = \begin{pmatrix} 0.677 \\ 0.710 \\ 0.194 \end{pmatrix}, \quad v_2 = \begin{pmatrix} 0 \\ -0.707 \\ 0.707 \end{pmatrix}, \quad v_3 = \begin{pmatrix} -0.792 \\ 0.566 \\ 0.226 \end{pmatrix}. \tag{4.80}$$

Since the components of the total state $x = (x_1, x_2, x_3) \in X$ are the probabilities of finding an animal of a given species in the respective region I, II, or III, of the area where it lives, the state space is

$$X = \{x; x \geqslant 0, x_1 + x_2 + x_3 = 1\} \subset E^3. \tag{4.81}$$

This is the simplex cut by the co-ordinate planes from the plane $x_1 + x_2 + x_3 = 1$ in the Euclidean space E^3. A vector perpendicular to this plane is $n = (1, 1, 1)$, and we have the inner products

$$n.v_1 \neq 0, \quad n \cdot v_2 = n.v_3 = 0. \tag{4.82}$$

Thus, only v_2 and v_3 are parallel to the plane X. It follows that any point $x \in X$ can be expressed as

$$x = c_2 v_2 + c_3 v_3 + x_0, \quad c_2 \in R, \quad c_3 \in R. \tag{4.83}$$

This gives the process

$$xt = A^t x = c_2 \rho_2^t v_2 + c_3 \rho_3^t v_3 + x_0 \to x_0 \text{ with } t \to +\infty. \qquad (4.84)$$

Since $\rho_2 > 0$ but $\rho_3 < 0$, we have an approach of x_0 through rapid transversal vibrations on the plane X.

For instance, from the initial state $(0.1, 0.2, 0.7)$ we get the following process:

x0:	x1:	x2:	x3:	x4:	x5:	x6:	x7:	x8:	x9:	x10:	x11:	x12:
0.100	0.675	0.244	0.567	0.325	0.506	0.370	0.472	0.395	0.453	0.410	0.443	0.418
0.200	0.213	0.573	0.349	0.523	0.394	0.490	0.417	0.472	0.431	0.462	0.439	0.456
0.700	0.113	0.183	0.084	0.152	0.100	0.139	0.110	0.132	0.115	0.128	0.118	0.125

x13:	x14:	x15:	x16:	x17:	x18:	x19:	x20:	x21:
0.436	0.422	0.433	0.425	0.431	0.426	0.430	0.427	0.429
0.443	0.453	0.445	0.451	0.447	0.450	0.447	0.450	0.448
0.120	0.124	0.121	0.123	0.122	0.123	0.122	0.123	0.122

Thus, the 21st step gives, within the accuracy of computation, the equilibrium state x_0, through the rapid vibrations illustrated in Fig. 34.

In fact, rapid vibrations are often met in biological context, which suggests that the reducible discontinuous dynamical systems derivable from a causal recursion but not from a differential equation are commonplace rather than exceptional in biology. We shall meet more of them in the ecological applications of Chapters 6 and 7.

Before leaving Ashby's Markovian process, let it be noted that ρ_1 is the Frobenius root, and $A: X \to X$ conveys a continuous one-to-one mapping of X on to itself. In fact, x_0 is the invariant point predicted by Brouwer's fixed-point theorem. It is the eigenvector to the eigenvalue $\rho_1 = 1$, with the norm $\|x_0\| = x_{01} + x_{02} + x_{03} = 1$. We shall see more Frobenius roots in the applications to the theory of economic growth in Chapter 8.

4.7. The Degeneracies of Causal Suction, Torsion, and Vortex, and of Rapid Vibrations, into Satellites and Comets

The various modes of asymptotic approach discussed above, all concern, of course, either self-regulating or self-steering dynamical systems. But such a system may degenerate into a system that is steerable from outside, and even into a disintegrating system. This is why causal suction, vortex, and rapid vibrations may degenerate into satellites. The same form of degeneration of causal torsion and of the related vibratory modes results in 'comets'.

(1) Periodical or quasi-closed satellites

For $\rho = 1$ the equations of causal vortex on a plane, (4.63), give

$$xt = c_1(\cos \omega t - k \cos \beta \sin \omega t) + c_2(k \cos \beta \cos \omega t + \sin \omega t),$$

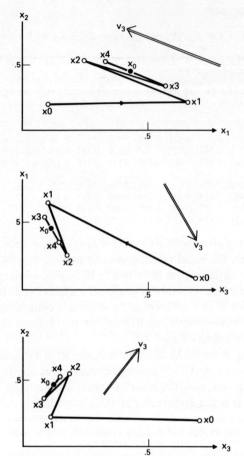

FIG. 34.   Transversal rapid vibrations toward the equilibrium state in Ashby's Markovian model, as projected on the three co-ordinate planes. (The direction of the relevant eigenvector v_3 is indicated on each plane.)

$$yt = k \sin \beta(-c_1 \sin \omega t + c_2 \cos \omega t). \qquad (4.85)$$

Here k, β, and ω are constants that depend on the causal recursion matrix A, while c_1 and c_2 determine the initial state. By varying this state we get a family of ovals symmetric with respect to the origin, and having the same axes that are orthogonal to one another (see Fig. 35a). This is the ideal form of *periodical satellites*. For $k = 1$ and $\beta = \pi/2$ it becomes a family of circles with the centre at the origin.

Other forms of periodical satellites are given by various nonlinear causal recursions. For a nonlinear recursion φ the ideal form appears only in the linear approximation

$$\varphi(z) = (\partial\varphi/\partial z)_{z_0}(z - z_0), \qquad (4.86)$$

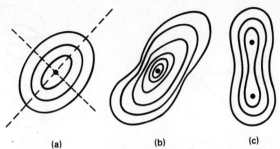

FIG. 35. Examples of closed trajectories of periodical satellite states: (a) the ideal form obtained for a linear causal recursion, and (b)–(c) its modifications in nonlinear systems.

which is valid in a small enough neighbourhood of the equilibrium state z_0. Figure 35b illustrates the gradual change of the linear ideal form to a nonlinear form with increasing distance from z_0.

Still another shape of periodical satellites is depicted in Fig. 35c, where the common satellites of two points of equilibrium are shown. But periodical satellites may also appear, say, around a limit cycle or self-steering trajectory.

Beginning with a three-dimensional state-space quasi-closed curves, too, may appear as satellites in a continuous dynamical system steerable from outside.

(2) Asymptotic satellites

The equations of causal suction on a plane E^2, which we can now consider in the form (4.6), take with the substitutions

$$\rho_1 = 1, \quad 0 < \rho_2 < 1, \tag{4.87}$$

the following form:

$$zt = c_1 v_1 + c_2 \rho_2^t v_2 \to c_1 v_1 \text{ with } t \to +\infty. \tag{4.88}$$

This can be called an *asymptotic satellite* (see Fig. 36a). For $c_1 \neq 0$ the origin is an *outsider point*, for $c_1 = 0$ the origin simply is the positive limit set of the asymptotic satellite in question.

With the eigenvalues

$$\rho_2 = 1, \quad 0 < \rho_1 < 1, \tag{4.89}$$

Eq. (4.6) gives

$$zt = c_1 \rho_1^t v_1 + c_2 v_2 \to c_2 v_2 \text{ with } t \to +\infty. \tag{4.90}$$

This too is an asymptotic satellite, but now its rectilinear trajectory is parallel to the eigenvector v_1 instead of v_2. The origin is an outsider point for $c_2 \neq 0$.

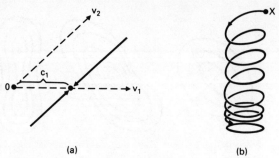

FIG. 36. Trajectories of asymptotic satellite states: (a) in a 2-dimensional state-space, (b) in a 3-dimensional state-space.

For $\rho_1 = \rho_2 = 1$ we have a case where the causal recursion is not locally linearizable at the equilibrium point: cf. Eq. (4.31).

Figure 36b shows an asymptotic satellite xR^+ from a three-dimensional pattern of trajectories. Note that while $\Lambda^+(x)$ defines a limit cycle, this limit cycle is not a goal, as the system obviously is not self-regulating but steerable from outside. This trajectory also belongs to a linear causal recursion $\varphi(z) = Az$, the three eigenvalues of the matrix A being the following: a pair of conjugate complex numbers, μ and μ^*, with the modulus one, and a real eigenvalue ρ_3 obeying $0 < \rho_3 < 1$.

Again the effects of nonlinearity will change the above ideal forms when we go farther from the origin.

(3) Rapidly vibrating satellites

With the substitutions

$$\rho_1 = 1, \qquad \rho_2 = -1, \quad \text{(Case 1)} \tag{4.91}$$

$$\rho_1 = -1, \quad \rho_2 = 1, \quad \text{(Case 2)} \tag{4.92}$$

$$\rho_1 = \rho_2 = -1, \qquad \text{(Case 3)} \tag{4.93}$$

the equations of causal suction on a plane, (4.6), take the forms

$$zt = c_1 v_1 + c_2(-1)^t v_2, \quad \text{(Case 1)} \tag{4.94}$$

$$zt = c_1(-1)^t v_1 + c_2 v_2, \quad \text{(Case 2)} \tag{4.95}$$

$$zt = (-1)^t(c_1 v_1 + c_2 v_2), \quad \text{(Case 3)} \tag{4.96}$$

respectively. All of them give *vibrating two-point satellites*, of the type shown in Fig. 37a. In Cases 1 and 2 the origin is an outsider point, while in Case 3 the vibrations go over the origin, making it the *centre* of the motion.

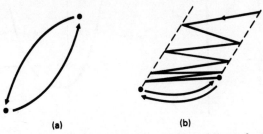

FIG. 37.   (a) A vibrating 2-point satellite state. (b) The trajectory of a zig-zagging satellite state.

With the substitutions

$$\rho_1 = -1, \quad 0 < \rho_2 < 1, \tag{4.97}$$

we get from Eq. (4.6):

$$zt = c_1(-1)^t v_1 + c_2 \rho_2^t v_2 \rightarrow c_1(-1)^t v_1 \text{ with } t \rightarrow +\infty. \tag{4.98}$$

This defines a *zig-zagging satellite* or vibrating asymptotic satellite shown in Fig. 37b. The state zt asymptotically approaches the two-point vibrating satellite that vibrates between the points $c_1 v_1$ and $-c_1 v_1$.

The substitutions

$$0 < \rho_1 < 1 \quad \text{and} \quad \rho_2 = -1 \tag{4.99}$$

give

$$zt = c_1 \rho_1^t v_1 + c_2(-1)^t v_2 \rightarrow c_2(-1)^t v_2 \text{ with } t \rightarrow +\infty. \tag{4.100}$$

This too is a zig-zagging satellite, with the vibrations now in the direction of v_2 instead of v_1.

All these vibrating satellites appear in their ideal forms indicated above only in the nearest vicinity of the equilibrium state z_0, if the causal recursion $\varphi(z)$ is nonlinear. Their form farther from z_0 depends on the kind of nonlinearity of the causal recursion.

(4) The comets derivable from causal torsion

If we make the n-fold eigenvalue associated with an n-dimensional causal torsion equal to one, we get a 'wandering state' that comes from infinity, passes by the equilibrium point, and recedes to infinity. That, obviously, can be called *comet*. An example is the rectilinear comet shown in Fig. 38a, which is obtained from the equations of two-dimensional causal torsion, (4.49), with the substitution $\rho = 1$, which gives:

$$xt = c_1, \quad yt = c_1 \alpha t + c_2. \tag{4.101}$$

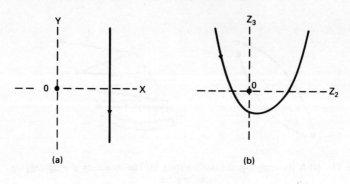

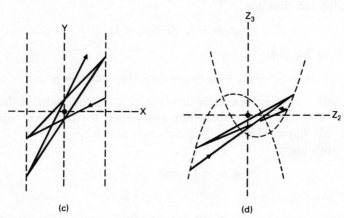

FIG. 38. A rectilinear (Picture a) and parabolic (Picture b) comet state passing the outsider point 0, and the corresponding zig-zagging comet states (Pictures c and d, respectively).

With the same substitution, the equations of three-dimensional causal torsion, (4.54), become

$$z_1 t = c_1, \quad z_2 t = c_1 \alpha t + c_2,$$

$$z_3 t = c_1 \frac{\alpha \gamma}{2} t^2 + \left[c_1 \frac{2\beta - \alpha \gamma}{2} + c_2 \gamma \right] t + c_3. \tag{4.102}$$

This gives a *parabolic comet* passing an outsider point, illustrated in Fig. 38b. Going to higher dimensionality brings out comets in whose equations polynomials of ever higher degree are involved.

Again, effects of nonlinearity appear and distort the above ideal types farther from the equilibrium state, provided that the causal recursion in question is nonlinear. And again, even points other than equilibrium points may appear as objects of asymptotic approach in the torsion fields whose degeneracies the comets are.

(5) Rapidly vibrating comets

If instead of $\rho = 1$ we make the substitution $\rho = -1$, we get a reducibly discontinuous system instead of a continuous one, and *rapidly vibrating comets*. An example is the vibratory version of the comet of 'first degree', Eq. (4.101), given by

$$xt = (-1)^t c_1, \quad yt = (-1)^t(c_1 \alpha t + c_2). \tag{4.103}$$

With $t \to +\infty$ the state $zt = (xt, yt)$ recedes to (positive or negative) infinity, while it keeps vibrating over the origin between the straight line $x = c_1$ and its mirror image $x = -c_1$ on the other side of the origin (see Fig. 38c).

The same substitution into the equations (4.54) of a three-dimensional torsion brings us, along with the vibrating comet of first degree we have already met, a vibrating comet of second degree, illustrated in Fig. 38d. Now the vibrations occur between the parabola of Fig. 38b and its mirror image on the other side of the origin, again over the origin. And again the amplitude of vibrations increases to infinity with $t \to +\infty$.

By continuing to torsions of higher dimensionality we get more complicated vibrations over the origin between curves of higher order, but the general picture remains the same.

(6) The combined attraction–repulsion field

Let us consider the linear system defined by $\varphi(z) = Az$, where the 2×2 matrix A has two single real eigenvalues obeying

$$\rho_1 > 1 > \rho_2 > 0. \tag{4.104}$$

With this substitution (4.6) gives $zt = (zt)_1 + (zt)_2$, such that

$$(zt)_1 = c_1 \rho_1^t v_1 \to \begin{cases} +\infty & \text{for } t \to +\infty, \\ 0 & \text{for } t \to -\infty, \end{cases}$$

$$(zt)_2 = c_2 \rho_2^t v_2 \to \begin{cases} 0 & \text{for } t \to +\infty, \\ +\infty & \text{for } t \to -\infty \end{cases} \tag{4.105}$$

in the positive sector defined by $c_1 > 0$ and $c_2 > 0$.

Seen from the *saddle point* $z = 0$, shown in Fig. 39a, these trajectories are *hyperbolic comets* that approach close to $z = 0$ and then pass it by receding to infinity. On the other hand, the same trajectories asymptotically approach the rectilinear trajectory of the point $zt = c_1 \rho_1^t v_1$ when $t \to +\infty$. Thus, seen from the perspective of the latter trajectory, those 'comets' are part of its suction field. The rectilinear unbounded attractor defined by $z = c_1 v_1$ is of

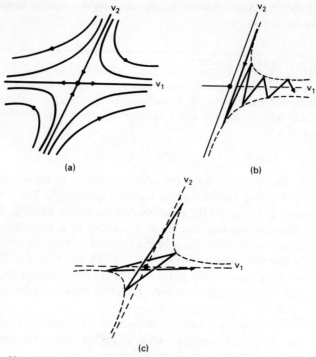

FIG. 39. (a) The curvilinear trajectories of hyperbolic comet states around a saddle point. (b) An example of the corresponding transversally zig-zagging comet states. (c) An example of the corresponding longitudinally zig-zagging comet states.

the type that in economics is called a *turnpike* (see Section 8.2). A turnpike system obviously is a linear self-steering system, and the mode of asymptotic approach in a two-dimensional turnpike system is, of course, causal suction. Accordingly, the continuous trajectories of a linear attraction–repulsion system can be alternatively interpreted either as comets passing by a saddle point or as self-steering trajectories approaching a stable turnpike.

But if one or both of the eigenvalues of our 2×2 matrix A are negative, while $|\rho_1| > 1$ and $|\rho_2| < 1$ are still true, we have rapidly vibrating comets. In the former case, the vibrations are *transversal* and pass the origin (Fig. 39b); in the latter case we have *longitudinal vibrations* over the origin (Fig. 39c). In both cases the amplitude increases without limit when $t \to +\infty$.

4.8. The General Connection Between Linear Difference and Differential Equations

By studying the modes of asymptotic approach of goals we have obtained, as by-products, solutions of the equations that define linear dynamical systems. In fact we have obtained the general solution of such systems. The linear system is a rare case in which the two mutually equivalent normal forms of causal recursion of a continuous system, (2.1) and (2.2), can indeed be derived from each other. Let us here consider the case, i.e. the connection between the equations

$$x(t + 1) = A(xt) \tag{4.106}$$

and

$$dx/dt = Bx, \tag{4.107}$$

where the matrices A and B are n-square and real.

Assuming that the matrix A has the single eigenvalues

$$\mu_j = \rho_j e^{i\omega_j}, \quad j = 1, 2, \ldots, n, \tag{4.107}$$

we can write, combining the results of the Sections 4.1 and 4.5, the general solution in the form

$$xt = \sum_{j=1}^{n} \rho_j^t [a_j(\cos \omega_j t \cdot v_j - \sin \omega_j t \cdot u_j)$$
$$+ b_j(\cos \omega_j t \cdot u_j + \sin \omega_j t \cdot v_j)]. \tag{4.108}$$

For real (and positive because of the continuity of the system) eigenvalues ρ_j we have $\omega_j = b_j = 0$.

From (4.108) we get, after some manipulation of terms:

$$dx/dt = \sum_{j=1}^{n} \rho_j^t [(a_j \log \rho_j - b_j \omega_j)(\cos \omega_j t \cdot v_j - \sin \omega_j t \cdot u_j)$$
$$+ (b_j \log \rho_j - a_j \omega_j)(\cos \omega_j t \cdot u_j + \sin \omega_j t \cdot v_j)], \tag{4.109}$$

$$x(t + 1) = \sum_{j=1}^{n} \rho_j^{t+1} [(a_j \cos \omega_j + b_j \sin \omega_j)(\cos \omega_j t \cdot v_j - \sin \omega_j t \cdot u_j)$$
$$+ (b_j \cos \omega_j - a_j \sin \omega_j)(\cos \omega_j t \cdot u_j + \sin \omega_j t \cdot v_j)]. \tag{4.110}$$

Comparing (4.110) with (4.108) we can see how the matrix A operates on the vectors v_j and u_j:

$$Av_j = \rho_j(\cos \omega_j \cdot v_j - \sin \omega_j \cdot u_j),$$
$$Au_j = \rho_j(\sin \omega_j \cdot v_j + \cos \omega_j \cdot u_j), \quad j = 1, 2, \ldots, n. \tag{4.111}$$

Similarly, by comparing (4.109) with (4.108) we find out how the matrix B operates on these vectors:

$$Bv_j = \log \rho_j \cdot v_j - \omega_j \cdot u_j,$$
$$Bu_j = \omega_j \cdot v_j + \log \rho_j \cdot u_j, \quad j = 1, 2, \ldots, n. \tag{4.112}$$

Thus, on each plane (v_j, u_j) corresponding to a conjugated couple of complex eigenvalues, the matrices A and B operate on the basis vectors v_j and u_j as the following 2×2 matrices:

$$A\begin{pmatrix} v_j \\ u_j \end{pmatrix} = \rho_j \begin{pmatrix} \cos \omega_j & -\sin \omega_j \\ \sin \omega_j & \cos \omega_j \end{pmatrix} \begin{pmatrix} v_j \\ u_j \end{pmatrix}.$$ (4.113)

$$B\begin{pmatrix} v_j \\ u_j \end{pmatrix} = \begin{pmatrix} \log \rho_j & -\omega_j \\ \omega_j & \log \rho_j \end{pmatrix} \begin{pmatrix} v_j \\ u_j \end{pmatrix}.$$ (4.114)

And on the eigenvectors v_j corresponding to the real eigenvalue ρ_j, they operate as follows:

$$Av_j = \rho_j v_j, \quad Bv_j = \log \rho_j \cdot v_j.$$ (4.115)

Therefore, using the notation

$$C \oplus D = \begin{pmatrix} C & 0 \\ 0 & D \end{pmatrix}$$ (4.116)

for the direct sum of any two square matrices C and D, we can represent the matrix A in the form

$$A = M \left\{ \bigoplus_{j=1}^{r} \rho_j \bigoplus_{k=r+1}^{n} \begin{pmatrix} \rho_k \cos \omega_k & -\rho_k \sin \omega_k \\ \rho_k \sin \omega_k & \rho_k \cos \omega_k \end{pmatrix} \right\} M^{-1}.$$ (4.117)

Here r is the number of real eigenvalues, and M is an n-square real non-singular matrix.

In a similar way, we can represent the matrix B as follows:

$$B = M \left\{ \bigoplus_{j=1}^{r} \log \rho_j \bigoplus_{k=r+1}^{n} \begin{pmatrix} \log \rho_k & -\omega_k \\ \omega_k & \log \rho_k \end{pmatrix} \right\} M^{-1}.$$ (4.118)

The formulae (4.117) and (4.118) are the best we can do by way of indicating the mutual correspondence of the matrices A and B, and thus of the linear difference and differential equations. By the discussion given above we can also see the mutual correspondence of their eigenvalues. To each eigenvalue μ of the matrix A there corresponds an eigenvalue λ of the matrix B, according to the following rule.

$$\mu = \rho e^{i\omega} \leftrightarrow \lambda = \log \rho + i\omega.$$ (4.119)

As for the case of multiple eigenvalues, it is sufficient here to refer to the method of study that was followed in Section 4.4.

Note on the Inhomogeneous Linear Equations. An inhomogeneous linear difference equation,

$$y(t + 1) = A(yt) + b,$$ (4.120)

can always be reduced to the homogeneous form (4.106) with the transformation of variables,

$$y = x - (I - A)^{-1}b. \tag{4.121a}$$

If the inverse $(1 - A)^{-1}$ does not exist, there are redundant dimensions in the state-space: dropping them, one gets a non-singular matrix $I - \hat{A}$ to be substituted into (4.121) in place of the matrix $I - A$.

An inhomogeneous linear differential equation,

$$dy/dt = By + b, \tag{4.121b}$$

can be likewise reduced to the homogeneous form (4.107) with the transformation

$$y = x - B^{-1}b. \tag{4.122}$$

Again, if B is singular there are redundant dimensions in the state-space; by removing them you end up with a non-singular matrix $\hat{B}$ to be substituted into (4.122) in place of B.

CHAPTER 5

The Variety of Dynamical Systems Generated by a Single Nonlinear Map (An Example)

In the preceding chapters we have considered the properties of dynamical systems starting with their uniquely determined causal recursions. Dynamical systems, in other words, have been the focus of interest. This corresponds to the ultimate purpose of this book, which is to consider the behaviour of real-world objects.

From a purely mathematical point of view we could have chosen another approach: starting with a given difference or differential equation, and studying the set of different dynamical systems it defines. Some lines of approach to system dynamics, notably those connected with the concepts of structural stability and instability (and, as an outcome of the latter property, the theory of bifurcations of equilibrium states), have indeed followed the latter method.

Since the latter approach has produced some results that are interesting from our present point of view as well, it will be discussed here in the form of an illustrative example. For this example a rather simple nonlinear difference equation is chosen – a difference equation rather than a differential one, since the variety of different types of dynamical systems generated by a nonlinear difference equation (a 'nonlinear map') is usually much richer than that generated by a nonlinear differential equation of comparable structure.

By way of example, we shall study the local behaviour (near its equilibrium state) of the following two-parameter map representing causal recursions defined on a Euclidean plane:

$$X(t + 1) = \varphi(Xt, \lambda), \quad X = (x, y) \in R^2, \quad \lambda = (a, b) \in R^2,$$
$$\varphi_x(X, \lambda) = a + (1 - b)x - xy^2, \quad \varphi_y(X, \lambda) = bx + xy^2. \tag{5.1}$$

This is an appropriate choice, as it generates in a neighbourhood of the equilibrium state $X_0 = (x_0, y_0)$,

$$x_0 = a/(b + a^2), \quad y_0 = a, \tag{5.2}$$

a system that is locally either self-regulating, or steerable from outside, or disintegrating, or a system with nilpotent causal recursion, depending on the values of parameters. Thus, all the categories of systems are represented except for the self-steering system.

Only local behaviour near X_0 will be investigated, as the *structure of the parameter space* is most pronounced there. Equation (5.1) is locally linearizable, and gives

$$(\partial\varphi/\partial X)_{X_0} = \begin{pmatrix} 1 - b - a^2 & -2a^2/(b + a^2) \\ b + a^2 & 2a^2/(b + a^2) \end{pmatrix} = \begin{pmatrix} 1 - p & -q \\ p & q \end{pmatrix}. \quad (5.3)$$

Here the notations $p = b + a^2$ and $q = 2a^2/p$ have been used. Thus, only a^2 and b matter. For the sake of convenience we shall use in the parameter space the co-ordinates $s = a^2$ and p. The eigenvalues of the matrix (5.3) are

$$\mu = (q - p + 1)/2 \pm \{[(q - p + 1)/2]^2 - q\}^{1/2}. \quad (5.4)$$

5.1. The Separatrices and Regions

The structure of the parameter half-plane $(s > 0, p)$ is shown in Fig. 40. The two separatrices

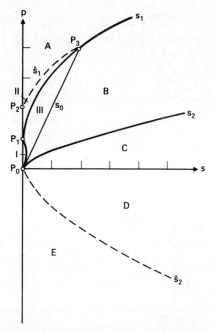

FIG. 40. The separatrices and regions in the parameter space of our example. The system is locally stable (self-regulating) in the equilibrium state for parameter values in Regions I, II, and III, with causal suction for Region I, causal vortex in Region III and rapid vibrations in Region II.

$$s_1 = p(p+1)/2 - p^{3/2} \quad \text{and} \quad s_2 = p(p+1)/2 + p^{3/2} \qquad (5.5)$$

are obtained directly from the condition $(q - p + 1)^2 = 4q$, on which (5.4) gives a double eigenvalue. Between these curves, i.e. in Regions III and B, we have a complex conjugate pair of eigenvalues, and outside them, in Regions I, II, A, C, D, and E a pair of single real eigenvalues.

The total region of complex eigenvalues is split by the straight line

$$s_0 = p/2 \text{ for } 0 < p < 4, \qquad (5.6)$$

obtained from the condition $q = 1$ of modulus one: $|\mu| = 1$. Thus, on the 'satellite line' s_0 we have, in a neighbourhood of the equilibrium state X_0, the periodical oval-shaped satellite states (see Fig. 41a) of a continuous dynamical system, and the system is (locally) steerable from outside.

Inside Region III we have $|\mu| < 1$, so that the system is locally stable near X_0, and the asymptotic approach to X_0 takes the form of oval-shaped spirals that has been called *causal vortex* (Fig. 41b). The system is (locally) continuous and self-regulating.

In Region B the modulus is larger than one, and the states recede from

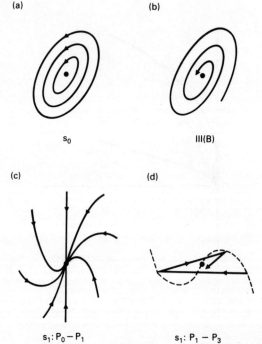

(a)

(b)

s_0

III(B)

(c)

(d)

$s_1 : P_0 - P_1$

$s_1 : P_1 - P_3$

FIG. 41.   The form of trajectories near the equilibrium state: (a) for the satellite line s_0 of Fig. 40, (b) for the vortex regions III and B, (c) for the separatrice s_1 between the points P_0 and P_1, (d) for the same separatrice between the points P_1 and P_3.

X_0 along the trajectories of causal vortex shown in Fig. 41b. The system is (locally) continuous and disintegrating.

Along the 'torsion lines' s_1 and s_2 we have a double real eigenvalue, and the pattern of trajectories near X_0 is generated by *causal torsion* shown in Fig. 41c. Between the points P_0 and P_1 on s_1 (Fig. 40) we have $0 < \mu < 1$, so that the system represented by Eq. (5.1) is for the respective parameter values (locally) continuous and self-regulating near X_0. Between P_1 and P_3 along s_1 there is $-1 < \mu < 0$. Here the states are rapidly vibrating (with period 2) over the point of equilibrium X_0, from one of the torsion lines of Fig. 41c to its mirror image on the other side of the point X_0 and back (Fig. 41d). The system is (locally) reducibly discontinuous and self-regulating near X_0.

From point P_3 to infinity along the curve s_1, and from P_0 to infinity on s_2, the system is (locally) disintegrating near X_0. In the former case we have $\mu < -1$ and thus divergent vibrations, i.e. the vibrations of Fig. 41d, run the other way round. In the latter case $\mu > 1$, and we have the continuous torsion system with the trajectories of Fig. 41c running in the other direction.

Coming to the areas of two single real eigenvalues in the parameter space, we still find two separatrices, viz. (see Fig. 40):

$$\hat{s}_1 = p(p - 2)/4 \text{ for } 2 < p < 4 \quad \text{and} \quad \hat{s}_2 = p(p - 2)/4 \text{ for } p < 0. \quad (5.7)$$

Both of them are segments of the parabola $s = p(p - 2)/4$. They are obtained directly from the condition that $\mu_- = -1$, where μ_- is the eigenvalue (5.4) with the minus sign. It is just as easy to verify that $\mu_+ > 1$ on $\hat{s}_2$, while $-1 < \mu_+ < 0$ is true on $\hat{s}_1$. On the segment $\hat{s}_1$ we accordingly get what was called 'zig-zagging satellite states' ending up with rapid vibrations over the point of equilibrium X_0 (see Fig. 42a). The system is (locally) reducibly discontinuous and steerable from outside. On the separatrice $\hat{s}_2$ we have 'zig-zagging states of disintegration' with transversal vibrations (Fig. 42b). The system is (locally) reducibly discontinuous and disintegrating.

By trivial but elaborate computation one can ascertain the following further results:

$$0 < \mu_+ < 1 \text{ and } 0 < \mu_- < 1 \text{ in Region I}, \quad (5.8)$$

$$-1 < \mu_+ < 0 \text{ and } -1 < \mu_- < 0 \text{ in Region II}, \quad (5.9)$$

$$-1 < \mu_+ < 0 \text{ and } \mu_- < -1 \text{ in Region A}, \quad (5.10)$$

$$\mu_+ > 1 \text{ and } \mu_- > 1 \text{ in Region C}, \quad (5.11)$$

$$\mu_+ > 1 \text{ and } \mu_- < -1 \text{ in Region D}, \quad (5.12)$$

$$\mu_+ > 1 \text{ and } -1 < \mu_- < 0 \text{ in Region E}. \quad (5.13)$$

Thus, the system is, in Region I, (locally) continuous and self-regulating, and the state of equilibrium X_0 is approached along trajectories of the type of *causal suction* illustrated in Fig. 42c. In Region II it is still (locally)

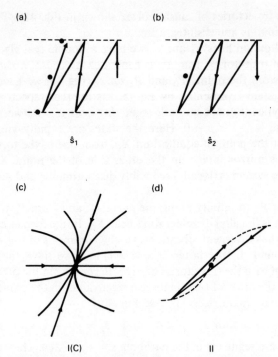

FIG. 42. (a)–(b): The types of zig-zagging satellite states (Picture a) and zig-zagging states of disintegration (Picture b) obtained for the separatrices $\hat{s}_1$ and $\hat{s}_2$, respectively. (c)–(d): The causal suction (Picture c) and longitudinal vibrations (Picture d) toward the equilibrium state obtained for Region I and Region II, respectively.

self-regulating but reducibly discontinuous, with longitudinal convergent vibrations of state over the point X_0 (between two trajectories of causal suction, cf. Fig. 42d).

In Regions A and E the system is (locally) reducibly discontinuous and disintegrating, the mode of disintegration being, in both cases, that of rapid vibrations between two trajectories of a combined attraction and repulsion field. But in Region A the vibrations are longitudinal, i.e. run over the equilibrium point X_0, while in Region E they are transversal and run past X_0.

In the remaining Regions C and D we still have a (locally) disintegrating system, but the mode of disintegration is now that of causal suction. In Region C the system is (locally) continuous, and the state recedes from X_0 along suction trajectories (Fig. 42c). In Region D the system is (locally) reducibly discontinuous, with divergent transversal vibrations of state between two suction trajectories.

To sum up the main results so far, we have found that (5.1) defines (cf. Fig. 40), locally, a continuous system inside the region whose boundaries are the co-ordinate axis $p = 0$ (for $s \geqslant 0$), the co-ordinate axis $s = 0$ (*for* $0 \leqslant p \leqslant 1$), and the torsion line s_1 (from the point P_1 onwards). Above the torsion-line boundary and below the s-axis it defines, locally, a reducibly discontinuous system.

Furthermore, as for the system categories of Fig. 8 the system (5.1) is locally self-regulating inside the Region I + II + III, steerable from outside on its boundaries consisting of the satellite lines s_0 and $\hat{s}_1$, and disintegrating elsewhere.

We have also found some results pertaining to the mode of asymptotic approach of the equilibrium state for self-regulating systems, and observed some types of satellite states associated with systems that are steerable from outside.

5.2. The Special Points

It remains to study the local behaviour of the causal recursion (5.1) at the special points P_0, P_1, P_2, and P_3 (see Fig. 40) of the parameter space.

(1) The nilpotent causal recursion at P_1

The parameter values $s = 0$ and $p = 1$ belonging to the point P_1 give

$$X_0 = (0,0), \quad (\partial\varphi/\partial X)_{X_0} = \begin{pmatrix} 0 & 0 \\ 1 & 0 \end{pmatrix} = A_1, \quad \text{and } \mu_+ = \mu_- = 0. \quad (5.14)$$

The matrix is nilpotent of index 2. Thus, for whatever initial state $x \in R^2$ we have $x2 = A_1^2 x = (0,0)$, and the causal recursion is also nilpotent. It is interesting to see that a nilpotent causal recursion appears at the point of the parameter space where the three regions of stability I, II, and III meet each other, as if the nilpotent recursion were an intensification of the stability properties of goal-directed systems.

(2) P_2 as a continuation of the satellite line $\hat{s}_1$

For P_2 we have $s = 0$ and $p = 2$, and thus

$$X_0 = (0,0), \quad (\partial\varphi/\partial X)_{X_0} = \begin{pmatrix} -1 & 0 \\ 2 & 0 \end{pmatrix}$$

$$= A_2, \quad \mu_+ = 0, \quad \text{and } \mu_- = -1. \quad (5.15)$$

This eliminates one of the dimensions of the state-space, leaving a two-point vibrating satellite state in the remaining dimension: for an initial state $X = (x,y)$ we have $Xt = (-x,2x)$ for an odd t, and $Xt = (x,-2x)$ for an even $t > 0$. The matrix is periodical, there being $A_2^3 = A_2$, and the remaining dimension of state-space is that determined by the straight line $y = -2x$.

The point P_2 can be conceived of as a continuation of the satellite line $\hat{s}_1$ that also gives vibrating satellite states.

(3) P_3 as a continuation of both the satellite line $\hat{s}_1$ and the torsion line s_1

For P_3 we have $s = 2$ and $p = 4$, and thus

$$X_0 = (2^{1/2}/4, 2^{1/2}), \quad (\partial\varphi/\partial X)_{X_0} = \begin{pmatrix} -3 & -1 \\ 4 & 1 \end{pmatrix},$$

and $\mu_+ = \mu_- = -1$.

$$(5.16)$$

This gives the following equations of the state $Xt = (xt, yt)$, with an initial state $X = (x, y)$:

$$xt = (-1)^t[x + t(2x + y)], \quad yt = (-1)^t[y - 2t(2x + y)]. \quad (5.17)$$

Hence, we see that if the initial state is on the straight line $y = -2x$ we get a vibrating two-point satellite state of the same kind as appeared at the point P_2. Thus, both P_2 and P_3, the two vertices of the satellite line $\hat{s}_1$, are faithful continuations of this line. On the other hand, if the initial state lies outside the straight line $y = -2x$, (5.17) defines divergent vibrations of period 2 in a system that is reducibly discontinuous. This shows that the point P_3 is also a continuation of the torsion line s_1, which for $p > 4$ also gives such vibrations.

(4) P_0 as a continuation of all the regions of the parameter space that meet at this point

P_0 is the origin of the parameter space, and for $s = 0$, $p = 0$ we get two real eigenvalues of $(\partial\varphi/\partial X)_{X_0}$, viz.

$$\mu_+ = 2s/p \quad \text{and} \quad \mu_- = 1. \quad (5.18)$$

Thus, the magnitude of μ_+ depends on the direction in which we approach the point P_0. Naming the different directions according to the regions or lines in question (Fig. 40) we have the following results:

$$I\text{-direction:} \qquad s/p < 1/2 \Rightarrow 0 < \mu_+ < 1, \qquad (5.19)$$

$$s_0\text{-direction:} \qquad s/p = 1/2 \Rightarrow \mu_+ = 1, \qquad (5.20)$$

$$C\text{-direction:} \qquad s/p > 1/2 \Rightarrow \mu_+ > 1, \qquad (5.21)$$

$$D\text{-direction:}\ p < 0, |s/p| > 1/2 \Rightarrow -1 < \mu_+ < 0, \qquad (5.22)$$

$$\hat{s}_2\text{-direction:}\ p < 0, |s/p| = 1/2 \Rightarrow \mu_+ = -1, \qquad (5.23)$$

$$E\text{-direction:}\ p < 0, |s/p| < 1/2 \Rightarrow \mu_+ < -1. \qquad (5.24)$$

Thus, for different directions we have different types of patterns of trajectories at P_0, reflecting those of the regions and lines of approach: a continuous approach of state to a one-point satellite state (from I-direction), continuous periodical satellite states as in Fig. 41a (from s_0-direction), continuous rectilinear satellite states (from D- and $\hat{s}_2$-directions), or vibrating states of disintegration with transversal vibrations (from E-direction).

As a summary of the special points we can say that all of these points behave dynamically as if they were continuations of the adjoining regions or lines of the parameter space: the local behaviour of the system parameters reflects the behaviour for those regions and lines.

5.3. Structural Stability and Instability

Definition. A dynamical system having causal recursion is *structurally stable* at a given point λ of the parameter space, if the type of trajectories of the system is invariant for a small enough change of λ.

There is no generally accepted formalization of the concept of the type of trajectories, but here we can take that concept to refer to the types of causal systems distinguished from each other in Sections 2.2 and 2.3 and in Chapter 3, plus the modes of trajectory fields investigated in Chapter 4.

Otherwise the formalization of the concept of structural stability is trivial. We are considering the causal recursion φ as a function

$$\varphi: X \times \Lambda \to X, \quad x(t + 1) = \varphi(xt, \lambda t), \tag{5.25}$$

where X is the state-space and Λ the parameter space. Obviously, a change of parameter induces a change of causal recursion, generally:

$$\lambda \to \lambda' \Rightarrow \varphi \to \varphi'. \tag{5.26}$$

But if the system is structurally stable at the point a small enough change $\Delta\lambda = \lambda' - \lambda$ retains the type, so that the different fields of trajectories generated by φ and φ' in X are of the same type.

Example. In our present example of parameter space, shown in Fig. 40, the system (more precisely, the field of trajectories in a close enough neighbourhood of the equilibrium state X_0) is structurally stable at every inner point of the Regions I, II, III, A, B, C, D, and E, but structurally unstable on the boundaries of these regions.

Note. 'Structural stability' is a rather odd notion: it has no connection at all with the concept of dynamical stability. For instance, our present model system is (locally at least) disintegrating within the region A. Nevertheless, it is 'structurally stable' there, according to the above definition. (For mathematical ideas behind structural stability and for recent developments in the field, see Steve Smale's fascinating essays in Smale, 1980.)

As a matter of fact so far, from the point of view of dynamics, structural instability has proved to be a more interesting object of study than has structural stability. Let us consider, for instance, the passage over the separatrice s_1 from Region I to Region III in Fig. 40. We get a succession of different types of the fields of trajectories, each of which defines a certain continuous mode of asymptotic approach of the equilibrium state X_0:

$$
\begin{array}{ccc}
\text{Causal suction} & \text{Causal torsion} & \text{Causal vortex} \\
\text{(for } s < s_1) & \text{(for } s = s_1) & \text{(for } s > s_1)
\end{array} \xrightarrow{\qquad} \xrightarrow{\qquad} \tag{5.27}
$$

Passing from left to right, the self-regulation of the system can be said to 'weaken', in the sense that first, for $s < s_1$, you go straight to the equilibrium state, then, for $s = s_1$, you turn around it by 180°, and finally for $s > s_1$, you revolve around it an infinite number of times while approaching it. But all the time the system is a self-regulating equilibrium system.

Let us then pass from Region III over the separatrice s_0 to Region B in Fig. 40. Now we first have a self-regulating equilibrium system with a vortex approach of X_0, then, for $s = s_0$, the system becomes (locally) steerable from outside and we have a field of satellites, and finally the system becomes (locally at least) disintegrating with a 'negative vortex':

$$
\begin{array}{ccc}
\text{Causal vortex} & \text{Periodical} & \text{Causal vortex} \\
\text{towards } X_0 \rightarrow & \text{satellites} \rightarrow & \text{off } X_0 \\
\text{(for } s < s_0) & \text{(for } s = s_0) & \text{(for } s > s_0)
\end{array} \tag{5.28}
$$

Our discussion has been local, i.e. restricted to a close enough neighbourhood of the equilibrium point X_0 such that the linear Lyapounov approximation is (approximately) valid. Suppose now that a global discussion of trajectories would indicate a limit cycle in the passage (5.28) for $s > s_0$. Then we would have, instead of (5.28):

$$
\begin{array}{ccc}
\text{Causal vortex} & \text{Periodical} & \text{Causal suction} \\
\text{towards the state} & \text{satellites} & \text{to a limit} \\
\text{of equilibrium } X_0 \rightarrow & \text{around } X_0 \rightarrow & \text{cycle} \\
\text{(for } s < s_0) & \text{(for } s = s_0) & \text{(for } s > s_0)
\end{array} \tag{5.29}
$$

This is what is called *Hopf bifurcation* (see Fig. 43). We shall soon study a case of Hopf bifurcation related to our example of parameter space.

The corresponding phenomenon of bifurcation in a one-dimensional state-space is *Feigenbaum bifurcation*. There we have:

$$
\begin{array}{ccc}
\text{Causal suction} & \text{Vibrating} & \text{Rapid vibrations} \\
\text{towards a state} & \text{two-point} & \text{towards a two-point} \\
\text{of equilibrium} \rightarrow & \text{satellites} \rightarrow & \text{limit cycle} \\
\text{(for } a < a_0) & \text{(for } a = a_0) & \text{(for } a > a_0)
\end{array} \tag{5.30}
$$

Here a is a parameter of the system.

There is, as yet, no general theory of bifurcations, just as there is no general theory of the related problems of self-organization. There are

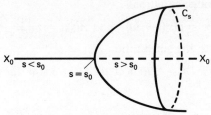

FIG. 43. Hopf bifurcation of the state of equilibrium X_0 in the dynamical system defined by the nonlinear map (5.1). The picture is to be seen as 3-dimensional, the line X_0—X_0 standing both for the straight line $p = 1$ of the parameter space (cf. Fig. 40) and for the point X_0 of the state-plane, which is depicted as orthogonal to this line. For parameter values $s < s_0$ the equilibrium X_0 is stable, while for $s > s_0$ a stable limit cycle C_s appears in the state-space.

interesting results concerning Feigenbaum bifurcations (Feigenbaum, 1978) and Hopf bifurcations, and some results about the reduction of bifurcations occurring in state-spaces of higher dimensions to those occurring in one- or two-dimensional spaces (Iooss, 1979). Examples of Feigenbaum bifurcations will be studied in Part 2, in connection with ecosystems. As for Hopf bifurcations, the most interesting question for a beginner is how to find out when they exist.

5.4. Hopf Bifurcation: an Algorithm for the Detection of its Existence

Let us return to our example of nonlinear map defined by (5.1). Its fixed point or, in the language of dynamical systems, its equilibrium state X_0, was given by (5.2). In the neighbourhood of this state the behaviour of the system is determined by the matrix $(\partial\varphi/\partial X)_{X_0} = A$ given by (5.3).

Our map had two real-valued parameters s and p. The line $s_0 = p/2$ in the parameter space (see Fig. 40) separates Region III, on which the map (5.1) defines a self-regulating equilibrium system with a causal-vortex approach to equilibrium, from Region B, where the map (5.1) gives a locally disintegrating system. In other words, with the parameter values belonging to Region B, the map defines a repulsion field around X_0, the motion of states receding from X_0 along vortex trajectories. Here, obviously, a Hopf bifurcation may occur: if, for parameter values $s > s_0$ a limit cycle occurs, the system is transformed, when passing over the line s_0, from a self-regulating equilibrium system to a periodically pulsating self-regulating system.

We can always study the case by keeping the value of one of the parameters constant: let this parameter be p. To study the possible existence of Hopf bifurcation, we then have to concentrate upon the dependence on the parameter s of the matrix A, its eigenvalues μ, and the vectors v and u. For the sake of uniqueness, let $\mu(s)$ be the complex eigenvalue with the plus sign before its imaginary part.

From the eigenvalue equations

$$A(s)(v + iu) = \mu(s)(v + iu) \quad \text{and} \quad A(s)(v - iu) = \mu^*(s)(v - iu) \qquad (5.31)$$

we first deduce, with the substitution $\mu(s) = \rho e^{i\omega}$:

$$(A - \rho \cos \omega \cdot I)v = -\rho \sin \omega \cdot u, (A - \rho \cos \omega \cdot I)u = \rho \sin \omega \cdot v. \qquad (5.32)$$

This gives

$$(A - \rho \cos \omega \cdot I)^2 v = -\rho^2 \sin^2 \omega \cdot v, \quad (A - \rho \cos \omega \cdot I)^2 u = -\rho^2 \sin^2 \omega \cdot u. \qquad (5.33)$$

The matrix $(A - \rho \cos \omega \cdot I)^2$ accordingly has a double eigenvalue $-\rho^2 \sin^2 \omega$. Hence, we can always (cf. Section 4.4) put

$$u = \begin{pmatrix} 1 \\ 0 \end{pmatrix}. \qquad (5.34)$$

Henceforth, the successive steps of calculation are as follows:

(1) Find the vector v by the formula (cf. (5.32))

$$v = (1/\rho \sin \omega)(A - \rho \cos \omega \cdot I)u. \qquad (5.35)$$

(2) Compute the variables $h = x - x_0$ and $k = y - y_0$ from the equation

$$\begin{pmatrix} h \\ k \end{pmatrix} = c_1 v + c_2(-u). \qquad (5.36)$$

(3) Give the result in the form

$$\begin{pmatrix} h \\ k \end{pmatrix} = M \begin{pmatrix} c_1 \\ c_2 \end{pmatrix}, \qquad (5.37)$$

where M is a 2×2 matrix, and check the computations so far made by the formula

$$M^{-1}AM = \rho \begin{pmatrix} \cos \omega & -\sin \omega \\ \sin \omega & \cos \omega \end{pmatrix}. \qquad (5.38)$$

(4) Expand the functions $\varphi_1(x, y)$ and $\varphi_2(x, y)$ in Taylorian series, at the point X_0, in terms of the variables h and k.

(5) Substitute, into the so-obtained series for $h1$ and $k1$, the complex numbers $Z = c_1 + ic_2$ and $Z^* = c_1 - ic_2$ by means of the formula (5.37).

(6) Eliminate, from the so-obtained series, the variable Z^*1, and compose the series $Z1 = \hat{\phi}(Z, Z^*)$. Check the computations by ascertaining that it obeys the formula derived by Iooss (1979):

$$Z1 = \hat{\phi}(Z, Z^*) = \mu(s)Z + R(Z, Z^*, s), \qquad (5.39)$$

where R is a function containing only second-order and higher terms in Z and Z^*.

(7) Compute the Iooss coefficients ξ_{pq} from the $Z1$-series, where they appear as coefficients of the respective $Z^p(Z^*)^q$ terms.

(8) Substitute $s = s_0$ into the Iooss coefficients, and then compute the coefficients $\alpha(s_0)$ and α by means of the formulae

$$\alpha(s_0) = \xi_{21} + \frac{2|\xi_{02}|^2}{\mu^2 - \mu^*} + \frac{|\xi_{11}|^2}{1 - \mu^*} + \frac{\xi_{11}\xi_{20}(1 - 2\mu)}{\mu^2 - \mu}, \tag{5.40}$$

$$\alpha = -\operatorname{Re}[\alpha(s_0)\mu^*]. \tag{5.41}$$

Here $\mu = \mu(s_0)$.

(10) Compute the derivative $(d\rho(s)/ds)_{s=s_0}$ and the potences $\mu^k(s_0)$ for $k = 2, 3, 4$.

Then a Hopf bifurcation exists on the conditions that

H.1. $\mu^k(s_0) \neq 1$ for $k = 1, 2, 3, 4$, $\qquad\qquad\qquad\qquad$ (5.42)

H.2. $(d\rho(s)/ds)_{s=s_0} > 0$, and $\qquad\qquad\qquad\qquad\qquad$ (5.43)

H.3. $\alpha > 0$. $\qquad\qquad\qquad\qquad\qquad\qquad\qquad\qquad\qquad$ (5.44)

Of course, we must also have $|\mu(s_0)| = 1$, which is necessary for having, at $s = s_0$, a passage from an attractive vortex field to a repulsive one (as is evident from Fig. 43, Hopf bifurcation may appear only when crossing over such a separatrice). The method given above is quite general and can always be applied to ascertain whether a Hopf bifurcation takes place or does not take place when crossing over such a separatrice in the parameter space.

To return to the parameter space of our example (Fig. 43), we can cross over the separatrice s_0 along the straight line $p = 1$. The crossing point then is $s = \frac{1}{2}$. We find, after some calculation, all the conditions of Hopf bifurcation met as follows:

H.1. $\mu(\frac{1}{2}) = (1 - i\sqrt{3})/2$, hence $\omega(\frac{1}{2}) = 60°$ so that $2\omega(\frac{1}{2}) = 120°$, $3\omega(\frac{1}{2}) = 180°$, $4\omega(\frac{1}{2}) = 240°$, all of them obeying the required $\omega(s_0) \neq 360°$.

H.2. $\rho(s) = (2s)^{\frac{1}{2}}$, thus $(d\rho/ds)_{s=\frac{1}{2}} = \frac{1}{2} > 0$.

H.3. $\alpha = 5/36 > 0$.

This kind of Hopf bifurcation is called *supercritical*. If $\alpha < 0$, with the other conditions unchanged, the Hopf bifurcation is *subcritical*: now the limit cycle appears for $s < s_0$ and is a repulsive one, sending the trajectories spiralling toward the equilibrium state X_0.

Methodological Note on Universal Causality

Part 1 has been entirely devoted to the study of the general properties of causal recursion. The reason for the approach taken here can be seen in the following methodological rule, which has never been violated in the exact sciences:

The Principle of Universal Causality:
  Every dynamical system has causal recursion in a complete state-description.

This principle has two important consequences.

First, it follows from it that causal recursion is the ultimate conceptual foundation of all sciences dealing with dynamical systems of any kind.

Second, the classification given in Part 1 of dynamical systems having causal recursion is exhaustive in the general case of dynamical systems as well – provided only that we are dealing with macroscopic systems. (When moving into mathematical dynamics in Chapter 2 we excluded the micro-world of quantum particles, but only that.)

Thus, any foundational study, for instance of the behavioural, social, or systems sciences, should also ultimately lead to the types of dynamical systems that have been investigated in Part 1. In the following Parts 2 and 3 we shall see examples of such theoretical reductions in some behavioural or social sciences dealing with *goal-seeking* dynamical systems, viz. population dynamics, economic growth theory, and the theory of social development.

In the applications given in Part 2, where causal recursions are explicitly stated, it is not implied, however, that a complete state-description has actually been reached. Such a conclusion would be premature in each of the cases discussed. The systems analysed in Part 2 should be considered as fairly comprehensive models involving causal recursion. No doubt the cases discussed there – the Verhulst approach of population dynamics, which with a slight modification extends (as we shall see) from competitive to predator–prey ecologies, and the balanced-growth systems of economics – are all more or less attempts at a complete state-description. But mathematical theory in these fields is not yet developed to permit categorical claims.

In Part 3, on other hand, only the *existence* of complete state-description, and thus of causal recursion, is assumed, not any particular form of recursion. In fact it is suggested that the topic discussed, viz. the development of human societies as wholes, belongs to the realm of 'objective complexity', where a comprehensive list of all the details of complete state-description is impossible to give, for practical, 'objective', reasons (man's life is too short for listing all the details, etc.). Nevertheless, even the mere postulate of the existence of causal recursion, when combined with the application of cybernetic entropy laws, leads to some general laws and principles concerning qualitative social development, as will be seen in Part 3.

Part 2

Simple Applications: Causal Recursion in Population Dynamics and Economic Growth Theory

(Measurable Systems)

Note: Here, *simple applications* of fundamental dynamics means applications in which causal recursion can be explicitly given, and the motion of state thus followed in detail. From the explicit construction of causal recursion it follows that the dynamical systems in question are *'measurable'* in the sense that the variables, which are essential to their behaviour, can be measured, in principle at least.

Measurable self-regulating systems appear, for instance, in population dynamics, and measurable self-steering systems in economic growth theory. First we shall consider in some detail the time-honoured Verhulstian ecosystems, again the focus of interest because of the Feigenbaum bifurcations (Feigenbaum, 1978) related to this type of models.

CHAPTER 6

Verhulstian Ecosystems and the Feigenbaum Bifurcations of their Equilibrium States

6.1. The Verhulst Axioms

Verhulstian population dynamics (Verhulst, 1838, 1845, 1847) deals with animal populations competing for food in a given area. In the basic model we have only one animal (or human) population, the magnitude of which is x, living off food whose amount in the area is a. We make the following assumptions.

(I) The more food there is, and the more animals there are, the greater the number of births of animals:

$$(\Delta x)_{\text{birth}} = kax, \qquad (6.1)$$

where k is a positive constant (the birth rate weighted by $1/a$).

(II) The more animals there are, the more deaths of animals there are:

$$(\Delta x)_{\text{death}} = -cx, \qquad (6.2)$$

where the death rate c is a positive constant.

(III) The more animals there are, and the more food there is, the greater the amount of food being consumed, there being

$$(\Delta a)_{\text{cons}} = -kax, \qquad (6.3)$$

so that the rate of consumption of food weighted by the number of animals is the same as the rate of birth of animals weighted by the amount of food.

(IV) The food arises entirely from decayed animals, the amount of food being entirely recycled into the system through the deaths of animals:

$$(\Delta a)_{\text{prod}} = cx. \qquad (6.4)$$

It follows from these assumptions that the total changes of x and a are given by

$$\Delta x = kax - cx \quad \text{and} \quad \Delta a = -kax + cx, \qquad (6.5)$$

respectively, so that

$$\Delta x + \Delta a = 0. \tag{6.6}$$

Thus, the 'total biomass' $N = x + a$ is constant, and the total state of the ecosystem can be represented by x alone, the state space being

$$X = \{x; 0 \leqslant x \leqslant N\} \subset R. \tag{6.7}$$

By the substitution $a = N - x$ we get from (6.5) the final equation of the 'motion' of state:

$$\Delta xt = (kN - c - kxt)xt. \tag{6.8}$$

This is often interpreted (cf. Nicolis and Prigogine, 1977, pp. 450–451) to mean the differential equation

$$dx/dt = (kN - c - kx)x, \tag{6.9}$$

which is known as the Verhulst equation for logistic growth. But this interpretation implies the further assumption that the ecosystem has to be a continuous dynamical system.

If that additional assumption is dropped, and the possible continuous or discontinuous nature of the dynamical system involved is left to be decided by the axioms (I)–(IV) alone, we have instead of Eq. (6.9) a richer mathematical structure with the equation

$$x(t + 1) - xt = (kN - c - kxt)xt, \tag{6.10}$$

and with the causal recursion

$$\varphi(x) = (1 + kN - c - kx)x. \tag{6.11}$$

6.2. The Preliminary (Local) Stability Analysis

Both the differential and difference equations give, along with the trivial but unstable rest state $x = 0$, the non-trivial equilibrium state

$$x_0 = N - c/k \in X \text{ for } N > c/k. \tag{6.12}$$

The 1×1 matrix $\partial\varphi/\partial x$ gives the 'eigenvalue'

$$\rho = (\partial\varphi/\partial x)_{x_0} = 1 - kN + c = k\left(\frac{c + 1}{k} - N\right). \tag{6.13}$$

For $\rho \neq 0$, i.e. for $N \neq (c + 1)/k$, the causal recursion (6.11) is locally linearizable at x_0. The formal stability analysis using the linear approximation near x_0 gives the following results:

(1) Small total biomass: $c/k < N < (c + 1)/k$. We have $0 < \rho < 1$, so that the system is locally stable and self-regulating at x_0, with a continuous suction approach to the final magnitude of the animal population x_0 that is between zero and $1/k$: $0 < x_0 < 1/k$.

(2) Medium total biomass: $(c + 1)/k < N < (c + 2)/k$. We have $-1 < \rho < 0$, and the system is still locally stable and self-regulating at x_0, but the asymptotic approach of x_0 is that of rapid vibrations, there being $1/k < x_0 < 2/k$ now.

(3) The special value $N = (c + 2)/k$. This gives $\rho = -1$, and the linearized equation of motion is

$$xt - x_0 = (-1)^t(x - x_0), \tag{6.14}$$

which holds good in an approximate sense for a small enough initial state-displacement $|x - x_0|$. This is a vibrating 2-point satellite, consisting of rapid vibrations between the points x and $x_0 + (x_0 - x)$. In this case the ecosystem is locally steerable from outside at x_0, where $x_0 = 2/k$.

(4) Large total biomass: $N > (c + 2)/k$. This corresponds to the case that $\rho < -1$. Thus, the ecosystem is locally disintegrating at x_0, through diverging rapid vibrations near x_0, from this state that is larger than $2/k$.

The special case $N = (c + 1)/k$, in which $\rho = 0$ and the causal recursion accordingly is not locally linearizable at x_0, must be treated separately. We can deal with it as follows. The causal recursion and the rest point are, in this case,

$$\varphi(x) = (2 - kx)x \text{ and } x_0 = 1/k, \tag{6.15}$$

respectively. It follows that

$$\varphi(x_0 + b) = \varphi(x_0 - b) = (1 - b^2k^2)x_0 \quad for \ 0 < b < 1/k, \tag{6.16}$$

so that

$$x1 = (1 - b^2k^2)x_0 \text{ for both } x = x_0 + b \text{ and } x_0 = x - b. \tag{6.17}$$

If the initial state is $x = x_0 + b$, the step $x \rightarrow x1$ means jumping over to the other side of x_0. For $x = x_0 - b$ no jump over x_0 takes place.

After the possible first jump over x_0 the state xt approaches x_0 monotonously, there being

$$xt = [1 - (bk)^{2^t}]x_0 \rightarrow x_0 \text{ with } t \rightarrow +\infty \quad for \ b < 1/k. \tag{6.18}$$

Hence, we can add the following result:

(5) The special value $N = (c + 1)/k$, with $\rho = 0$. This is the value of the total biomass that separates the domains of N where the system approaches x_0 continuously or through rapid vibrations, respectively (see Fig. 44), and indeed the asymptotic approach of x_0 is a curious combination of rapid vibrations and continuous approach: for an initial state $x = x_0 + b$ (with $0 < b < 1/k$) first a jump over x_0 takes place, after which the approach to x_0 is continuous suction; for an initial state $x = x_0 - b$ the approach is that of continuous suction from the very beginning.

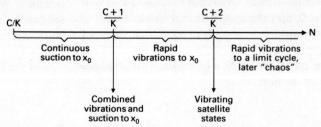

FIG. 44. The continuum of total biomass in the 1-dimensional Verhulstian ecosystem. For $N \leqslant (c + 2)/k$ the picture indicates the local behaviour of state near the equilibrium state x_0. For $n > (c + 2)/k$ it anticipates some later results concerning the divergence of state from x_0.

Surveying the local results (cf. Fig. 44) we observe that the causal-recursion approach confirms the assumption of continuous system only for a small total biomass $N < (c + 1)/k$, which corresponds to a low level of final animal population, viz. $x_0 < 1/k$.

With increasing biomass the asymptotic approach to the equilibrium state x_0 becomes, after the curious interlude encountered for the special value $N = (c + 1)/k$, that of rapid vibrations. The ecosystem thus becomes reducibly discontinuous, but for $(c + 1)/k < N < (c + 2)/k$ it is still locally stable and survives small disturbances from the equilibrium state x_0.

The amount of food that is available in the equilibrium state x_0 is always the same, since it is defined by

$$a_0 = N - x_0 = c/k, \qquad (6.19)$$

which is constant if the death rate c and the birth rate k weighted by the amount of food are constants. But with a larger total biomass an individual animal has, on average, less to eat, since the number of animals then increases. This is reflected in the vibration of their number until, at $N = (c + 2)/k$, the equilibrium state x_0 becomes unstable.

First, for the special value $N = (c + 2)/k$, the ecosystem is locally steerable from outside at x_0; any small 'external disturbance' pushes the ecosystem from x_0 to another precarious equilibrium defined by a satellite state, until another disturbance throws it again into another satellite state. When the total biomass further increases, the ecosystem becomes locally disintegrating at x_0, and x_0 becomes a repelling state, like $x = 0$. In this domain only a global analysis can tell what will become of the system.

The analysis of global behaviour of the Verhulstian ecosystems $S(N, c, k)$, to be undertaken now, will much improve on the results of local analysis even in other respects.

6.3. Verhulstian Ecosystems as Self-regulating Equilibrium Systems: a Small or Medium Total Biomass

(1) Some general rules of the motion of state

By rewriting the basic difference equation (6.10) in the form

$$x(t + 1) - xt = k(x_0 - xt)xt \qquad (6.20)$$

we get the rules

$$xt < x_0 \Rightarrow x(t + 1) > xt \quad \text{(Rule a)},$$
$$xt > x_0 \Rightarrow x(t + 1) < xt \quad \text{(Rule b)}$$

illustrated in Fig. 45. Thus, the state xt always moves in the direction of the vector pointing from xt to x_0.

In Case (a) we get from Eq. (6.20) the further rules

$$x_0 - xt > 0 \Rightarrow \frac{x(t + 1) - xt}{x_0 - xt} = kxt \begin{cases} > 1 \text{ for } xt > 1/k & \text{(Rule a1)}, \\ = 1 \text{ for } xt = 1/k & \text{(Rule a2)}, \\ < 1 \text{ for } xt < 1/k & \text{(Rule a3)}. \end{cases}$$

For Rule a3 we get the two sub-rules $a3_1$ and $a3_2$ shown in Fig. 46, depending on whether $x_0 > 1/k$ or $x_0 \leqslant 1/k$.

In case (b) the corresponding rules are

$$xt - x_0 > 0 \Rightarrow \frac{xt - x(t + 1)}{xt - x_0} = kxt \begin{cases} > 1 \text{ for } xt > 1/k & \text{(Rule b1)}, \\ = 1 \text{ for } xt = 1/k & \text{(Rule b2)}, \\ < 1 \text{ for } xt < 1/k & \text{(Rule b3)}. \end{cases}$$

Now there are the two sub-rules $b1_1$ and $b1_2$ shown in Fig. 47, according to whether $x_0 > 1/k$ or $x_0 \leqslant 1/k$.

These simple rules help us to solve entirely the global behaviour of small Verhulstian ecosystems, and they also make a good beginning for a study of those of medium size.

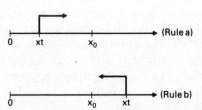

FIG. 45. The rules that indicate the direction of the motion of state in a Verhulstian ecosystem.

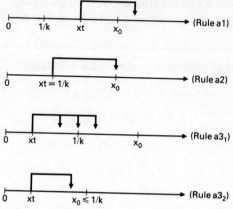

FIG. 46.   The rules that indicate the next step taken by a state $xt < x_0$ in a Verhulstian ecosystem.

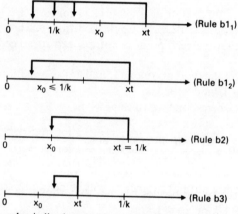

FIG. 47.   The rules indicating the next step taken by a state $xt > x_0$ in a Verhulstian ecosystem.

(2) Quasi-continuous small ecosystems: N ⩽ (c + 1)/k

These are systems that have a small total biomass N, and that are locally continuous near the equilibrium magnitude of animal population x_0 that obeys $x_0 \leqslant 1/k$. By applying the above rules we can easily show that these small ecosystems survive any disturbance of state, $d = x - x_0$, where the perturbed state x may be any state of X except for the state $x = 0$.

For an initial population $x < x_0$ we get, by a repeated application of Rule $(a3_2)$ of Fig. 46, a monotonous growth of xt towards x_0, which it reaches with $t \to +\infty$. The change function

$$f(x) = \varphi(x) - x = kx(x_0 - x) \tag{6.21}$$

obeys

$$f(x) = 0 \text{ for } x = 0 \text{ and } x = x_0, \quad df/dx = 0 \text{ for } x = x_0/2, \qquad (6.22)$$

there being $f(x_0/2) = kx_0^2/4 > 0$. Thus, the growth of the animal population towards the equilibrium value x_0 has the familiar form of 'logistic growth' (Fig. 48) known from the Verhulst differential equation (6.9).

When the initial state obeys $x_0 < x < 1/k$, we get, by a repeated application of Rule (b3) of Fig. 47, a monotonous decrease of the animal population toward the equilibrium value x_0, which is again reached with $t \to +\infty$. Thus, in both of the cases so far discussed, the global behaviour of the Verhulstian ecosystem as defined by the causal recursion (6.11) is of the same type as that obtained from the Verhulst differential equation. In other words, the system behaves as if it were globally continuous.

There are only two remainders of the fact that these systems, as generated by the causal recursion (6.11), are globally discontinuous. First, if the initial state is $x = 1/k$, while $x_0 < 1/k$, the system jumps right away to equilibrium: Rule (b2) of Fig. 47 gives $x = 1/k \to x1 = x_0$ (see Fig. 49).

Second, for an initial state $x > 1/k$ (but $x \leqslant N$) we first get, by Rule (b1$_2$) of Fig. 47, a jump over the equilibrium point: $x \to x1 < x_0$. Here we must stop to ensure that the zero point is not reached or exceeded, i.e. that

$$x1 = x + kx(x_0 - x) > 0. \qquad (6.23)$$

This is equivalent to

$$x < x_0 + 1/k. \qquad (6.24)$$

But this is always true for all $x \leqslant N$, since $x_0 + 1/k = N - c/k + 1/k > N$ for $c < 1$, and in any non-trivial ecosystem the rate of death of the animals, c, must be smaller than one: the extreme value $c = 1$ means that all the individuals that are alive at moment t will be dead at moment $t + 1$.

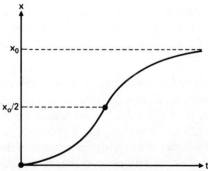

FIG. 48. The 'logistic' curve showing the approach of the equilibrium state x_0 in a quasi-continuous Verhulstian ecosystem.

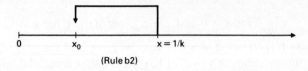

(Rule b2)

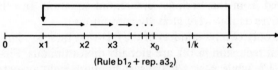

(Rule b1$_2$ + rep. a3$_2$)

FIG. 49. The anomalies of the motion of state encountered in a quasi-continuous Verhulstian ecosystem.

From the state $x1$ onwards we get, even for $x > 1/k$, by a repeated application of Rule (a3$_2$) of Fig. 46, a monotonous growth of the animal population towards the equilibrium value x_0, which is reached with $t \to +\infty$ (Fig. 49).

Thus, we have shown that quasi-continuous small ecosystems are globally stable, i.e. self-regulating, everywhere in the state-space $X = [0, N]$ except for the point $x = 0$. The domain of stability (or survival) of the Verhulstian systems $S(N, c, k)$ is

$$D_0 = (0, N] \text{ for } c/k < N \leqslant (c + 1)/k. \tag{6.25}$$

Had we started with the Verhulst differential equation (6.9), we would have found the domain of stability D_0 to be valid for all the values of the total biomass N. This is not true for the causal recursion (6.11) and the resulting difference equation (6.10), which we can consider as the complete mathematical representation of the Verhulstian ecosystems $S(N, c, k)$. The results concerning Eq. (6.10), reported above and in the following sections, come from Aulin (1987c).

Note on Ecocatastrophes Related to Hunger.   The two sudden decreases of the animal population observed above for a large initial population, i.e. the recorded jumps from the values $x = 1/k$ and $x > 1/k$, can be given an ecological interpretation. Obviously, for a constant total biomass, N, the amount of food, a, becomes scarce with the increase of the animal population (since $at = N - xt$). The scarcity of food makes a sudden reduction of the population necessary. In the domain of vibratory systems, i.e. for $N > (c + 1)/k$ (to be considered later), such hunger catastrophes become commonplace, as is evident from the vibratory nature of the development of populations. In the following discussions we shall not consider this phenomenon separately any more, as it is obvious enough.

(3) Vibrating ecosystems of medium size: $(c + 1)/k < N \leqslant (c + 2)/k$

This is the medium range of the total biomass, N, which locally, i.e. for small enough displacements $d = x - x_0$ of state, displayed rapid (period 2) vibrations converging to x_0 or, for the special value $N = (c + 2)/k$, a vibrating 2-point satellite state. The equilibrium point now obeys $1/k < x_0 \leqslant 2/k$.

The global behaviour is again determined by the rules of Figs 46 and 47. For an initial state $x = 1/k$ Rule (a2) gives a jump $x \to x1 = x_0$, so that the equilibrium state is reached immediately. For $x < 1/k$ a repeated application of Rule (a3$_1$) brings the state xt, at some moment t, either to the point $1/k$ or over it so that $xt > 1/k$. In the former case x_0 is again reached immediately, by Rule (a2). In the latter case we apply Rule (a1) to get the state $x(t + 1)$ right of x_0: $x(t + 1) > x_0$. Then we can apply Rule (b1$_1$), and then again either Rule (a1) or Rule (a2), or a repeated Rule (a3$_1$) followed by either (a2) or (a1), as the case may be. Thus, the state either jumps to x_0 or keeps vibrating over it (see Fig. 50).

The initial states x obeying $1/k < x < x_0$ or $1/k < x_0 < x$ bring nothing new to the above processes. In the former case we begin with Rule (a1), go on with Rule (b1$_1$), and then apply either Rule (a1) or (a2) or a repeated

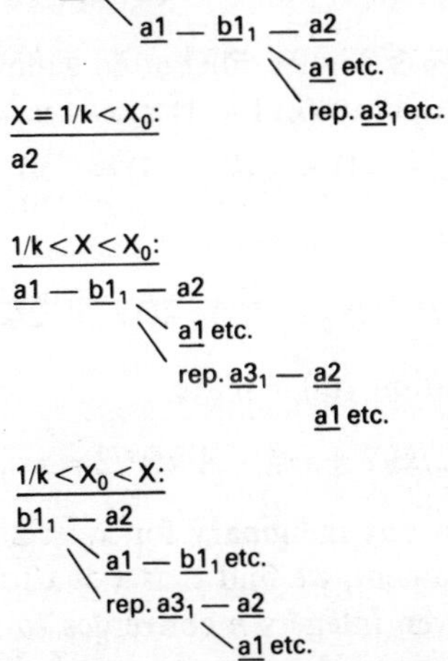

FIG. 50. The algorithms for the determination of global behaviour in a Verhulstian ecosystem for the medium range of total biomass.

Rule $(a3_1)$ followed by either (a1) or (a2), etc., as above. In the latter case we begin with Rule $(b1_1)$, and continue as before. The state again either, at some stage of the process, jumps directly to x_0 or keeps vibrating over it (cf. Fig. 50).

(4) The convergence of state vibrations in the medium range of biomass

Let us now study the convergence vs divergence of the vibrations of state over the point of equilibrium x_0. For $xt < x_0$, Eq. (6.20) gives

$$x(t + 1) - x_0 = (x_0 - xt)(kxt - 1) < x_0 - xt \Leftrightarrow xt < 2/k, \qquad (6.26)$$

which in the medium range is always true, since now $xt < x_0 \leqslant 2/k$. For $xt > x_0$ we similarly get

$$x_0 - x(t + 1) = (xt - x_0)(kxt - 1) \leqslant xt - x_0 \Leftrightarrow xt \leqslant 2/k. \qquad (6.27)$$

Thus, the vibrations over x_0 converge, provided that the initial population is small enough, viz. $x \leqslant 2/k$.

But we can enlarge the potential domain of stability to initial populations larger than $2/k$. Let us start with $x > 2/k$. By (6.27) the first step is a 'divergent' one: $x_0 - x1 > x - x_0$. But the next step is, by (6.26), a 'convergent' one: $x2 - x_0 < x_0 - x1$. Under which condition does the latter step win, so that $x2 < x$ for $x > 2/k \geqslant x_0$? The answer is obtained by the following argument.

Since

$$x2 = \varphi(x1) = x1 + kx1(x_0 - x1), \qquad (6.28)$$

we get successively for $x - x_0 > 0$ and $x_0 - x1 > 0$:

$$x2 \leqslant x \Leftrightarrow x2 - x_0 \leqslant x - x_0 \Leftrightarrow x1 - x_0 + kx1(x_0 - x1)$$
$$\leqslant x - x_0 \Leftrightarrow (x_0 - x1)(kx1 - 1) \leqslant x - x_0 \Leftrightarrow kx1 - 1$$
$$\leqslant (x - x_0)/(x_0 - x1) = 1/(kx - 1) \Leftrightarrow (kx1 - 1)(kx - 1) \leqslant 1, \quad (6.29)$$

which is equivalent to

$$y(x) = x^2 - (x_0 + 2/k)x + (x_0 + 2/k)/k \geqslant 0. \qquad (6.30)$$

The roots of the equation $y(u) = 0$ are

$$u_{1,2} = (x_0 + 2/k)/2 \pm (x_0^2 - 4/k^2)^{\frac{1}{2}}/2 = (x_0 + 2/k)/2 \pm v. \qquad (6.31)$$

But here the expression v is imaginary for $x_0 < 2/k$, which makes $x2 < x$ true. Repeating the argument, we find that $x > x2 > x4 > x6 > \ldots$, so that the sequence xn with even integers n converges to $x_0 < 2/k$ (provided that the limit $x = N$ is not exceeded in the course of vibrations, which we have to check later).

To prove the convergence of the sequence xn for odd integers n as well, it suffices to study the derivative of the function

$$g(xt) = x_0 - x(t+1) = (xt - x_0)(kxt - 1), \qquad (6.32)$$

for which we get:

$$dg/dxt = 2kxt - kx_0 - 1 > 0 \text{ for } xt > (x_0 + 1/k)/2. \qquad (6.33)$$

But xt is surely larger than the average of $1/k$ and x_0 for $xt > x_0 > 1/k$, which is now the case. Thus, it is valid for all the states in the sequence $x2$, $x4$, $x6$,... Then it follows from (6.33) that $x1 < x3 < x5 < ...$, so that the sequence xn converges to the equilibrium point x_0 for odd integers n as well. This completes the proof of convergence of the vibrations in the medium range of N for $x > 2/k$, provided that the limit $x = N$ is not exceeded in these vibrations (which we shall soon check).

In the remaining case, that $x_0 = 2/k$, the roots u_1 and u_2 of Eq. (6.30) reduce to the double root $u = 2/k$. Thus, for an initial population $x = u = 2/k$ we have, of course, $xt = 2/k = x_0$, since this is the equilibrium state now. For $x \neq 2/k$, while $x_0 = 2/k$, we again get the convergence of vibrations by the above argument.

(5) The full domain of stability in the medium range of N

Let us now take up the limitations to the convergence of vibrations due to the possible exceeding of the limits of state-space in the course of these vibrations. For the limit $x = 0$, the formulae (6.23) and (6.24) apply here as well, and the lower limit of state-space is never even reached. For the upper limit $x = N$, we have to satisfy the following condition:

$$x1 = x + kx(x_0 - x) \leqslant N \qquad (6.34)$$

there being, in the medium range of N,

$$1/k < x < x_0 \leqslant 2/k. \qquad (6.35)$$

In view of $x_0 = N - c/k$ the condition (6.34) is equivalent to

$$z(x) = x^2 - (x_0 + 1/k)x + (x_0 + c/k)/k \geqslant 0. \qquad (6.36)$$

The roots of the equation $z(x) = 0$ are

$$x_{\pm} = (x_0 + 1/k)/2 \pm [(x_0 - 1/k)^2 - 4c/k^2]^{\frac{1}{2}}/2. \qquad (6.37)$$

They are imaginary for

$$x_0 < (1 + 2c^{\frac{1}{2}})/k, \text{ i.e. for } N < M = (c + 1 + 2c^{\frac{1}{2}})/k, \qquad (6.38)$$

and reduce to a double real root for $N = M$. In both cases the condition $x1 \leqslant N$ is fulfilled.

For $c \geqslant 1/4$ we have $M \geqslant (c + 2)/k$, so that the condition $N \leqslant M$ is, in this case, automatically met in the medium range of total biomass. Thus, for a death rate larger than or equal to $1/4$ the domain of stability (or survival) of the ecosystem $S(N, c, k)$ of medium size comprises the whole state-space except for the unstable equilibrium state $x = 0$:

$$D_0 = (0, N] \text{ for } (c + 1)/k < N \leqslant (c + 2)/k \text{ and } c \geqslant 1/4. \qquad (6.39)$$

For $c < 1/4$ the critical value M is smaller than $(c + 2)/k$, and thus may already impose limitations on the domain of stability in the medium range of N. But for $N \leqslant M$ we have the unreduced domain of stability:

$$D_0 = (0, N] \text{ for } (c + 1)/k < N \leqslant M \text{ and } c < 1/4. \qquad (6.40)$$

(6) The reduced domain of stability in the medium range of N

As soon as the actual total biomass N is between the critical value M and $(c + 2)/k$, which is possible for $c < 1/4$, we get a reduced domain of stability for an ecosystem $S(N, c, k)$ of medium size. In this case we have two positive roots x_+ and x_-, located symmetrically on different sides of the 'middle point'

$$N_0/2 = (x_0 + 1/k)/2 \qquad (6.41)$$

of the causal recursion diagram (Fig. 51). They obey

$$1/k < x_- < N_0/2 < x_+ < x_0 < N. \qquad (6.42)$$

If the initial state x is between x_- and x_+, the condition $x1 \leqslant N$ is not met, but the ecosystem perishes. To study the so-reduced domain of survival we can first easily prove, using (6.11), that

$$\varphi[(1 - c)N] \leqslant N \text{ for } N \leqslant \hat{N} = \frac{(2 - c)/(1 - c)}{k}. \qquad (6.43)$$

But for $N < (c + 2)/k$ the condition $N \leqslant \hat{N}$ is always met (see Fig. 54). It follows from $\varphi[(1 - c)N] < N$ that (see Fig. 51) $(1 - c)N > x_+$, i.e. that $\varphi(N) > x_+$. By using this we can see from Fig. 51 that the vibrations of state never diverge, if they start from an initial state $x \in [x_+, N]$.

By means of Fig. 51 we can also see that the whole domain of stability (or survival) can be written as the infinite set-theoretical sum

$$D_1 = [x_+, N] + [z_1, x_-] + [z_2, y_1] + [z_3, y_2] + \dots \qquad (6.44)$$

for $M < N \leqslant (c + 2)/k$ and $c < 1/4$. Here the interval limits obey

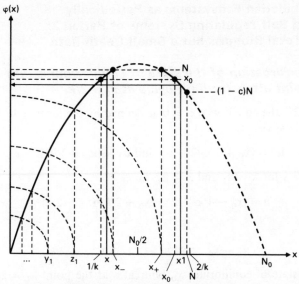

FIG. 51. The causal-recursion function of the one-dimensional Verhulstian ecosystem. The forbidden zone (x_-, x_+) and the first derived forbidden zone (y_1, z_1) are shown. The 'cap' and 'tail' are indicated by dashed lines on the curve of the φ-function. The picture also illustrates how a state $x < x_-$ in its motion $x \rightarrow x1$ jumps over the forbidden zone.

$$\varphi(y_1) = x_-, \quad \varphi(y_2) = y_1, \quad \varphi(y_3) = y_2, \ldots,$$
$$\varphi(z_1) = x_+, \quad \varphi(z_2) = z_1, \quad \varphi(z_3) = z_2, \ldots \quad (6.45)$$

It follows that the survival of the ecosystem $S(N, c, k)$ for the values of parameters mentioned is precarious, as it depends on whether the initial state is within some of the intervals (6.44). But if it is, the vibrations of state converge asymptotically to the equilibrium state x_0. If it is not, the vibrations of state diverge over the upper limit of state-space, and the system disintegrates and perishes.

Note on the Chaotic Regimes. When x approaches zero the 'forbidden zones' of Fig. 51 become dense everywhere. Hence, if the accuracy of measurement of the population x is limited, there is a *chaotic regime* in some domain $(0, x_{\min})$. If the initial population falls into this domain we cannot tell whether the ecosystem will survive or perish, as we cannot distinguish, within the limits of the accuracy of measurement, whether x is in the domain of stability D_1 or outside it. Similar chaotic regimes appear in connection with Feigenbaum bifurcations, to be considered later (see Section 6.5) before every limit N_∞ of a series of bifurcations. This note on the appearance of chaotic regimes will not be repeated there.

6.4 Verhulstian Ecosystems as Periodically Pulsating Self-regulating Systems of Period 2: a Large Total Biomass but a Small Death Rate

(1) The bifurcation of the state of equilibrium x_0 into a two-point attractor limit cycle at $x_0 = 2/k$

For $x_0 > 2/k$ the equation $y(x) = 0$ as defined by the formula (6.30) has two positive roots, viz.

$$u_1 = (x_0 + 2/k)/2 + v \quad \text{and} \quad u_2 = (x_0 + 2/k)/2 - v. \tag{6.46}$$

Thus, $x2 = x$ for $x = u_1$ and $x = u_2$. Since

$$v = (x_0^2 - 4/k^2)^{\frac{1}{2}}/2 = (x_0 + 2/k)^{\frac{1}{2}}(x_0 - 2/k)^{\frac{1}{2}}/2 \tag{6.47}$$

we have

$$u_1 \to x_0 \quad \text{and} \quad u_2 \to x_0 \quad \text{for } x_0 \to 2/k. \tag{6.48}$$

Thus, the state of equilibrium x_0 bifurcates at the point $x_0 = 2/k$ into the two points u_1 *and* u_2 (see Fig. 52).

We can easily check, by using (6.11), that

$$\varphi(u_1) = u_2 \quad \text{and} \quad \varphi(u_2) = u_1. \tag{6.49}$$

Thus, the process $u_1 \to u_2 \to u_1 \to \cdots$ defines a 2-point cycle. The mechanism of the bifurcation of this cycle from the state x_0 is illustrated in Fig. 52.

Next we have to check two things: first, for which values of parameters do the points $x = u_1$ and $x = u_2$ belong to the state-space $[0, N]$, so that we have a real limit cycle? Second, for which values of parameters is it an attractor, locally at least?

Since $v > (x_0 - 2/k)/2$ and

$$u_1 \leqslant N \Leftrightarrow x_0 \leqslant [2 + c^2/(1 - c)]/k \Leftrightarrow N \leqslant \frac{(2 - c)/(1 - c)}{k} = \hat{N} \tag{6.50}$$

we have $0 < u_2 < 2/k < x_0 < u_1 \leqslant$ N for $N \leqslant \hat{N}$. Thus, the function $\hat{N}(c)$ determines the limit in the parameter space (see Fig. 54), below which the 2-point limit cycle keeps within the state-space $X = [0, N]$.

Turning to the second question we first observe, by applying the argument (6.29)–(6.31) again for $x > x_0$, the following rule:

$$x2 > x \text{ for } x_0 < x < u_1, \quad \text{while } x2 < x \text{ for } x_0 < u_1 < x. \tag{6.51}$$

Modifying the argument (6.29)–(6.31) for the case $x < x_0$ we can see that

$$x2 > x \text{ for } 0 < x < u_2, \quad \text{while } x2 < x \text{ for } u_2 < x < x_0. \tag{6.52}$$

The rules (6.51) and (6.52) are illustrated in Fig. 53.

What we need for proving existence of an attractor limit cycle, however,

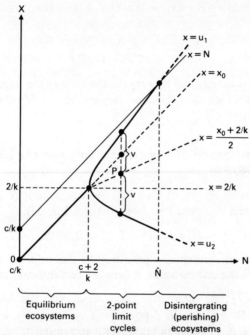

FIG. 52. The mechanism of the first Feigenbaum bifurcation leading to a 2-point limit cycle. At the value $\hat{N}$ of total biomass the 2-point limit cycle becomes unstable in the way shown in the picture.

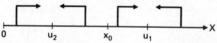

FIG. 53. The rules indicating the directions of rapid vibrations toward the 2-point limit cycle (u_1, u_2) in a Verhulstian ecosystem.

is to prove that

$$x < x2 < u_1 \text{ for } x_0 < x < u_1, \quad u_1 < x2 < x \text{ for } x_0 < u_1 < x, \quad (6.53)$$

and the corresponding rules for u_2. We can confine ourselves to a local discussion here (the global behaviour and the full or reduced domains of stability of the system $S(N, c, k)$ for $N > (c + 2)/k$ can be analysed just as in Section 6.3).

As for the local stability of our 2-point limit cycle, we can study the derivative

$$dx2/dx = (dx2/dx1)(dx1/dx) = (1 + kx_0 - 2kx1)(1 + kx_0 - 2kx). (6.54)$$

Hence we get:

$$(dx2/dx)_{u_1} = (1 + kx_0 - 2ku_1)(1 + kx_0 - 2ku_2) = (dx2/dx)_{u_2}. \quad (6.55)$$

This gives

$$(dx2/dx)_{u_1,u_2} > 0 \text{ for } x_0 < 5^{\frac{1}{2}}/k, \text{ i.e. for } N < (c + 5^{\frac{1}{2}})/k = \tilde{N}. \quad (6.56)$$

Thus, it is the function $\tilde{N}(c)$ that marks the limit in the parameter space (see Fig. 54), below which the bifurcated 2-point limit cycle is locally stable, and thus an attractor.

(2) The significance of the critical value of the death rate: $\hat{c} = 0.381966\ldots$

The next question is: on which condition is the limit $\hat{N}$ smaller than the limit $\tilde{N}$? A trivial computation shows that (cf. Fig. 54):

$$\hat{N} \leqslant \tilde{N} \Leftrightarrow c \leqslant (3 - 5^{\frac{1}{2}})/2 = 0.38196.. = \hat{c}. \quad (6.57)$$

Thus, we have to distinguish, according to the value of the death rate c compared with the critical value $\hat{c}$, between two different cases:

(i) $c \leqslant \hat{c}$: *only one bifurcation may occur*, viz. the bifurcation of the equilibrium state x_0 into the 2-point limit cycle (u_1, u_2), which obtains for the values $N \leqslant \hat{N}$ of total biomass. With increasing total biomass, i.e. for $N > \hat{N}$, the ecosystem loses its stability and perishes (cf. Figs 52 and 54).

(ii) $c > \hat{c}$: *further bifurcations may occur*, according to the Feigenbaum scheme (Feigenbaum, 1978). This case will be investigated in Section 6.5.

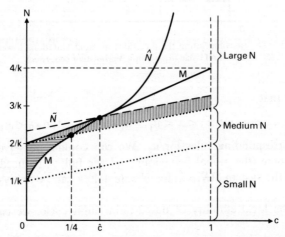

FIG. 54. The separatrices and regions in the parameter space of a Verhulstian ecosystem with a constant birth-parameter k. The region of stable 2-point limit cycles is shaded with vertical lines, that of the reduced domain of stability for $N < (c + 2)/k$ with horizontal lines.

But there is something more to say about the critical value $\hat{c}$. In fact, there is a 'triple coincidence' of the functions $\hat{N}$, $\tilde{N}$, and M for this value:

$$\hat{N}(\hat{c}) = \tilde{N}(\hat{c}) = M(\hat{c}). \tag{6.58}$$

The curves $N = \hat{N}(c)$ and $N = M(c)$ touch each other at this point of the parameter space, there being (see Fig. 54)

$$\hat{N}(c) > M(c) \text{ for } c \neq \hat{c}. \tag{6.59}$$

The mutual order of magnitude of M, $\hat{N}$, and $\tilde{N}$ is (cf. Fig. 54)

$$M < \hat{N} < \tilde{N} \text{ for } c < \hat{c}, \quad \text{while } \tilde{N} < M < \hat{N} \text{ for } c > \hat{c}. \tag{6.60}$$

All the functions $M(c)$, $\hat{N}(c)$, and $\tilde{N}(c)$ are monotonously increasing, their ranges being (cf. Fig. 54)

$$1/k = M(0) \leqslant M(c) \leqslant M(1) = 4/k, \tag{6.61}$$

$$2/k = \hat{N}(0) \leqslant \hat{N}(c) \leqslant \hat{N}(1) = \infty, \tag{6.62}$$

$$5^{\frac{1}{2}} = \tilde{N}(0) \leqslant \tilde{N}(c) \leqslant \tilde{N}(1) = 1 + 5^{\frac{1}{2}}. \tag{6.63}$$

(3) The domains of stability for $c \leqslant \hat{c}$

In the triangle-shaped area of Fig. 54 bordered by the curves $N = M$, $N = \hat{N}$, and the straight line $N = (c + 2)/k$, we again have, since $N > M$ there, the forbidden zones. From the condition

$$x1 \leqslant N \tag{6.64}$$

it follows (by the argument that we used previously to justify the reduced domain of stability but to be applied here to the now-relevant modification of Fig. 51) that the domain of stability of the 2-point limit cycle (u_1, u_2) is given, by the same infinite set-theoretical sum as before, the reduced domain of stability of the equilibrium point x_0, viz.

$$D_1 = [x_+, N] + [z_1, x_-] + [z_2, y_1] + [z_3, y_2] + \cdots$$
$$\text{for } (c + 2)/k < N \leqslant \hat{N} \text{ and } c < 1/4, \tag{6.65}$$

the intervals being those defined by (6.45).

But the value $1/4$ of the death rate c is also the value on which the curve $N = M(c)$ reaches the line $N = (c + 2)/k$ in the parameter space, there being (see Fig. 54)

$$M(c) < (c + 2)/k \text{ for } c < 1/4,$$
$$M(c) = (2 + c)/k \text{ for } c = 1/4, \quad \text{and}$$
$$M(c) > (c + 2)/k \text{ for } c > 1/4. \tag{6.66}$$

It follows that we have the full domain of stability of the 2-point limit cycle

$$D_0 = (0, N] \text{ for } 1/4 < c \leqslant \hat{c} \text{ and } (c + 2)/k < N \leqslant M. \tag{6.67}$$

The system is disintegrating and perishes for $N > \hat{N}$, so that we have an empty domain of stability in this case

$$D = \varnothing \text{ for } c \leqslant \hat{c} \text{ and } N > \hat{N}. \tag{6.68}$$

6.5. The Feigenbaum Bifurcations and their Interruptions in Verhulstian Ecosystems in the General Case: a Large Death Rate and Total Biomass

(1) 'Cap effect' and 'tail effect'

We have seen above that the scheme of Feigenbaum bifurcations (Feigenbaum, 1978) may be interrupted by the condition that the magnitude of the animal population, x, cannot exceed the upper limit, N. If the limit is exceeded, which for a small death rate $c \leqslant \hat{c}$ happens for $N > \hat{N}(c)$, the system is no longer self-regulating but is disintegrating, and perishes.

In fact Feigenbaum's assumptions (Feigenbaum, 1978) would involve, when translated into our terms, that $\varphi(x) \to 0$ with $x \to N$. But we have followed the Verhulstian axioms of population dynamics, which lead to the causal recursion (6.11), and thus to (see Fig. 55)

$$\varphi(x) \to 0 \text{ for } x \to x_0 + 1/k = N + (1 - c)/k > N \text{ for } 0 < c < 1, \tag{6.69}$$

there being

$$\varphi(N) = (1 - c)N > 0 \text{ for } c < 1. \tag{6.70}$$

The Verhulstian ecosystems, therefore, only satisfy the Feigenbaum conditions in the limit case, where the rate of mortality c approaches the extreme value one. For $c = 1$ all the individuals that are alive at moment t would be

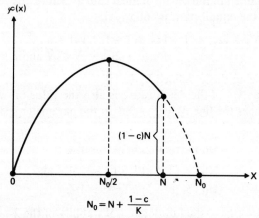

$$N_0 = N + \frac{1 - c}{K}$$

FIG. 55. An illustration showing that the causal recursion of a Verhulstian ecosystem belongs to the class of Feigenbaum functions only in the limit $c \to 1$.

dead at moment $t + 1$. Because of a positive birth rate this does not necessarily mean the extinction of the total animal population, and the value $c = 1$ may be possible for some populations met in microbiology. But for normal animal populations we have $0 < c < 1$, and the scheme of Feigenbaum bifurcations is interrupted at some value of total biomass, say $N_F(c)$, because of the condition that $x < N$. This rule, as we have already seen, is valid for $c \leqslant \hat{c}$ with the value

$$N_F(c) = \hat{N}(c) \quad \text{for } c \leqslant \hat{c}. \tag{6.71}$$

Let us consider the mechanism of interruption in the case $c \leqslant \hat{c}$ in some more detail. In this case the interruption of the Feigenbaum scheme of bifurcations was due to the fact that the point u_1, belonging to the 2-point limit cycle that emerged from the first bifurcation, exceeded the upper limit of state-space, $x = N$ (see Fig. 52). But this implies that the other point u_2 belongs to the 'forbidden zone':

$$\varphi(u_2) = u_1 > N \Leftrightarrow u_2 \in (x_-, x_+). \tag{6.72}$$

Obviously this can be generalized to any m-point limit cycle produced by the Feigenbaum scheme of bifurcations, m being any integer larger than one. In other words, as soon as one of the points of the limit cycle, say u, exceeds the upper limit $x = N$ of state-space, there is another point of the same limit cycle, the *predecessor* u^p of u in this limit cycle, for which

$$\varphi(u^p) = u > N, \quad \text{and thus } u^p \in (x_-, x_+), \tag{6.73}$$

hold good.

Conversely, if one of the points of the limit cycle, u, enters the 'forbidden zone' (x_-, x_+) of state-space (see Fig. 52), its *follower* u^f in this limit cycle exceeds the limit $x = N$, since

$$\varphi(u) = u^f > N \quad \text{for } u \in (x_-, x_+). \tag{6.74}$$

Thus, we have the following general rule (Aulin, 1987b):

> The 'cap effect' and the 'tail effect', i.e. that some of the points of a limit cycle belonging to the Feigenbaum scheme of bifurcations enters the forbidden zone (x_-, x_+) or exceeds the limit $x = N$, respectively, always occur simultaneously, i.e. for the same values of the parameters c and N, and together they lead to the interruption of the scheme of bifurcations at the respective point $(c, N_F(c))$ of the parameter space.

It follows from the rule that we can study the interruptions of bifurcations, which take place in the normal case that $c < 1$, by considering the exceeding of the upper limit of state-space, $x = N$, and forget the 'forbidden zones'. The latter are needed only for a study of the global domains of stability, such as was undertaken in earlier sections but will not be done here. First, however, we have to discuss the ideal case of uninterrupted Feigenbaum bifurcations, i.e. the case that $c = 1$.

(2) The ideal case of uninterrupted Feigenbaum bifurcations: $c = 1$ and $N > (c + 2)/k = 3/k$

In this case, where a total renewal of the whole population takes place from any moment t to moment $t + 1$, there are no limitations for all Feigenbaum bifurcations to occur. Thus, the qualitative picture of the different kinds of systems obtained for different values of the parameter N is as follows (Feigenbaum, 1978).

Using the notations

$$N_1 = (c + 2)/k = 3/k \quad \text{for } c = 1, \tag{6.75}$$

$$N_2 = N = (c + 5^{\frac{1}{2}})/k = (3.2360679..)/k \text{ for } c = 1, \tag{6.76}$$

we have the first bifurcation leading to the 2-point attractor limit cycle (u_1, u_2) at $N = N_1$. At the value $N = N_2$ the second bifurcation takes place: each of the points u_1 and u_2 bifurcates into two new points, which together make up a 4-point attractor limit cycle (see Fig. 56).

With increasing value of the total biomass, N, the first cascade of bifurcations goes on as illustrated in Fig. 56. There is the mth bifurcation leading to a 2^m-point attractor limit cycle at a value $N = N_m$, until we come to the limit $N = N_\infty^{(2)}$, where the 'set of harmonics' of the 2-point limit cycle

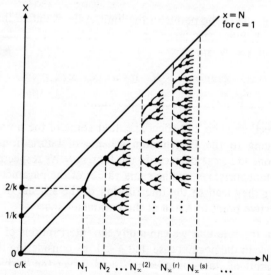

FIG. 56. The full scheme of Feigenbaum bifurcations obtained in Verhulstian population dynamics for the death rate $c = 1$.

(u_1, u_2) ends:

$$N_1 < N_2 < N_3 < \cdots < N_\infty^{(2)}. \tag{6.77}$$

By denoting here

$$\Delta N_m = N_m - N_{m-1} \tag{6.78}$$

we have (Feigenbaum, 1978) the rule of convergence,

$$\lim_{m \to \infty} \Delta N_{m+1}/\Delta N_m = 1/\delta, \tag{6.79}$$

with the Feigenbaum number $\delta = 4.66920\ldots$ Thus, the convergence of the sequence (6.77) is very strong.

Following the Feigenbaum scheme (see Fig. 56), a new set of harmonics is started at $N = N_\infty^{(2)} = N_1^{(r)}$, where an r-point limit cycle is bufurcated from x_0. At some value $N = N_2^{(r)} > N_1^{(r)}$ it gives way to a $4r$-point limit cycle, and so on, there being the bifurcation of an $r \cdot 2^m$-point limit cycle at some value $N = N_m^{(r)}$. Again, a limit value $N = N_\infty^{(r)}$ exists, and the convergence obeys the formula (6.79) with the same universal constant δ, as has been shown by Feigenbaum (1978). (*Note*: these limit cycles may be repelling.)

A new set of harmonics, say that of an s-point limit cycle, is again started at $N = N_\infty^{(r)} = N_1^{(s)}$, and so we get an endless scheme of sets of harmonics, whose mutual order, it seems (see Feigenbaum, 1978), depends on the precise form of the causal recursion, the first set of harmonics, viz. that of the 2-point cycle, being the same for all.

All the bifurcated points in these sets of harmonics of various basic limit cycles $2, r, s, \ldots$ are situated, for $c = 1$, below the limit $N = x_0 + c/k = x_0 + 1/k$, as is illustrated in Fig. 56. Thus, no interruption of the scheme of bifurcations occurs in this ideal case.

(3) The normal case of interrupted bifurcations for $1 > c > \hat{c} = 0.381966\ldots$ and $N > (c + 2)/k$

We already know that the interruption of the first bifurcated limit cycle (u_1, u_2) for a small death rate $c < \hat{c}$ takes place for the value

$$N_F(c) = \hat{N}(c) = \frac{(2 - c)/(1 - c)}{k} \quad \text{(valid for } c < \hat{c}). \tag{6.80}$$

For these small death rates only the first bifurcation actually takes place: when the total biomass, N, increases, the ecosystem perishes.

For a larger death rate, c, the scheme of bifurcations continues, and we have at least the second bifurcation occurring a the value $N = N(c) > \hat{N}(c)$ (see Fig. 54). However, the bifurcations are interrupted at the *critical value of total biomass* $N_F(c)$, whose magnitude depends on the death rate, c (see Fig. 57). We get the following general rule (Aulin, 1987b):

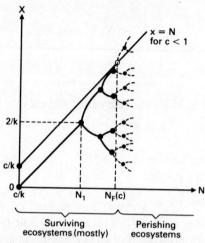

FIG. 57. The mechanism of interruption of the scheme of Feigenbaum bifurcations for a normal death rate $\hat{c} < c < 1$ in Verhulstian population dynamics. The interruption of the scheme takes place when the uppermost bifurcated branch exceeds the limit $x = x_0 + c/k = N$, which for $\hat{c} < c < 1$ is bound to happen at some value $N_F(c)$.

If, for a given value of the death rate, c, the total biomass exceeds the critical value $N_F(c)$, the rapid vibrations of state exceed the upper limit $x = N$ of state-space, and the ecosystem perishes.

We have, by now, shown that in each case the Verhulst axioms lead to the destruction of the ecosystem for large enough total biomass. (Is this the story of Sahara and the Middle-Eastern deserts, briefly told?) The result is more interesting, as the Verhulstian nonlinear model was initially constructed in order to avoid the difficulties connected with the linear Malthusian model, where the growth of population led to the destruction of the ecosystem in the end.

Note. From the universality proved by Feigenbaum (1978) it follows that the results obtained here are not restricted to the special φ-function used here: they remain true for any de-capped and de-tailed Feigenbaum function.

CHAPTER 7

The Role of Self-regulation in Biological Evolution and in Different Ecologies

7.1. The Survival of the Fittest

If two species of animals, with the respective populations x_1 and x_2, are living off the same kind of food, whose amount in the area is a, the total biomass will be $N = x_1 + x_2 + a$. Instead of Eq. (6.10), the Verhulst axioms now give the two-component causal recursion

$$x(t + 1) = \varphi(xt), \quad \text{with} \tag{7.1}$$

$$\varphi_i(x) = (1 - c_i)x_i + k_i(N - x_1 - x_2)x_i \quad \text{for } i = 1, 2. \tag{7.2}$$

For the total state $x = (x_1, x_2)$ we get the equilibrium states $x = (0,0)$,

$$x' = (n_1, 0), \quad \text{and} \quad x'' = (0, n_2). \tag{7.3}$$

Here the notation $n_i = N - c_i/k_i$ is used. Now we have:

$$\partial \varphi / \partial x = \begin{pmatrix} 1 + k_1 n_1 - k_1(2x_1 + x_2) & -k_1 x_1 \\ -k_2 x_2 & 1 + k_2 n_2 - k_2(2x_2 + x_1) \end{pmatrix}. \tag{7.4}$$

This gives, for the steady states, the eigenvalues

$$\rho_1(0,0) = 1 + k_1 n_1, \qquad \rho_2(0,0) = 1 + k_2 n_2, \tag{7.5}$$

$$\rho_1(x') = 1 - k_1 n_1, \qquad \rho_2(x') = 1 + k_2(n_2 - n_1), \tag{7.6}$$

$$\rho_1(x'') = 1 + k_1(n_1 - n_2), \quad \rho_2(x'') = 1 - k_2 n_2. \tag{7.7}$$

Thus, only x' and x'' can be equilibrium goals, the former only if $n_1 > n_2$, the latter if the reverse is true. This means that only the 'fittest' species survives. The same result is obtained from the Verhulst differential equations of the case.

But again the causal recursion (7.1)–(7.2) gives the additional k-conditions on local stability (and self-regulation):

$$\text{for } x': \quad 0 < n_1 < 2/k_1 \quad \text{and} \quad 0 < n_1 - n_2 < 2/k_2, \tag{7.8}$$

$$\text{for } x'': \quad 0 < n_2 < 2/k_2 \quad \text{and} \quad 0 < n_2 - n_1 < 2/k_1. \tag{7.9}$$

The complete result concerning the asymptotic approach to the state x' is given in Fig. 58, which shows the parameter space associated with this state. As a curiosity, even a nilpotent causal recursion (of index 2) appears, corresponding to the special point $(n_1 = 1/k_1, n_1 - n_2 = 1/k_2)$ of the parameter space.

If $n_1 = n_2$, while the k-conditions are met, our ecosystem is steerable from outside in both its goal states, x' and x'': various satellite states occur, in which the two species may enjoy a precarious coexistence. If the k-conditions are not met, bifurcations may appear first, but the fittest as well as the unfit perish, for a large enough N.

7.2. The Coexistence of Different Species

The proper solution for a permanent coexistence of different species is based on their using different kinds of food. Let us consider two biomasses,

$$N_1 = a_1 + x_1 + \beta x_2 \quad \text{and} \quad N_2 = a_2 + x_2 + \beta x_1, \qquad (7.10)$$

where a_1 is the food mainly used by species 1 but also, in a proportion indicated by the coefficient β, by the other species (a_2 being interpreted in a similar way). This gives the causal recursion

$$\varphi: \begin{cases} \varphi_1(x) = (1 - c_1)x_1 + k_1(N_1 - x_1 - \beta x_2)x_1, \\ \varphi_2(x) = (1 - c_2)x_2 + k_2(N_2 - x_2 - \beta x_1)x_2, \end{cases} \qquad (7.11)$$

or, if a differential equation is used, the continuous system determined by

$$dx_1/dt = k_1(n_1 - x_1 - \beta x_2)x_1, \quad dx_2/dt = k_2(n_2 - x_2 - \beta x_1)x_2, \quad (7.12)$$

where the notations $n_i = N_i - c_i/k_i$ $(i = 1, 2)$ have been used.

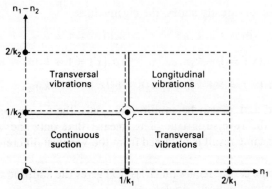

FIG. 58. The parameter space associated with the equilibrium state x' in a Verhulstian ecosystem of two competitive species. A dashed line marks a boundary whose points are not included in the region in question. The special point $(1/k_1, 1/k_2)$ gives a nilpotent causal recursion.

The closer to zero is β, the less the two species are using the same food, and with $\beta \to 0$ the two ecosystems do not interact, and we are back in the case where Eq. (6.10) applies to each of them. With $\beta \to 1$ the food resources of the different species become the same, and we are back in the zero-sum game studied in Section 7.1. For $0 < \beta < 1$ we get, in addition to the trivial solution $x = (0,0)$, three other steady-state solutions:

$$x' = (n_1, 0), \ x'' = (0, n_2), \text{ and } x_0 = \left(\frac{n_1 - \beta n_2}{1 - \beta^2}, \frac{n_2 - \beta n_1}{1 - \beta^2} \right). \quad (7.13)$$

By applying again the method of local linearization, we observe that the system is self-regulating in each of these states on the following conditions:

for x': $\qquad \beta n_1 > n_2 \text{ and } n_1 < \min\left\{ \frac{2}{k_1}, \frac{1}{\beta}\left(\frac{2}{k_2} + n_2 \right) \right\},$ $\qquad$ (7.15)

for x'': $\qquad n_2 > n_1/\beta \text{ and } n_2 < \min\left\{ \frac{2}{k_2}, \frac{1}{\beta}\left(\frac{2}{k_1} + n_1 \right) \right\},$ $\qquad$ (7.14)

for x_0: $\beta n_1 < n_2 < n_1/\beta \text{ and } (n_1 - \beta n_2)(n_2 - \beta n_1) < \frac{1 - \beta^2}{\beta^2} \cdot \frac{2}{k_1 k_2}.$ $\qquad$ (7.16)

If a continuous (suction) approach to the steady-state goal is required, we must replace $2/k_i$ by $1/k_i$ every time it appears above, and $2/k_1 k_2$ by $1/k_1 k_2$. Otherwise the approach happens through rapid fluctuations.

If we pay attention only to the first conditions mentioned above, we again get the same results from both the difference and differential equations. This result is illustrated in Fig. 59. Hence, we see that the introduction of the auxiliary parameter β has cleared a space between the domains of self-regulation of x' and x'' for the third domain of self-regulation, viz. that of the state x_0, where both species are coexisting. For the special points $n_2 = \beta n_1$ and $n_2 = n_1/\beta$ we have:

$$n_2 = \beta n_1 \ \Rightarrow x_0 = x' \ \text{ and } \rho_1 = 1, \quad (7.17)$$

$$n_2 = n_1/\beta \Rightarrow x_0 = x'' \ \text{ and } \rho_2 = 1. \quad (7.18)$$

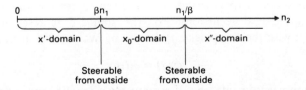

FIG. 59. The regions of validity of different equilibrium states, indicated in terms of the animal population n_2, in a Verhulstian ecosystem of two species using partly the same, partly different food: a domain of co-existence appears in the medium range of (both) animal populations.

With these critical values of the parameter n_2 the system becomes steerable from outside: any causal change of populations will affect their mutual relations.

The total animal population in the area will be the largest when both species are living in it permanently:

$$x_{0_1} + x_{0_2} = \frac{n_1 + n_2}{1 - \beta} > \max\{n_1, n_2\}. \tag{7.19}$$

Likewise, if one of the species wins, and the other one becomes extinct, the population of the winning species is again larger than if the other one had won. Thus, 'evolution leads to the steadily growing exploitation of each niche' (Nicolis and Prigogine, 1977, p. 457; their analysis was based on differential equations).

But again, the difference equation gives further conditions, and the total biomass becomes unsupportable by the area if its size increases over the limit corresponding to the value $N_F(c)$ in the ecosystems $S(N, c, k)$.

Note on Ecological Self-organization. If the ecosystem of our examples is self-organizing, i.e. capable of changing the values of the parameters of its own causal recursion (such as the c_i, k_i, n_i, N_i, and β above), the system may move from one of the domains of stability to another. Self-organization might be accomplished by mutations or other 'innovations' on some other level of state-description from that used above. Such ecological self-organization would be of the 'Prigogine type' (see Section 2.4), as distinct from the 'Ashby type', in which the changes in parameter values are caused on the same level of state-description on which the causal recursion is defined.

Note on Age-dependent Population Dynamics. If predictions concerning the development of age distributtion in a given population are required, we must use a model, where the total state x is the age distribution. In the 'general model' developed by G. F. Webb (1985), a continuous dynamical system composed of positive half-trajectories xR^+ in a Banach space B is investigated. Due to the mathematical nature of the states $x \in B$ as distributions, the problem setting and the underlying assumptions now take the form of integral rather than of differential equations. But again, a continuous dynamical system emerges, and can be given representations in terms of discrete-time causal recursions. For the standard representation φ the connection with the mappings $S(t)$ defined by Webb (1985, p. 73) is

$$\varphi^t(x) = S(t)x \tag{7.20}$$

for any non-negative integer t. If a reducibly discontinuous system, instead of a continuous one, is introduced by starting with a discrete-time formulation, rapid vibrations will again appear, but this possible extension of age-dependent population dynamics will not be discussed in this book.

7.3. General Remarks on Self-regulation in Competitive Ecology

Let us again consider two species, with the respective populations x and y, living in a certain area. For the variables of change, Δx and Δy, we can write

$$\Delta x = X(x, y)x, \quad \Delta y = Y(x, y)y, \tag{7.21a}$$

where X and Y are the *rates of growth* of x and y respectively. Following the current terminology, we have in this ecosystem

(1) *competitive ecology*, if

$$\partial X/\partial y < 0 \quad \text{and} \quad \partial Y/\partial x < 0, \tag{7.21b}$$

(2) *predator–prey ecology*, with the predator x and prey y, if

$$\partial X/\partial y > 0 \quad \text{but} \quad \partial Y/\partial x < 0, \tag{7.22}$$

(3) *symbiosis ecology*, if

$$\partial X/\partial y > 0 \quad \text{and} \quad \partial Y/\partial x > 0. \tag{7.23}$$

The complete Verhulst axioms, when formulated for an ecosystem of two species, obviously define competitive ecology. They give (cf. (7.11) or (7.12))

$$X(x, y) = k_1(n_1 - x - \beta y), \quad Y(x, y) = k_2(n_2 - y - \beta x), \tag{7.24}$$

so that

$$\partial X/\partial y = -k_1\beta < 0 \quad \text{and} \quad \partial Y/\partial x = -k_2\beta < 0. \tag{7.25}$$

The 'curves' defined by the equations

$$X(x, y) = 0 \quad \text{and} \quad Y(x, y) = 0, \tag{7.26}$$

denoted in Fig. 60a by ξ and η, respectively, are the straight lines

$$y = (n_1 - x)/\beta \quad (\xi)$$

and

$$y = n_2 - \beta x \quad (\eta), \tag{7.27}$$

as $0 < \beta < 1$. Their intersection is the point of equilibrium, where both species coexist:

$$(\hat{x}, \hat{y}) = \left(\frac{n_1 - \beta n_2}{1 - \beta^2}, \frac{n_2 - \beta n_1}{1 - \beta^2} \right). \tag{7.28}$$

Above the line ξ we have $X < 0$, and below it $X > 0$. Similarly, above η there is $Y < 0$ and below it $Y > 0$. Hence, we can draw the arrows, shown in Fig. 60a, indicating the prevalent direction of the motion of state in each of the areas separated from each other by the lines ξ and η, and by the co-

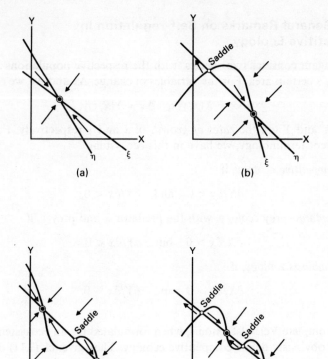

FIG. 60. The alternation of equilibrium and saddle points in a continuous ecosystem of two competing species, as illustrated in the cases where the rate of growth $X(x, y)$ of the first population is (a) a linear function, (b) a polynomial of second degree, (c) a polynomial of third degree, and (d) a polynomial of fourth degree, while the rate of growth $Y(x, y)$ is linear in each case.

ordinate axes. For continuous systems the arrows suggest that the point of intersection of ξ and η is a 'sink', i.e. an equilibrium goal of the respective differential-equations system.

The same simple method also serves us in more complicated cases, as shown by Figs 60b, 60c, and 60d (Hirsch and Smale, 1974), where the rate of growth, X, has been supposed to be a a polynomial of second, third, and fourth degree, respectively. This rough geometric analysis (to be secured by analytic means) reveals at the points of intersection of ξ and η both equilibrium goal states and saddle points, as shown in the figures. Moreover, their appearance is quite independent of the actual form of the curves ξ and η, but depends only on their mutual positions. The only restrictions as to the form of these curves are dictated by the conditions of competitiveness, (7.21), and by the following conditions that we can set:

$$X(x,0) > 0 \text{ for } 0 \leqslant x < a \quad \text{and} \quad Y(0,y) > 0 \text{ for } 0 \leqslant y < b, \quad (7.29)$$

and

$$X(x,0) < 0 \text{ for } x > a \quad \text{and} \quad Y(0,y) < 0 \text{ for } y > b, \quad (7.30)$$

a and b being finite positive numbers. It follows from (7.21), (7.29), and (7.30) that every straight line $x = x_0$ for $0 \leqslant x_0 < a$ intersects ξ exactly once in the positive sector, and not at all if $x_0 > a$. Similarly every straight line $y = y_0$ for $0 \leqslant y_0 < b$ intersects η exactly once in this sector, and not at all if $y_0 > b$.

The topologies in Fig. 60 suggest that every *continuous* ecosystem of two competing species is what we have called a self-regulating equilibrium system, possibly having several distinct equilibrium goal states with their own domains of stability. This can indeed by proved (Hirsch and Smale, 1974, Chapter 12, Section 3) for differentiable systems.

Another conclusion from Fig. 60 concerns the frequent appearance of saddle points in these ecosystems. Take, for instance, that one in Fig. 60b, situated between a 'sink' and the point $(0, b)$. Suppose that the curve α shown in Fig. 61 is the right-hand branch of the 'inset' that a saddle point always has (cf. Fig. 39a). Then α is always the separatrice between the trajectories, like the one through the point P, that tend to $(0,b)$, and those, like the one through the point Q, tending to the sink. If the ecosystem, by some external disturbance, is pushed the small distance from Q to P, it is set on a course where it is going to lose one of its species in the end. This is what is called the *bifurcation of behaviour* in an ecosystem, where a small 'environmental disturbance' may engender a great change.

Hirsch and Smale (1974) call such a change an 'ecological catastrophe' (p. 273). But it is not! It is only the loss of one species in an ecosystem, which otherwise goes on unharmed. This may be a deplorable loss, but it hardly deserves the name of catastrophe, at least not in comparison with the

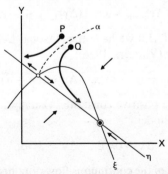

FIG. 61. The 'bifurcation of behviour' (Hirsch and Smale, 1974) in a continuous ecosystem of two competitive species: the two trajectories starting from the two neighbouring points P and Q, respectively, diverge from each other giving rise to widely different behaviours of the system.

real catastrophes occurring in discontinuous ecosystems, derived from the Verhulst axioms, when total biomass exceeds a limit of survival (Chapter 6): the ecosystem as a whole is destroyed. No doubt the bifurcations of behaviour discussed by Hirsch and Smale appear in discontinuous ecosystems too, along with total catastrophes.

7.4. On Self-regulation in Predator–Prey Ecology

(1) The modifications of Verhulst axioms for a predator–prey ecology

Let us consider two animal populations x and y, the prey x being the food of the predator y, while the food of the prey is abundantly available all the time, and can be considered as constant. Then the Verhulst axioms (I)–(IV), when applied to the prey population, reduce to the following:

(I) (for the prey): the more animals there are, the amount of food remaining the same, the more births there are of these animals;

(II) (for the prey): the prey being the food of the predators, the more predators there are and the more prey animals there are, the more prey animals are being eaten by predators, and thus the more deaths of prey animals there are.

Thus, the positive change of x due to the births of prey animals is proportional to x, while the negative change due to the deaths of prey animals is proportional to both x and y:

$$(\Delta x)_{birth} \sim x, \quad (\Delta x)_{death} \sim -xy. \tag{7.31}$$

For predators the Verhulst axioms (I) and (II) apply unchanged. The prey, x, now being the food of predators, y, we get:

$$(\Delta y)_{birth} \sim xy, \quad (\Delta y)_{death} \sim -y. \tag{7.32}$$

Combining (7.31) and (7.32) we have the classical Lotka–Volterra model of predator–prey ecology (Volterra, 1936):

$$\Delta x = (a - by)x, \quad \Delta y = (cx - d)y, \tag{7.33}$$

where a, b, c, and d are positive constants. Usually it has been applied as a differential-equations model:

$$\Delta x = dx/dt, \quad \Delta y = dy/dt.$$

A geometric analysis of the continuous flow obtained, if (7.33) is interpreted as a differential equation, gives the diagram of Fig. 62a, that suggests continuously oscillating states, either periodical satellites or causal vortex. The former possibility is borne out by the observation that (7.33), interpreted

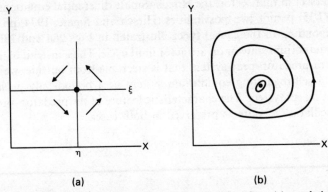

FIG. 62. The characteristic fluctuations of prey (x) and predator (y) populations, as obtained from the Lotka–Volterra model: (a) the approximate directions of trajectories given by the 'geometric method', (b) the shapes of trajectories given by a global analysis.

as differential equations, are Hamiltonian equations with the Hamiltonian function

$$H(q, p) = \mathrm{d}q + \mathrm{a}p - c\mathrm{e}^q - b\mathrm{e}^p, \quad x = \mathrm{e}^q, y = \mathrm{e}^p. \tag{7.34}$$

The closed curves $H = \text{const.}$, illustrated in Fig. 62b, are satellites, and the Lotka–Volterra differential equations thus define a periodically pulsating system that is steerable from outside.

(2) The Hirsch–Smale correction to the Lotka–Volterra equations

An improvement of the Lotka–Volterra model was suggested by Hirsch and Smale (1974). They introduced, along with the assumptions leading to (7.33) and (7.34), the further axiom that the growth of both predator and prey populations must be limited.

Instead of (7.33) we now get

$$\Delta x = (a - by - ux)x, \quad \Delta y = (cx - d - vy)y, \tag{7.35}$$

where u and v are positive limited-growth constants. The straight lines ξ (i.e. $X = 0$) and η (i.e. $Y = 0$) now are given by

$$y = (a - ux)/b \quad \text{with } u/b > 0 \quad (\xi), \tag{7.36}$$

$$y = (cx - d)/v \quad \text{with } c/v > 0 \quad (\eta), \tag{7.37}$$

respectively.

If ξ and η do not intersect in the positive sector at all, predators die out and the prey population converges to the final magnitude $\hat{x} = a/u$. If they do intersect in that sector, the Hirsch–Smale differential equations obtained from (7.35) permit two possibilities (Hirsch and Smale, 1974, pp. 264–265) corresponding to the causal types illustrated in Figs 28a and 28b: either a final equilibrium state or an attractor limit cycle. Thus, instead of the Lotka–Volterra predator–prey system that is steerable from outside, we now have either a self-regulating equilibrium system or a periodically pulsating self-regulating ecosystem. The characteristic feature of the predator–prey system, viz. oscillation of state, is preserved in both cases.

CHAPTER 8

The Conditions of Self-steering of Economic Development

8.1. A Heuristic Model of Self-steering and Self-organization of Economic Development in the Very Long Run: the Generalized Steinmann–Komlos Model

The current neoclassical theory of economic growth, given in the textbooks of economics, can best be considered as a rather rough approximation of reality. Its fundamental state variables are not exactly measurable, although they are considered as quantitative variables for which causal recursions of definite functional form can be suggested. They are theoretical constructs intended to help the understanding of the great outlines of economic development.

This being the case, one could expect that the very first question to be answered by that theory should be: does the theory offer an explanation of the main historical facts concerning the development of economic systems through the ages of social history? Surprisingly this question seems to have been broached, within the framework of neoclassical theory, only quite recently in a model (Steinmann and Komlos, 1988) whose mathematical formulation in a slightly generalized form will be given here.

(1) The generalized Steinmann–Komlos model

Let the total state of the production system be given by

$$x = (Q, R, K, L), \qquad (8.1)$$

where

Q = the total output of the system,
R = the natural resources available,
K = the capital available,
L = the labour force available,

each of them being represented by a positive number.

For a society living in a geographically limited area we can assume R to be constant, and write the equations of motion of the state in their characteristic neoclassical form as follows:

$$\dot{Q}/Q = \sigma + \beta \dot{K}/K + \gamma \dot{L}/L, \quad \text{i.e. } Q_t = Ae^{\sigma t}R^\alpha K_t^\beta L_t^\gamma, \tag{8.2}$$

$$\dot{K} = sQ - \delta K, \quad \dot{L} = gL, \quad \alpha + \beta + \gamma = 1, \tag{8.3}$$

where $\exp(\sigma t)$ stands for technological development, and all the constants are positive, s being the rate of saving, δ the rate of depreciation of capital (we have completed the model by the addition of this term: depreciation did not feature in the original Steinmann–Komlos model), and g the rate of growth of the labour force (i.e. population).

The labour equation (8.3) of course gives immediately

$$L_t = L_0 e^{gt}. \tag{8.4}$$

The capital equation (8.3) can be written as

$$\dot{K} = sAR^\alpha L_0^\gamma e^{(\sigma + \gamma g)t}K_t^\beta - \delta K_t. \tag{8.5}$$

This is a Bernoulli equation, which can easily be integrated (for computational details see Section 8.3) to give

$$K_t = [Ce^{-(1-\beta)\delta t} + De^{(\sigma + \gamma g)t}]^{\frac{1}{1-\beta}}, \tag{8.6}$$

$$C = K_0^{1-\beta} - D, \quad D = \frac{sAR^\alpha L_0^\gamma(1-\beta)}{\sigma + \gamma g + (1-\beta)\delta}. \tag{8.7}$$

Hence, by (8.2), we get the output Q_t:

$$Q_t = AR^\alpha L_0^\gamma e^{(\sigma + \gamma g)t}[Ce^{-(1-\beta)\delta t} + De^{(\sigma + \gamma g)t}]^{\frac{\beta}{1-\beta}}. \tag{8.8}$$

The equations (8.4), (8.6), and (8.8) together give the general solution of the equations of motion (8.2)–(8.3).

By putting $C = 0$ we get the special solution

$$\hat{K}_t = \hat{K}_0 e^{\lambda t}, \quad \hat{Q}_t = \hat{Q}_0 e^{\lambda t}, \quad \hat{L}_t = L_t, \tag{8.9}$$

$$\hat{K}_0 = D^{1/(1-\beta)}, \quad \hat{Q}_0 = AR^\alpha L_0^\gamma D^{\beta/(1-\beta)}, \tag{8.10}$$

$$\lambda = \frac{\sigma + \gamma g}{1 - \beta}. \tag{8.11}$$

This is what, in the neoclassical theory of economic growth, is called the *golden age path*.

(2) Self-steering to the golden age path

It is readily seen from (8.6) and (8.8) that the convergence of the general solution to the special solution, viz. to the golden age path,

$$K_t \to \hat{K}_t, \quad Q_t \to \hat{Q}_t, \quad \text{for } t \to \infty, \tag{8.12}$$

holds good if, and only if

$$0 < \lambda/\delta < \beta/(\alpha + \gamma). \tag{8.13}$$

Thus, we have the following result*:

Theorem 8.1.1. The self-steering of economic development to the golden age path takes place if, and only if, the rate of economic growth in proportion to the rate of depreciation of capital is smaller than the ratio $\beta/(\alpha + \gamma)$ of elasticities.

(For a discussion of the condition (8.13) see Section 8.3.2.)

Other asymptotic properties of the model are listed below:

$$\lim_{t \to \infty} Q_t/K_t = \lim_{t \to \infty} \hat{Q}_t/\hat{K}_t = \hat{Q}_0/\hat{K}_0 = (\lambda + \delta)/s, \qquad (8.14)$$

$$\lim_{t \to \infty} K_t/L_t = \lim_{t \to \infty} Q_t/L_t = 0 \quad \text{for } \sigma < \alpha g, \qquad (8.15)$$

$$\lim_{t \to \infty} L_t/K_t = \lim_{t \to \infty} L_t/Q_t = 0 \quad \text{for } \sigma > \alpha g, \qquad (8.16)$$

$$\lim_{t \to \infty} K_t/L_t = (1/L_0^\gamma)D^{\frac{1}{1-\beta}} = \frac{\lambda + \delta}{s}\left(\lim_{t \to \infty} Q_t/L_t\right) \quad \text{for } \sigma = \alpha g. \qquad (8.17)$$

Obviously, we have to distinguish between three cases:

Case a: $\sigma < \alpha g$ (Self-steering to economic misery). This is the case of very slow technological progress or stagnation. In this case the *Malthusian trap* is realized: the population L grows faster than total output, so that there is an ever-smaller amount of output for each individual.

Case b: $\sigma > \alpha g$ (Self-steering to wealth). This is the case of a technologically advancing society. The Malthusian trap is avoided, and the ratio Q/L grows, meaning increasing wealth.

Case c: $\sigma = \alpha g$. This case represents a theoretical possibility and an ideal, where the magnitudes σ and αg coincide. In such a case we have

$$\lambda = g$$

so that the golden age path reduces to a *balanced growth path* (this is always the case for $\sigma = \alpha = 0$, of course),

$$\hat{K}_t = \hat{K}_0 e^{gt}, \quad \hat{Q}_t = \hat{Q}_0 e^{gt}, \quad L_t = L_0 e^{gt}, \qquad (8.18)$$

represented in the state space (Q,K,L) by a straight line through the origin. Since in this case the ratio Q_t/L_t approaches a constant with $t \to \infty$, the wealth per capita tends to a constant value. Since, on the other hand, the rate σ of technical progress in this case is balanced by the elasticity α of exploitation of natural resources and by the rate g of growth of population, the case may be of some interest as an objective of economy.

---

*It is easy to show that, in view of the general definitions of dynamical systems with a full causal recursion given in Section 2.3, the economic system discussed here is steerable from outside for $\lambda/\delta = \beta/(\alpha + \gamma)$, disintegrating for $\lambda/\delta > \beta/(\alpha + \gamma)$, and displays the degeneration of self-steering to self-regulation for zero growth or negative growth, i.e. for $\lambda \leqslant 0$.

In all the cases *a–c*, as can be seen from (8.14), the ratio of savings, sQ, (all of which is invested) to capital, K, approaches $\lambda + \delta$ whether the golden age path is reached or not.

(3) The curves of equivalence of output

Let us consider the constant value of production function at a fixed point of time $\hat{t}$. By (8.2)–(8.3) we have:

$$Q = Ae^{\sigma\hat{t}}R^{1-\beta-\gamma}K^{\beta}L^{\gamma}, \tag{8.19}$$

or, equivalently,

$$\log Q = B + \beta\log(K/R) + \gamma\log(L/R), \tag{8.20}$$

$$B = \log A + \log R + \sigma\hat{t}. \tag{8.21}$$

In terms of the variables β and γ the curve of equivalence of $\log Q$ is a straight line. Its slope is given by

$$\frac{d\gamma}{d\beta} = -\frac{\log(K/R)}{\log(L/R)}, \tag{8.22}$$

there being

$$\frac{\partial \log Q}{\partial \beta} = \log(K/R) \quad \text{and} \quad \frac{\partial \log Q}{\partial \gamma} = \log(L/R). \tag{8.23}$$

Let us study the different kinds of economies according to the mutual magnitudes of the hypothetical factors of production, R, K, and L.

(i) *Hunting and gathering economy*: $R > L > K$

At this stage of economic development natural resources are the most important factor of production, labour the second in the order of importance, while capital is still very scarce. Thus we have, according to (8.22) and (8.23):

$$\frac{d\gamma}{d\beta} < -1, \quad \frac{\partial \log Q}{\partial \beta} < 0, \quad \frac{\partial \log Q}{\partial \gamma} < 0. \tag{8.24}$$

The resulting 'curves' of equivalence of $\log Q$ and the direction of growth of Q are shown in Fig. 63.1.

(ii) *Agricultural economy*: $L > R > K$

Now labour has passed natural resources as the most important factor of production, while capital remains the least important. Hence we get:

$$\frac{d\gamma}{d\beta} > 0, \quad \frac{\partial \log Q}{\partial \beta} < 0, \quad \frac{\partial \log Q}{\partial \gamma} > 0. \tag{8.25}$$

The curves of equivalence and the direction of growth of $\log Q$ now are those indicated in Fig. 63.2.

(iii) *Industrial economy*: $L > K > R$

Capital has started its rise as an important factor of production, passing natural resources in this first phase of 'capitalism'. We have now:

$$-1 < \frac{d\gamma}{d\beta} < 0, \quad \frac{\partial \log Q}{\partial \beta} > 0, \quad \frac{\partial \log Q}{\partial \gamma} > 0. \tag{8.26}$$

The corresponding curves of equivalence and the direction of growth of $\log Q$ are shown in Fig. 63.3.

(iv) *Post-industrial economy*: $K > L > R$

Now the position of capital as the most important factor of production has been established, while labour has returned to the position of second largest factor. We get:

$$\frac{d\gamma}{d\beta} < -1, \quad \frac{\partial \log Q}{\partial \beta} > 0, \quad \frac{\partial \log Q}{\partial \gamma} > 0. \tag{8.27}$$

This is illustrated in Fig. 63.4. Let it be noted that our conception of capital includes the intellectual capital invested in production.

(4) The three technological revolutions as self-organizations of the economic system

For the sake of simulation of technological revolutions, let us make the convention that the most important factor of production is given the exponent (i.e. the value of α, β, or γ, as the case may be) 0.6, the second largest factor the exponent 0.3, and the smallest factor the exponent 0.1. Then the four stages of economic development that were above distinguished from each other get the following numbers:

		α	β	γ
(i)	$R > L > K$	0.6	0.1	0.3
(ii)	$L > R > K$	0.3	0.1	0.6
(iii)	$L > K > R$	0.1	0.3	0.6
(iv)	$K > L > R$	0.1	0.6	0.3

The three technological revolutions, or 'structural changes of economy', if you like, can now be understood in view of two principles of self-organization:

(I)   The elasticity (β or γ) of the fastest-growing factor of production (K or L) tends to increase on each stage of development.

(II)   The total output Q is maximized, within the limits of each form of economy.

The structural changes of economy are then derived as follows.

(i) → (ii):   *Agricultural revolution*

In a hunting and gathering economy (Fig. 63.1), with the gradual growth of population, labour is the fastest-growing factor of production. Thus, the exponent γ of L tends to grow from its original value 0.3 toward the maximum 0.6, as indicated in Fig. 63.1 by the thick arrow. The principle of maximization of Q keeps the value of β simultaneously at its minimum 0.1. Thus, the agricultural economy (ii) follows the hunting and gathering economy (i).

Let us assume, for the sake of simplicity, a constant rate of growth, g, of population of the magnitude 0.25%. With the value $\alpha = 0.6$ holding good for a hunting and gathering economy, the threshold of the avoidance of the Malthusian trap in this kind of society is

$$\sigma_{min} = \alpha g = 0.6 \cdot 0.0025 = 0.00150. \tag{8.28}$$

In the development (i) → (ii) towards an agricultural society the threshold is lowered to half:

$$\sigma_{min} \text{ (ii)} = 0.3 \cdot 0.0025 = 0.00075. \tag{8.29}$$

Thus, the Malthusian trap – the decrease of the average share of total output available to each individual – becomes a little easier to avoid than it was originally, in a hunting and gathering society. However, no essential change is yet taking place, and it is safer to assume that both agricultural and hunting–gathering economies belong to *Case a* of the preceding section: the Malthusian trap still prevails.

(ii) → (iii): *Industrial revolution*

Now capital, K, and its elasticity, β, start to grow within an agricultural society (Fig. 63.2). On the other hand, the principle of maximization of the output, Q, keeps the value of γ, according to Fig. 63.2, at its maximum 0.6. Thus, β can only reach the value 0.3. The resulting route to industrial revolution is indicated in Fig. 63.2 by the thick arrow.

In this structural change of economy the threshold of avoidance of the Malthusian trap is essentially lowered, being now 1/6 of what it was originally:

$$\sigma_{min}\text{(iii)} = 0.1 \cdot 0.0025 = 0.00025. \tag{8.30}$$

Accordingly, the society most probably now avoids the Malthusian trap, and we have the *Case b* of self-steering to wealth.

(iii) → (iv): *Post-industrial revolution.*

Capital continues to grow within an industrial society, and its elasticity β reaches the maximum value 0.6 (see Fig. 63.3). Thus, labour loses its position as the most important factor of production, and its elasticity γ has to decrease. Because of the principle of maximization of output; however, it remains at the next largest value, 0.3, so that the resulting change from an

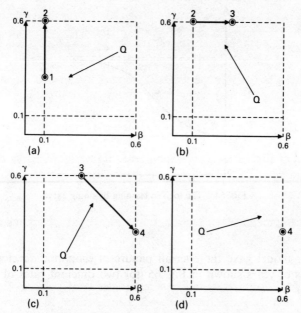

FIG. 63.   The lines of equivalence of the logarithm of total output, log Q, in the four successive stages of economic development are orthogonal to the growth of Q indicated by an arrow. The thick arrows show the three technological revolutions.

industrial to a post-industrial economy is that indicated by the thick arrow in Fig. 63.3. The bulk of capital now probably consists of *human capital*, i.e. of acquired skills and accumulated knowledge.

As to the Malthusian trap, the threshold of its avoidance preserves its minimum value:

$$\sigma_{\min}(\text{iv}) = 0.1 \cdot 0.0025 = 0.00025. \tag{8.31}$$

(5) The learning-by-doing effect

This effect, which speeds up the economic growth at the beginning of a new technological revolution, was described by Steinmann and Komlos (1988) by making the constant A in the production function dependent on time:

$$A_t = A_T + \frac{A_\infty - A_T}{1 + k/(t - T)}. \tag{8.32}$$

Here T is the point of time when a new technological revolution starts, and A_∞ is the final level of A reached with that technology. Thus, learning-by-doing obeys a logistic curve (Fig. 64). Taking into account the learning-by-doing effect, the computer simulation performed by Steinmann and Komlos

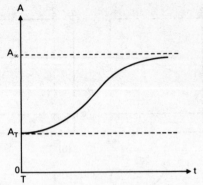

FIG. 64. The logistic learning-by-doing curve.

with their model gave the over-all picture of economic development over thousands of years shown in Fig. 65 (for two different rates of saving, 5% and 10%).

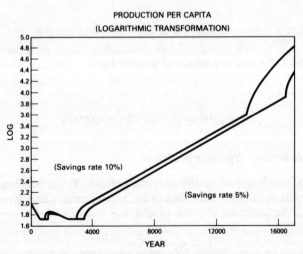

FIG. 65. A simulation of the development of $\log(Q/L)$ in the very long run, as given by the Steinmann-Komlos model for the savings rates 10% (upper curve) and 5% (lower curve). The first fall in the Malthusian trap near the left margin occurs in a hunting-and-gathering society, the next one, after a learning-by-doing effect, in an agricultural one. The steady escape from the trap characterizes an industrial society, the wealth of which grows further with the coming of a post-industrial society near the right margin. (Source: Steinmann and Komlos, 1988.)

(6) The Goodwin and Kondratieff cycles

Let us consider an industrial society (i.e. $\sigma > \alpha g$), advanced enough along the Steinmann–Komlos scale of development (cf. Fig. 65) for us to consider the value of t in (8.6) and (8.8) sufficiently large to make the ratio Q/K very nearly a constant already, say c, while Q/L, of course, is growing exponentially, as shown in Fig. 65.

Making a departure from neoclassical growth theory, we now distinguish between the total labour force, L, and the employed labour force, L_{eff}. We can also make the ratio, Q/L_{eff}, equal to $\exp(at)$ multiplied by a positive constant, the rate of growth of Q/L_{eff}, a, also being a positive constant. By introducing another new independent variable (along with L_{eff}), viz. the wage rate, w, at which L_{eff} is paid, ignoring the depreciation of capital, and assuming that the investment is now equal to the profit, we have

$$dK/dt = Q(1 - u), \quad \text{with } u = wL_{eff}/Q, \tag{8.33}$$

supposed to hold true since this point of development.

Hence, we get successively:

$$\dot{L}_{eff}/L_{eff} = \dot{Q}/Q - a = \dot{K}/K - a = c(1 - u) - a, \tag{8.34}$$

$$\dot{v}/v = \dot{L}_{eff}/L_{eff} - \dot{L}/L = c(1 - u) - g - a, \tag{8.35}$$

where the rate of employment has been denoted by v:

$$v = L_{eff}/L. \tag{8.36}$$

If we then make the special Goodwin assumption (Goodwin, 1967) that the wage rate, w, is a function of the rate of employment, v, such that

$$\dot{w}/w = bv - d, \tag{8.37}$$

where b and d are two further positive constants, then we obtain for the share of labour bill, $u = wL_{eff}/Q$, the equation

$$\dot{u}/u = \dot{w}/w + \dot{L}_{eff}/L_{eff} - \dot{Q}/Q = bv - d - a. \tag{8.38}$$

Thus we have, on the plane (u,v) of the state-space, a predator–prey subsystem obeying the Lotka–Volterra equations (8.35)–(8.38). They define an oscillating Hamiltonian system that is steerable from outside. However, by replacing the investment sQ, which we made earlier, by the investment $(1 - u)Q$ made in the Goodwin model, one can incorporate Goodwin's cycles in an extended Solow–Swan model, and show that the cycles then become damped Akerlof and Stiglitz, 1969.

Due to the oblique location of the plane (u, v) with respect to the Q-dimension in the five-dimensional total state-space, the Goodwin cycles are projected on Q as variations of growth along the non-golden-age-path trajectories. They are supposed to be short compared with the long cycles of, say, 50–70 years, predicted by Kondratieff (1935), and superimposed as slower variations on the Goodwin cycles.

For an approach to Goodwin cycles in terms of a Verhulst difference equation (for $c = N = 1$) see Pohjola (1981). The Kondratieff cycles have recently been shown much interest: see e.g. Korpinen (1987).

8.2. A Down-to-earth Approach to Economic Growth: the Dynamic Input–Output Model

If the Steinmann–Komlos model represents a rather abstract form of general theorizing using concepts that are not meant to be exactly measurable, the dynamic input–output model developed by Wassily Leontief (1953) and Andras Bródy (1970) is without doubt the most down-to-earth approach to economic growth ever made: it employs only such variables and parameters whose empirical values can be immediately computed on the basis of the statistics concerning national economy.

The model is linear and is therefore best conceived as a short-term approximation of economic growth, or even as a cross-section of the state of national economy and of the state of economic growth at a given point of time.

(1) The condition of stability of the 'turnpike'

The basic equation of motion of state in a dynamic input–output model can be written in the form

$$(I - B)xt = C[x(t + 1) - xt], \qquad (8.39)$$

where:

B = the $n \times n$ matrix of input coefficients indicating the economic inputs into each sector of the economy, per one unit of product of this sector, in the period t;

C = the $n \times n$ matrix giving the amounts of stock of each commodity in each sector, per one unit of product of this sector, at the end of the production period t;

xt = the vector of outputs of each sector at the period t.

The following assumptions can be made about the matrices:

$$\det(I - B) \neq 0, \quad \det C \neq 0, \qquad (8.40)$$

$$(I - B)^{-1} \geqslant 0, \quad D = (I - B)^{-1}C \geqslant 0, \qquad (8.41)$$

$$D = \text{non-decomposable.} \qquad (8.42)$$

It follows that D is a Frobenius matrix having a positive eigenvalue v_1 with a positive eigenvector h; v_1 being larger than or equal to the absolute value of any other eigenvalue of D. Hence, the causal recursion matrix

$$A = I + D^{-1} \qquad (8.43)$$

that appears in the recursion formula

$$x(t + 1) = A \cdot xt, \qquad (8.44)$$

has a positive eigenvalue ρ_1 obeying

$$\rho_1 = 1 + v_1^{-1} > 1, \quad Ah = \rho_1 h. \qquad (8.45)$$

There is, accordingly, a special solution of (8.44) given by

$$\hat{x}t = c_1 \rho_1^t h > 0, \quad \rho_1 > 1, \qquad (8.46)$$

with the initial state

$$\hat{x} = c_1 h, \qquad (8.47)$$

where c_1 is a positive real number.

The trajectory (8.46) is a straight line through the origin; it is called the *turnpike* of the national economy in question and is illustrated in Fig. 66a. By separating the special solution (8.47) in the general solution of (8.44) we can write the latter in the form

$$xt = c_1 \rho_1^t h + \sum_{j=2}^{n} c_j \rho_j^t [p_j(t) v_j \cos \omega_j t + q_j(t) u_j \sin \omega_j t]. \qquad (8.48)$$

Here the other eigenvalues of the matrix A have been denoted by

$$\mu_j = \rho_j \exp(i\omega_j), \qquad (8.49)$$

where ρ_j is the modulus and ω_j the phase. The expressions $p_j(t)$ and $q_j(t)$ are polynomials of t, which are not otherwise interesting in this connection.

Obviously,

$$\lim_{t \to \infty} (xt - c_i \rho_1^t h) = 0 \qquad (8.50)$$

if, and only if, all the other eigenvalues have an absolute value smaller than one:

$$\rho_j < 1 \quad \text{for } j = 2, 3, \ldots, n. \qquad (8.51)$$

Thus, we have the following theorem.

Theorem 8.2.1. The other trajectories asymptotically approach the turnpike if, and only if, the other eigenvalues of the causal-recursion matrix, except the one that is associated with the eigenvector defining the turnpike, have an absolute value smaller than one.*

---

*Note that the system is dissipative only on the further condition that ρ_1 is between 1 and $1/\rho_2 \rho_3 \cdots \rho_n$.

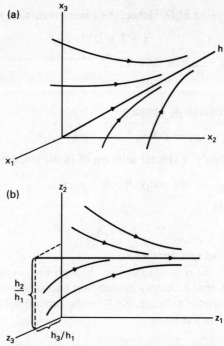

FIG. 66. The fields of self-steering state-trajectories in a stable (Picture a) and relatively stable (Picture b) turnpike system. Causal suction has been assumed, but vortex, torsion, or rapid vibrations might appear as well.

If the condition were fulfilled, there would already be self-steering of economic development to the momentary, or short-term, balanced growth path represented by the turnpike. However, the condition is hardly ever satisfied by any national economy. Thus, in the short-term development, where the dynamic input–output model applies, the economic system is not self-steering.

(2) The 'relative stability' of turnpike

If the condition of stability of the turnpike (8.51), is relaxed by replacing it with the condition that

$$\rho_j < \rho_1 \quad \text{for all } j = 2, 3, \ldots, n, \tag{8.52}$$

we have the case that in economic literature is referred to as 'relative stability'.

It follows from (8.52) that we can now write the general solution of the basic equation of the model in its causal-recursion form, (8.44), as follows:

$$xt = \rho_1^t [c_1 h + \sum_{j=2}^{n} \langle (\rho_j/\rho_1)^t \rangle. \tag{8.53}$$

Here the bracketed expression obeys

$$\langle (\rho_j/\rho_1)^t \rangle \to 0 \quad \text{when } t \to \infty. \tag{8.54}$$

[Warning! One should not jump from the formulaes (8.53) and (8.54) to the conclusion that the general solution xt converges to the turnpike solution $c_1\rho_1^t h$ when t goes to infinity: we know by Theorem 8.2.1 that for this conclusion the relaxed condition (8.52) is insufficient. For the 'strong', not 'relative', convergence you can only trust sums and differences, not products or proportions: from $xt - yt \to 0$ you can conclude the convergence $xt \to yt$, but from the product $xt = (yt)(zt)$ or the proportion $xt/yt = zt$, together with the known asymptotic behaviour of zt, you can only tell the asymptotic behaviour of the ratio xt/yt, not that of the difference $xt - yt$!]

What we can conclude from (8.53) and (8.54) is the convergence of the ratios of the components of state:

$$\frac{x_i t}{x_k t} = \frac{\rho_1^t \left[c_1 h_i + \sum_{j=2}^{n} \langle (\rho_j/\rho_1)^t \rangle_i \right]}{\rho_1^t \left[c_1 h_k + \sum_{j=2}^{n} \langle (\rho_j/\rho_1)^t \rangle_k \right]}$$

$$= \frac{c_1 h_i + \sum_{j=2}^{n} \langle (\rho_j/\rho_1)^t \rangle_i}{c_1 h_k + \sum_{j=2}^{n} \langle (\rho_j/\rho_1)^t \rangle_k} \to \frac{c_1 h_i}{c_1 h_k} = h_i/h_k \tag{8.55}$$

when $t \to \infty$. Hence the name 'relative stability' for the case where the relaxed condition (8.52) is used.

Theorem 8.2.2. In a relatively stable turnpike system the orbital convergence of other trajectories to a rectilinear trajectory can be achieved by a transformation of state description, where all but one of the original state components are replaced by their proportions to the unchanged one.

Proof. In a transformation of state, say, (we could choose any x_i as the unchanged component)

$$x \to z = (x_1, x_2/x_1, x_3/x_1, \ldots, x_n/x_1) \tag{8.56}$$

the linear causal recursion (8.44) is changed to

$$z(t + 1) = \varphi(zt), \tag{8.57}$$

the nonlinear new causal recursion, φ, being given by

$$z_1(t + 1) = a_{11} z_1(t) + \sum_{k=2}^{n} a_{1k} z_1(t) z_k(t), \tag{8.58}$$

$$z_j(t+1) = \frac{a_{j1} + \sum_{k=2}^{n} a_{jk}z_k(t)}{a_{11} + \sum_{k=2}^{n} a_{1k}z_k(t)} \quad \text{for } j = 2, 3, \ldots, n. \tag{8.59}$$

Because of the convergence (8.55) all the trajectories in the transformed state-space asymptotically approach the straight line (see Figure 66b) defined by

$$z_2 = h_2/h_1, \quad z_3 = h_3/h_1, \ldots, z_n = h_n/h_1, \tag{8.60}$$

which proves the theorem. $\square$

(3) The turnpike stabilities and the general theorem on orbital convergence

We have here an occasion to test again the predictions of Theorem 4.6.1 on orbital convergence concerning the eigenvalues of the matrix Ω. For a stable turnpike, h, the condition (8.51) clearly satisfies Theorem 4.6.1, since now the matrix Ω is a constant and we can put $\theta = 1$. For a relatively stable turnpike we can get the again constant matrix Ω by eliminating the dimension z_1 from the matrix $\partial\varphi/\partial z$. With the notations

$$\partial\varphi/\partial z = \begin{pmatrix} \partial\Phi_{11} & \partial\Phi_{1k} \\ \partial\Phi_{j1} & \partial\Phi_{jk} \end{pmatrix}, \quad A = \begin{pmatrix} a_{11} & a_{1k} \\ a_{j1} & a_{jk} \end{pmatrix}, \quad h = \begin{pmatrix} h_1 \\ h \end{pmatrix}, \tag{8.61}$$

where $\partial\varphi_{11} = \partial\varphi_1 \partial z_1$ and the rest is self-explanatory, we get from (8.59):

$$\Omega = (\partial\Phi_{jk})_{z(h)} = (1/\rho_1)A_{jk} - (1/\rho_1 h_1)\bar{h}A_{1k}. \tag{8.62}$$

Case $n = 2$. For $n = 2$, the matrix Ω reduces to the real number

$$\Omega = a_{22}/\rho_1 - a_{12}h_2/\rho_1 h_1. \tag{8.63}$$

By the eigenvalue equation $Ah = \rho_1 h$ we have

$$a_{12}h_2 = (\rho_1 - a_{11})h_1, \tag{8.64}$$

so that, since now $TrA = a_{11} + a_{22} = \rho_1 + \rho_2$,

$$\Omega = (a_{11} + a_{22} - \rho_1)/\rho_1 = \rho_2/\rho_1 < 1, \tag{8.65}$$

which was to be shown.

Case $n = 3$. For $n = 3$, the matrix Ω is 2-square, and we can use the identities

$$\text{Tr}\,\Omega = \beta_2 + \beta_3, \quad \det\Omega = \beta_2\beta_3, \quad \text{Tr}A = \rho_1 + \mu_2 + \mu_3, \det A = \rho_1\mu_2\mu_3, \tag{8.66}$$

valid for the eigenvalues β_2 and β_3 of the matrix Ω and the eigenvalues ρ_1,

μ_2, and μ_3 of the matrix A. From the eigenvalue equation $Ah = \rho_1 h$ it follows that we have, for the expression $\text{Tr}(\bar{h}A_{1k}) = A_{1k}\bar{h}$, the formula

$$A_{1k}\bar{h} = (\rho_1 - a_{11})h_1. \tag{8.67}$$

By substituting this into (8.62) we get:

$$\text{Tr}\,\Omega = (a_{22} + a_{33})/\rho_1 - (\rho_1 - a_{11})/\rho_1 = (\mu_2 + \mu_3)/\rho_1. \tag{8.68}$$

As to the determinant of Ω, a straightforward calculation from (8.62) gives

$$\det \Omega = (1/\rho_1^2 h_1) \cdot \det \begin{vmatrix} h_1 & a_{12} & a_{13} \\ h_2 & a_{22} & a_{23} \\ h_3 & a_{32} & a_{33} \end{vmatrix}. \tag{8.69}$$

Because of the eigenvalue equation $Ah = \rho_1 h$, written as

$$h_j = (1/\rho_1) \sum_{k=1}^{3} a_{jk}h_k, \quad j = 1, 2, 3, \tag{8.70}$$

the determinant in equation (8.69) can be decomposed into the sum of $(h_1/\rho_1)\det A$ and two vanishing determinants. Hence,

$$\det \Omega = (1/\rho_1^3)\det A = \mu_2\mu_3/\rho_1^2. \tag{8.71}$$

It follows from (8.68) and (8.71) that the characteristic equation of the matrix Ω is

$$\beta^2 - \beta[(\mu_2 + \mu_3)/\rho_1] + \mu_2\mu_3/\rho_1^2 = 0. \tag{8.72}$$

The solution gives

$$\beta_2 = \mu_2/\rho_1, \quad \beta_3 = \mu_3/\rho_1. \tag{8.73}$$

The relaxed condition (8.52) then gives the result that the absolute value of the eigenvalues β_2 and β_3 of the matrix Ω are smaller than one, as was to be shown.

(4) The complete dynamic input–output model and the problem of economic measurement

Not even the relative stability of the turnpike is guaranteed by the statistics concerning national economy. But there are turnpike theorems that suggest (see e.g. Tsukui and Murakami, 1980) that the trajectories of the system keep close to the turnpike most of the time for any conceivable planning period, and these suggestions are borne out by the more comprehensive empirical calculations based on statistical data for national economies (e.g. Aulin-Ahmavaara, 1987). It follows that the proportions h_i/h_k are good enough approximations of the empirical values of x_i/x_k (for the trajectory realized by a national economy) to make the dynamic input–output model, in its complete form, an important tool in the theory of economic measurement. For this end the model has to be completed in the following two ways:

(i) Balanced growth prices (Bródy, 1970).

A price vector, p, dual to the output vector, x, is introduced by writing

$$p_{t+1}(I + C - B) = (1 + r_t)p_t C. \tag{8.74}$$

Here r_t is the rate of interest. Because of (8.46) and (8.47), the turnpike solution, $\hat{x}t$, obeys the growth equation

$$\hat{x}(t + 1) = (1 + \lambda)\hat{x}t, \quad \lambda = v_1^{-1} \tag{8.75}$$

On a balanced growth path like this, the prices cannot change because demand and supply balance each other. Hence, balanced growth prices are given by substituting $\hat{p} = p_t = p_{t+1}$ into (8.74):

$$\hat{p} = \hat{p}(B + \hat{r}C), \tag{8.76}$$

Here, $\hat{r}$ is the balanced-growth interest rate. Due to (8.75) and (8.39) any point $\hat{x}$ of the turnpike obeys a similar equation:

$$\hat{x} = (B + \lambda C)\hat{x}. \tag{8.77}$$

By equating the scalar product $\hat{p}\hat{x}$ as given by (8.76) and (8.77) we get: $\hat{r} = \lambda$. Thus, the balanced-growth interest rate equals the (maximal) rate of growth of an economy reached on the turnpike.

Hence, the solution of the two eigenvalue equations

$$D\hat{x} = (1/\lambda)\hat{x}, \quad \hat{p}G = (1/\lambda)\hat{p}, \quad \text{with } G = C(I - B)^{-1}, \tag{8.78}$$

gives all that is needed for making the variables $x_1, x_2, \ldots, x_n$ commensurable, as measured on the turnpike: the terms of the sum

$$\hat{p}\hat{x} = \hat{p}_1\hat{x}_1 + \hat{p}_2\hat{x}_2 + \cdots + \hat{p}_n\hat{x}_n \tag{8.79}$$

can be considered to be measured by the same yardstick, and to indicate the ideal production obtainable with the *growth potential*, λ, of an economy that uses the technology defined by the matrices B and C (Aulin-Ahmavaara, 1987). Thus, the *technical change* from a given production period to another becomes strictly measurable in terms of the respective values of λ and of the respective decompositions (8.79). (For more details, see Aulin-Ahmavaara, 1987).

(ii) The production of human capital and labour.

Traditionally, the dynamic input–output model has not considered the production of human capital and of different kinds of labour. In view of its use as a strict measurement of technical change, one should, of course, be able to include the production of human capital and labour as well. Another problem has been connected with the long gestation periods of some capital goods (and of the production of human capital, of course). Recently, it has been shown (Aulin-Ahmavaara, 1987) that both problems have a common solution in terms of a procedure for increasing the number of 'sectors' included in the model, while preserving the form of the basic equations (8.39) and (8.74). The only change in the equations of motion is that B- and C-matrices now become functions of λ. Still a uniquely determined turnpike exists, and is easily computed. 'The yearly production process cannot be envisaged any more in the old fashion as producing just commodities and services. What we produce is, firstly and overwhelmingly: humans' (Bródy, 1988).

(5) The price–output duality and the problematic turnpike properties

From what was said in the preceding passage it is evident that the dynamic input–output model is again, in a new way, present in the analysis of the structure of a national economy. Therefore, let some notes be added on the price–output duality and the turnpike theorems in connection with Leontief systems. Let us start by rewriting (8.48), assuming single eigenvalues, in the form

$$x_t = \sum_{j=1}^{n} \mu_h^t h_j, \quad x_0 = \sum_{j=1}^{n} h_j, \quad \mu_j = \rho_j \exp(i\omega_j), \quad \mu_1 = \rho_1 > 1. \quad (8.48')$$

Here h_j is, in the general case, a complex eigenvector belonging to the, in the general case, complex eigenvalue μ_j, $\mu_1 = \rho_1$, however, being the positive eigenvalue associated with a positive eigenvector h_1.

We can call (8.48') the expansion of the state x_t, in terms of the eigenvectors h_j, according to the eigenvalues μ_j. In the same sense, the Frobenius matrix D, taken to be the causal recursion matrix of a variable s_t, generates an expansion of s_t in terms of the same eigenvectors h_j, but according to the eigenvalues v_j of D. The matrix G, given in (8.78), considered as the causal recursion matrix of a variable r_t, gives an expansion of r_t in terms of the eigenvectors $q_j = (I - B)h_j$ of G, according to its eigenvalues v_j. Obviously D and G have the same eigenvalues.

From (8.74) it follows that the causal recursion matrix of the price vector p_t is $\rho_1 \Psi^{-1}$, with $\Psi = I + G^{-1}$. In the theory of planning based on Leontief systems one can show (Tsukui and Murakami, 1980) that the vector u_t of discounted prices has, along the 'optimal paths', the causal recursion matrix Ψ^{-1}. The discounted prices u_t are the prices that one pays for wares that will be available at the point of time t in the future: obviously you cannot pay the full price p_t beforehand. These matrices all generate the expansions of the respective variables in terms of their eigenvectors, according to their eigenvalues, which are given in Table 1. (The price vectors p and u are row vectors, the other vectors are columns.)

The couple of positive eigenvectors (h_1, k_1) is called a pair of mutually dual turnpikes. From Table 1 we can at once read the following result.

Theorem 8.2.3. If one of a pair of dual turnpikes is stable, the other one is unstable. The reverse statement, deducing the stability of a turnpike from the instability of its dual, is valid only provided that the price turnpike is that of discounted prices.

We can learn more of the price–output duality by considering the properties of the eigenvalues μ_j and the related eigenvectors h_j. From the non-singularity of D, it follows that $v_j \neq 0$ and $v_j \neq \infty$ must hold true for $j = 2, 3, \ldots, n$. This implies that $\mu_j = 1 + 1/v_j \neq 1$ (and $\neq \infty$) also holds good. The matrix $\Phi (= A$ in our previous notation) is not necessarily non-singular, but if it is not we can always reduce its dimensionality and obtain a non-singular matrix. Thus, we can assume that $\mu_j \neq 0$ holds true for all j. (This in turn implies that we have $v_j \neq -1$.) In a similar way we can arrange, without loss of generality, for Ψ to be non-singular.

Since D is a (positive) Frobenius matrix, it may have, along with complex eigenvalues and eigenvectors, the following combinations of real eigenvalues and eigenvectors, in addition to the Frobenius root $v_1 > 0$ and its eigenvector $h_1 > 0$:

(1) $v_j > 0$, in which case $\rho_j > 1$ and $h_j \not> 0$ (not non-negative); or
(2) $v_j < 0$, in which case $\rho_j < 1$ and $h_j > 0$.

As a consequence of these restrictions, valid for real eigenvalues and real eigenvectors, we first get the classification of continuous Leontief systems involving no cycles (i.e. $\rho_j > 0$ for all j), shown in Table 2. The six categories of systems indicated there can be adapted to the discontinuous, rapidly vibrating systems too, by substituting for the eigenvalues ρ_j in Table 2 (in this case) their absolute values $|\rho_j|$. The same six categories can be also extended to include the cycles, i.e. the possibility that some pairs of complex conjugate numbers also appear among the eigenvalues μ_j of Φ. In this case,

TABLE 1
(Only the variables x, p, and u have significance in economic theory)

Variable	Recursion matrix	Eigenvalues	Eigenvectors
s_t	$D = (I - B)^{-1}C$	v_j	h_j
r_t	$G = C(I - B)^{-1}$	v_j	$q_j = (I - B)h_j$
x_t	$\Phi = I + D^{-1}$	$\mu_j = 1 + 1/v_j$	h_j
v_t	$\Psi = I + G^{-1}$	$\mu_j = 1 + 1/v_j$	$q_j = (I - B)h_j$
p_t	$\rho_1 \Psi^{-1}$	ρ_1/μ_j	k_j
u_t	Ψ^{-1}	$1/\mu_j$	k_j

TABLE 2
Continuous Leontief Systems Involving No Cycles

Type	Eigenvalues	x-system	The dual p-system	Turnpike h_1	The price equilibrium k_1	Trajectory field around the turnpike
1	$\rho_j > \rho_1 > 1 \land j \neq 1$	disintegrating, relatively unstable	self-regulating	unstable	stable	repulsion (x-system), attraction (p-system)
2	$\rho_1 < \rho_j \land 1 \neq 1$	disintegrating, relatively stable	disintegrating	unstable	unstable	repulsion (in both systems)
3	$\exists$ some ρ_j: $\rho_1 > \rho_j > 1$, otherwise $\rho_j > \rho_1 > 1, j \neq 1$	disintegrating, relatively unstable	disintegrating	unstable	unstable	repulsion (x-system), combined attraction–repulsion (p-system)
4	$\exists$ some ρ_j: $\rho_j > \rho_1 > 1$, otherwise $0 < \rho_j < 1, j \neq 1$	disintegrating, relatively unstable	disintegrating	unstable		combined attraction–repulsion (x-system), repulsion (p-system)
5	$\exists$ some ρ_j: $\rho_1 > \rho_j > 1$, otherwise $0 < \rho_j < 1, j \neq 1$	disintegrating, relatively stable	disintegrating	unstable	unstable	combined attraction–repulsion (x-system), repulsion (p-system)
6	$0 < \rho_j < 1 \land j \neq 1$	self-steering, relatively stable	disintegrating	stable	unstable	attraction (x-system), repulsion (p-system)

the symbols ρ_j in Table 2 are to be reinterpreted as the absolute values $|\mu_j|$. In both cases we have to add a seventh category to those given in Table 2, because we may have $\rho_j = -1$, i.e. $|\rho_j| = 1$, for some $j \neq 1$ in a discontinuous Leontief system, and $|\mu_j| = 1$, i.e. $\rho_j = 1$, for some $j \neq 1$ in a Leontief system involving cycles. But then we have, in fact, an exhaustive classification of Leontief systems.

A 'catenary movement', where the state first approaches the turnpike and then moves away from it, is possible only if a combined attraction–repulsion field (where e.g. $x_t \to \infty$ for both $t \to \infty$ and $t \to -\infty$) surrounds the turnpike in question: this is true in any linear system with a turnpike, in view of the discussion of linear systems given in Chapter 4. For instance, in Leontief systems such a trajector field is realized, for x-systems, in Types 4 and 5 and, for p-systems, only in Type 3, and never in the dual systems simultaneously (see Table 2). A similar result is obtained for any of the systems with either the turnpike h_1 or the turnpike k_1 (Table 1).

In the most detailed discussion of turnpike properties in Leontief systems so far (Tskui and Murakami, 1980) the reader is baffled by the assumption, obviously made by the authors, that the 'catenary movements' of state, with respect to the turnpike, are a general property of all Leontief systems and all systems having the same eigenvectors h_i or k_i (like those in Table 1). Further perplexity is caused by their discussion of the variables x_t/ρ_1^t and $\rho_1^t u_t$ (rather than of the variables x_t and u_t) having the causal recursion matrices Φ/ρ_1 and $\rho_1 \Psi^{-1}$, respectively.

What are the 'turnpike properties' of which we speak? Suppose, for instance, that you are given a fixed initial state, x_0, of output, a fixed price vector p, and a free choise between a number, K, of different production technologies, σ, each of them represented by the respective technical matrices B_σ and C_σ of a Leontief system. One of the technologies, say $\sigma = 1$, is defined to be optimal, in the sense that the rate of growth on its balanced growth path $h_1(1)$, henceforth called the proper turnpike, is largest among all the $h_1(\sigma)$'s. You are then given the task of maximizing a utility function, for instance the capital $pB_{\sigma(T)}x_T$, in a future period T over the variation of the $K(T + 1)$ different 'possible paths' $x_0, x_1, x_2, \ldots, x_T$, and finding out the 'optimal paths' that yield the maximum.

It is intuitively clear that the optimal paths should have a tendency to favour the optimal technology, and even the vicinity of the related optimal balanced growth path $h_1(1)$. This is the 'turnpike property', which in various specifications is expressed by various turnpike theorems. And it is natural to assume some kind of catenary movements, since the turnpike has first to be approached, starting with an initial state x_0 that is outside it, and possibly outside that part of state-space related to the optimal technology. Finally, an optimal path may have to leave the vicinity of the turnpike $h_1(1)$, since the prescribed price vector p determines the desired structure of capital, which is not necessarily realizable on the turnpike or its vicinity.

Let $x_t(1)$ be the component of the total state vector x_t in the part of state-space related to the optimal (or 'turnpike') technology, and let $\bar{x}_t(1)$ be the remaining component, related to the 'non-turnpike' technologies. By defining the norm, $\|x\|$, as the sum of the absolute values of all components of x, and the distance $d(x, y)$ between two state-vectors x and y by $\|x - y\|$, the interesting variables in connection with turnpike theorems in Leontief systems obviously are $\|\bar{x}_t(1)\|$ and $d(x_t(1), h_1(1))$. Instead of them, Tsukui and Murakami (1980, pp. 14–24) consider the variable $z_t = x_t/\rho_1^t$, proving that, for instance, given two positive numbers, ε and ω, *and a sufficiently long planning period*, $(0, T)$, you can always make $\|\bar{z}_t(1)\| < \varepsilon$ and $d(z_1(t), h_1(1) < \omega$ for $0 < T_1 \leqslant t \leqslant T_2 < T$, where T_1 and T_2 are two points in time.

While the non-existence of catenary movements in every Leontief system does not necessarily cripple the turnpike theorems as stated by Tsukui and Murakami (1980, Chapter 1) (although it reduces the generality of the intended simultaneous existence of the x- and u-turnpike properties), the use of the variable z_t obscures their meaning. This is because, when written for the state x_t, the above conditions read $\|\bar{x}_t(1)\| < \varepsilon\rho_1^T$ and $d(x_t(1), h_1(1)) < \omega$ ρ_1^T, respectively. Now, if T increases, the limits $\varepsilon\rho_1^T$ and $\omega\rho_1^T$ increase too, and both of them tend to infinity with T, thus making a mockery of the idea of a sufficiently long planning period $(0, T)$ as a condition of turnpike properties.

This is not to say that general turnpike properties known from abstract economic theory, or something like them, could not be valid in Leontief systems as well. But the above observations suggest that in the formulation of those properties the special dynamics of Leontief systems plays a special role, and that there is still room for improvement in the existing turnpike theorems by taking those dynamics more into account.

8.3. The Condition of Self-steering to the Golden Age Path in the Solow–Swan and Related Mainstream Models

Both of the models so far discussed, viz. that of Steinmann and Komlos, and the dynamic input–output model, are, from an orthodox neoclassical point of view, maverick approaches to economic growth. It is high time to see what is the current mainstream in economic growth theory.

(1) The basic Solow–Swan model

Without doubt this model has played a fundamental role in all neoclassical growth theory. It describes the production system as a continuous dynamical system based on the following assumptions:

(i) A production function $F(K, L)$, homogeneous of degree one:

FMSD—F

$$Q = F(K, L) \quad \text{(in the absence of technical progress)}, \quad (8.80)$$

$$F(aK, aL) = aQ \quad \text{for all } a \in R, Q = K\partial Q/\partial K + L\partial Q/\partial L. \quad (8.81)$$

(ii) The marginal productivity of both capital and labour:

$$r = \partial Q/\partial K - \delta = \text{interest rate}, \quad w = \partial Q/\partial L = \text{wage rate}, \quad (8.82)$$

so that the latter equation (8.81) gives the income distribution:

$$Q = (r + \delta)K + wL, \quad (r + \delta)K = \text{capital bill}, \quad wL = \text{labour bill}.$$
$$(8.83)$$

(iii) A constant rate, s, of saving, a constant rate, δ, of depreciation of capital, and all savings, sQ, invested:

$$dK/dt = sQ - \delta K. \quad (8.84)$$

(iv) A constant rate of growth of labour (population):

$$dL/dt = gL, \quad \text{i.e. } L_t = L_0 e^{gt}. \quad (8.85)$$

(v) A constant rate of technological progress, represented by some of the following forms of the production function:

$$Q = F(K, e^{\sigma_h t}L) \quad \text{(Harrod-neutral technical change)}, \quad (8.86)$$

$$Q = F(e^{\sigma_s t}K, L) \quad \text{(Solow-neutral technical change)}, \quad (8.87)$$

$$Q = e^{\sigma t}F(K, L) \quad \text{(Hicks-neutral technical change)}. \quad (8.88)$$

(vi) Constant shares of rental and labour bills:

$$(r + \delta)K/Q = \beta = \text{constant share of capital (rental) bill}, \quad (8.89)$$

$$wL/Q = \gamma = 1 - \beta = \text{constant share of labour bill}, \quad (8.90)$$

which gives, when combined with Hicks neutrality:

$$dQ = (\partial Q/\partial K)dK + (\partial Q/\partial L)dL + (\partial Q/\partial t)dt = (\beta dK/K + \gamma dL/L + \sigma)Q, \text{ i.e.}$$
$$(8.91)$$

$$Q = e^{\sigma t}K^\beta L^\gamma \text{ (with a constant of integration } A = 1\text{)}. \quad (8.92)$$

Thus, the Cobb–Douglas production function, which is obviously neutral in all the ways (8.86)–(8.88), results in $\sigma_h = \sigma/\gamma$, $\sigma_s = \sigma/\beta$.

Assumption (vi) is often replaced by an unspecified function $F(K, L) = Lf(K/L)$. In a discussion of the asymptotic behaviour the assumption may be defended, as one of Samuelson's 'stylized facts of capitalism' states that the relative shares of wages and profits are constant in the long run. On the other hand, one can study the redistribution of income even with a Cobb–Douglas production function by assuming time-dependent β and γ.

(2) The condition of self-steering in the Solow-Swan model

By substituting (8.85) and (8.92) into (8.84) we get

$$dK/dt = sL_0^{\gamma}e^{(\sigma + \gamma g)t}K^{\beta} - \delta K. \tag{8.93}$$

This is linearized by the Leibnitz substitution

$$z = K^{1-\beta}, \tag{8.94}$$

which gives

$$dz/dt = \gamma sL_0^{\gamma}e^{(\sigma + \gamma g)t} - \gamma \delta z, \text{ with } 1 - \beta = \gamma. \tag{8.95}$$

The homogeneous equation gives

$$dz_1/dt = -\gamma \delta z_1 \Rightarrow z_1 = Ce^{-\gamma \delta t}, \tag{8.96}$$

while a special solution of the total equation (8.95) is

$$z_2 = De^{(\sigma + \gamma g)t}, \text{ with } D = sL_0^{\gamma}/(g + \delta + \sigma/\gamma). \tag{8.97}$$

Thus, the general solution of (8.95) is given by

$$z_t = Ce^{-\gamma \delta t} + De^{(\sigma + \gamma g)t}. \tag{8.98}$$

By substituting this into (8.94) we get successively:

$$K_t = [Ce^{-\gamma \delta t} + De^{(\sigma + \gamma g)t}]^{1/\gamma}, \tag{8.99}$$

$$Q_t = L_0^{\gamma}e^{(\sigma + \gamma g)t}[Ce^{-\gamma \delta t} + De^{(\sigma + \gamma g)t}]^{\beta/\gamma}$$
$$= L_0^{\gamma}[Ce^{(\gamma/\beta)(\sigma + \gamma g - \beta \delta)t} + De^{(1/\beta)(\sigma + \gamma g)t}]^{\beta/\gamma}. \tag{8.100}$$

It follows from (8.99) and (8.100) that, corresponding to the value $C = 0$, there is a special solution

$$\hat{K}_t = \hat{K}_0 e^{\lambda t}, \quad \hat{Q}_t = \hat{Q}_0 e^{\lambda t}, \quad L_t = L_0 e^{gt}, \tag{8.101}$$

with the common rate of growth of $\hat{K}_t$ and $\hat{Q}_t$

$$\lambda = g + \sigma/\gamma, \tag{8.102}$$

the initial values being given by

$$\hat{K}_0 = D^{1/\gamma}, \quad \hat{Q}_0 = L_0^{\gamma}\hat{K}_0^{\beta}, \quad \text{and } L_0. \tag{8.103}$$

With both of the other initial values being determined by L_0 and by constant parameters, the curve in the state-space defined by (8.101) is uniquely determined. Since

$$\hat{Q}_t/\hat{K}_t = \hat{Q}_0/\hat{K}_0 = \text{const.}, \tag{8.104}$$

the curve is entirely on the plane (8.104). Furthermore, since

$$\lim_{t \to -\infty} \hat{Q}_t = \lim_{t \to -\infty} \hat{K}_t = 0, \qquad (8.105)$$

the curve starts from the origin. It is the golden age path of the Solow–Swan model illustrated in Fig. 67.

It can be seen from (8.99) and (8.100) [Notice the warning given in Section 8.2.2!] that the convergence

$$Q_t \to \hat{Q}_t, \quad K_t \to \hat{K}_t \quad \text{when } t \to \infty, \qquad (8.106)$$

is valid if, and only if,

$$0 < \lambda/\delta < \beta/\gamma \qquad (8.107)$$

is satisfied. This result could, of course, have been obtained directly from

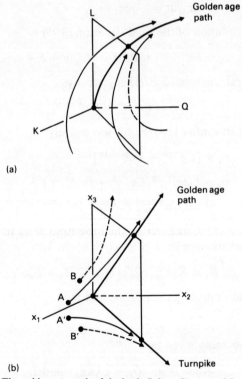

(a)

(b)

FIG. 67. (a) The golden age path of the basic Solow–Swan model with a Cobb–Douglas production function, and state-trajectories approaching it asymptotically. (b) The rectilinear golden age path in the space (x_1, x_2, x_3), with the A-trajectory converging to it and the C-trajectory diverging from it. Their projections on the (x_1, x_2)-plane starting from the points A' and B', respectively, both converge to the turnpike, which is the projection of the golden age path. The A- and B-trajectories belong to different values of the parameters of the system, the condition of self-steering being fulfilled in the former case, and not fulfilled in the latter case. Here $x_1 = e^{\sigma t} L^\gamma$, $x_2 = K^\gamma$, and $x_3 = Q^\gamma$.

our general theorem concerning the orbital convergence to the golden age path, Theorem 8.1.1, with the substitution $\alpha = 0$.

The condition (8.107) is, therefore, the condition of self-steering to the golden age path of the Solow–Swan model. It indicates that

> economic development is self-steering towards the goal one has chosen for it (by choosing the values of parameters), if the rate of economic growth in proportion to the rate of depreciation of capital is smaller than the share of capital bill (including human capital) in proportion to the share of labour bill of production.

In view of the results obtained in Section 8.1, economic self-steering, i.e. the achievement of the goals chosen for economy, is thus surest and quickest in the post-industrial period, and next-surest and quickest in the industrial period – always provided that economic growth is cautious, and accompanied by well-timed depreciation of obsolete capital tools.

Or, stated otherwise, we have:

> The General Rule of Self-steering in Different Types of Economic Systems: The threshold of self-steering to the golden age path is lowest in the post-industrial period, the second lowest in the industrial period, and highest in the agricultural and hunting-and-gathering periods of economic development.

Here the threshold is thought to be represented by the magnitude $(\alpha + \gamma)/\beta$, which the ratio δ/λ must exceed in order to satisfy the general condition of self-steering to the golden age path given by (8.13).

(3) The 'capital jelly' vintage model

Suppose that we want to consider capital tools from different years separately, indicating the greater efficiency of modern tools. Then we write

$$K_t = \int_{-\infty}^{t} K_v(t)\mathrm{d}v, \qquad (8.112)$$

$$Q_t = \int_{-\infty}^{t} Q_v(t)\mathrm{d}v, \qquad (8.113)$$

$$L_t = \int_{-\infty}^{t} L_v(t)\mathrm{d}v, \qquad (8.114)$$

where $K_v(t)$ stands for the capital tools coming from the year v, $Q_v(t)$ for the part of total output produced by these means, and $L_v(t)$ for the labour force working with these tools.

Writing the Cobb–Douglas production function (8.92) separately for each year v we have

$$Q_v(t) = e^{\sigma v} K_v(t)^\beta L_v(t)^\gamma = L_v(t)[J_v(t)/L_v(t)]^\beta, \qquad (8.115)$$

where the *efficient capital* has been denoted by

$$J_v(t) = e^{(\sigma/\beta)v} K_v(t). \qquad (8.116)$$

Assumption (iii) concerning saving, depreciation, and investment can now be incorporated in the forms

$$K_v(v) = I(v) = sQ_v \text{ for all } v \leqslant t \text{ (investment = saving)}, \qquad (8.117)$$

$$K_v(t) = K_v(v)e^{-\delta(t-v)} \qquad \text{(depreciation)}, \qquad (8.118)$$

respectively, the remaining assumptions (ii) and (iv) being unchanged.

We now make the additional (seventh) assumption that the efficient capital per worker has to be, in order for capital goods from different years to be competitive, the same for all years:

$$J_v(t)/L_v(t) = j(t) = \text{independent of } v. \qquad (8.119)$$

Hence, we get successively:

$$\int_{-\infty}^{t} J_v(t)\mathrm{d}v = j(t)\int_{-\infty}^{t} L_v(t)\mathrm{d}v, \quad \text{or } J_t = j(t)L_t, \qquad (8.120)$$

$$j(t) = J_v(t)/L_v(t) = J_t/L_t, \qquad (8.121)$$

$$Q_t = \int_{-\infty}^{t} Q_v(t)\mathrm{d}v = j(t)\int_{-\infty}^{t} L_v(t)\mathrm{d}v = (J_t/L_t)^\beta L_t = J_t^\beta L_t^\gamma, \qquad (8.122)$$

$$\frac{\mathrm{d}J}{\mathrm{d}t} = \frac{\mathrm{d}}{\mathrm{d}t}\int_{-\infty}^{t} J_v(t)\mathrm{d}v = s\frac{\mathrm{d}}{\mathrm{d}t}\int_{-\infty}^{t} Q_v \exp[(\sigma/\beta)v - \delta(t-v)]\mathrm{d}v$$

$$= sQ_t e^{(\sigma/\beta)t} - \delta s \int_{-\infty}^{t} Q_v \exp[(\sigma/\beta)v - \delta(t-v)]\mathrm{d}v$$

$$= sQ_t e^{(\sigma/\beta)t} - \delta J_t = sL_0^\gamma e^{(\gamma g + \sigma/\beta)t} J_t^\beta - \delta J_t. \qquad (8.123)$$

The equation so obtained for the *capital jelly*, J_t, is again a Bernoulli equation. Together with (8.122) it gives

$$J_t = [Ce^{-\gamma\delta t} + D_1 e^{(\sigma_s + \gamma g)t}]^{1/\gamma}, \quad \sigma_s = \sigma/\beta, \qquad (8.124)$$

$$Q_t = L_0^\gamma [Ce^{(\gamma/\beta)(\gamma g - \beta\delta)t} + D_1 e^{(1/\beta)(\sigma + \gamma g)t}]^{\beta/\gamma}, \qquad (8.125)$$

$$D_1 = sL_0^\gamma/(g + \delta + \sigma_s). \qquad (8.126)$$

Thus, the golden age path is now given by

$$\hat{J}_t = \hat{J}_0 e^{\lambda_1 t}, \qquad \hat{Q}_t = \hat{Q}_0 e^{\lambda_2 t}, \qquad L_t = L_0 e^{gt}, \qquad (8.127)$$

$$\lambda_1 = g + \sigma_s/\gamma, \quad \lambda_2 = g + \sigma/\gamma, \quad \hat{J}_0 = D_1^{1/\gamma}, \quad \hat{Q}_0 = L_0^\gamma D_1^{\beta/\gamma} \qquad (8.128)$$

and the condition of self-steering to this path is

$$0 < g/\delta < \beta/\gamma. \tag{8.129}$$

The condition does not differ much from that of the Solow–Swan model, (8.107): only the rate of growth of labour, g, has here replaced the rate of growth, λ, split in two parts.

8.4. Criticism of Neoclassical Growth Theory from the Point of View of Mathematical Dynamics

We have seen above that the basic Solow–Swan models of neoclassical growth theory are dynamically stable only on a condition to be imposed on their parameters, viz. that the ratio δ/λ or, in the case of the vintage model, the ratio δ/g, has to exceed the threshold γ/β.

This condition seems to have been neglected in neoclassical growth theory because of certain adopted practices, which have either directed attention away (point (1) below) or have been unnecessarily restrictive in scope (as in points (2) and (3)).

(1) The 'stability in the sense of Harrod'

Let us study, by way of an example, the basic Solow–Swan model with a Cobb–Douglas production function. The long-term predictions of the model are given by the asymptotic behaviour of its variables when $t \to \infty$. It follows immediately from the solution (8.99)–(8.100) that the asymptotic behaviour of some of the most important derived variables are as follows:

$$Q/K \to (g + \sigma_h + \delta)/s, \tag{8.130}$$

$$K/L \sim [s/(g + \sigma_h + \delta)]^{1/\gamma} e^{\sigma_h t}, \tag{8.131}$$

$$Q/L \sim [s/(g + \sigma_h + \delta)]^{\beta/\gamma} e^{\sigma_h t}, \tag{8.132}$$

$$r = (\beta Q/K) - \delta \to (\beta/s)(g + \sigma_h + \delta) - \delta, \tag{8.133}$$

$$w = \gamma Q/L \sim \gamma [s/(g + \sigma_h + \delta)]^{\beta/\gamma} e^{\sigma_h t}, \tag{8.134}$$

$$w/(r + \delta) = \gamma K/\beta L \sim (\gamma/\beta)[s/(g + \sigma_h + \delta)]^{1 + \beta/\gamma} e^{\sigma_h t}, \tag{8.135}$$

$$c = (1 - s)Q/L \sim (1 - s)[s/(g + \sigma_h + \delta)]^{\beta/\gamma} e^{\sigma_h t}. \tag{8.136}$$

It is a consequence of the exponential form of the functional expressions in the model that for all of these variables y, including the basic variables Q, K, and L as well, the rate of growth $\dot{y}/y$ approaches a constant value, positive or zero, when $t \to \infty$. This is what in older economic literature (Jorgenson, 1960) was called 'stability in the sense of Harrod' of the variables in question.

Obviously the Harrod stability so defined has nothing to do with dynamical stability: a dynamical system, where all the significant variables are Harrod-stable, can still be steerable from outside or disintegrating, thus dynamically unstable. The concept of Harrod stability is already out of fashion. I mention it here only since this concept seems, in its time, to have been a reason why dynamical stability was not investigated in neoclassical growth theory.

(2) The restriction of neoclassical growth theory to per capita variables

Another reason for shunning the study of dynamical stability in the state-space (Q, K, L) of the basic variables has been, and still seems to be, the restriction to per capita variables in the later formulations of neoclassical growth theory. This practice has been accompanied by the claim that economics is a science dealing with proportional variables only. While the claim is not entirely pointless, it nevertheless eliminates in advance the possible self-steering of the economic system. This is because such a practice makes a dogma of bounded (positive half-)trajectories, and thus restricts the possible stability of the system to self-regulation.

Indeed, if we consider the per capita variables (corrected for technical progress)

$$k = K/\bar{L} \quad \text{and} \quad q = Q/\bar{L}, \quad \text{where } \bar{L} = e^{\sigma_{h}t}L, \tag{8.137}$$

and apply the general solution (8.99)–(8.100), we have a dynamical system with a fixed point (k_0, q_0). All the trajectories in such a two-dimensional economic system converge to the point of equilibrium (k_0, p_0) without any condition. Such a truncated economic system is a self-regulating equilibrium system, the unbounded trajectories having been eliminated by the convention (8.137). A similar operation can of course be performed in any neoclassical growth model. In this way one can avoid studying the stability or instability of the original economic system.

However, a science of economics, where the outcome Q, for instance, cannot be considered as a basic state variable, seems curiously and unnecessarily restricted in scope. On the other hand, as soon as the outcome Q (for instance, the national income of a national economy) is included among the components of total state, we face the problem of dynamical stability in the form studied in Section 8.3.

(3) The current emphasis on the global saddle-point stability in Hamiltonian economics

The growth models discussed so far consider the economic system as a spontaneously developing, potentially self-steering dynamical system, whose internal laws of development are approximated by the equations of motion of the model. A different attitude to economic development is represented by the 'optimation school'. It starts by asking not what are the laws of a more or less spontaneous development, but how we can affect them by imposing on it some consciously chosen objectives.

Let us reconsider the Solow–Swan model in this vein. We first set up the objective:

$$\text{Maximize} \int_0^\infty e^{-\rho t} \frac{L}{1-a}[c^{1-a} - 1]dt \text{ with respect to } c, \qquad (8.138)$$

subject to the condition (now expressed on the 'demand side')

$$dK/dt = e^{\sigma t}K^\beta L^\gamma - cL - \delta K = Q - cL - \delta K. \qquad (8.139)$$

Here $\exp(-\rho t)$ is the discounting factor, and the rest of the integrand in (8.138) is our utility function, c being the consumption per capita and a the coefficient of risk aversion. The Malthusian growth is again assumed: $L = L_0 e^{gt}$.

The problem so stated is equivalent to the following one:

$$\text{Maximize } H(K, P; c) = \frac{L}{1-a}[c^{1-a} - 1] + P[Q(K, L) - cL - \delta K] \qquad (8.140)$$

with respect to c, the equations of motion having the Hamiltonian form with the 'perturbation term' ρP:

$$dK/dt = \partial H/\partial P, \quad dP/dt = -\partial H/\partial K + \rho P. \qquad (8.141)$$

The first of these equations reproduces (8.139), while the second one gives

$$dP/dt = -P(\partial Q/\partial K - \delta) + \rho P = P(\rho + \delta) - \beta e^{\sigma t}K^{\beta-1}L^{1-\beta}P. \qquad (8.142)$$

A further equation for the 'imputed price' P is obtained from the condition of maximization of H:

$$\partial H/\partial c = Lc^{-a} - PL = 0, \quad \text{thus } P = c^{-a}. \qquad (8.143)$$

The unique 'optimal path' is singled out from among the solutions $(K(t), P(t))$ of the two differential equations (8.139) and (8.142) by the transversality condition

$$e^{-\rho t}K(t)P(t) \to 0 \quad \text{when } t \to \infty. \qquad (8.144)$$

Let us study the golden age path (cf. Section 8.3) again. In this purpose, assume the existence of a special solution $(\hat{K}, \hat{P}; \hat{c})$, in which $(d\hat{K}/dt)/\hat{K} = g + \sigma_h = \lambda$, $(d\hat{c}/dt)/\hat{c} = \sigma_h$, thus $(d\hat{P}/dt)/\hat{P} = -a\sigma_h$. Hence we get, by (8.139) and (8.142):

$$(d\hat{Q}/dt)/\hat{Q} = \sigma + \beta(d\hat{K}/dt)/\hat{K} + (1 - \beta)g = \lambda, \qquad (8.145)$$

$$\partial \hat{Q}/\partial \hat{K} = \rho + \delta - (\mathrm{d}\hat{P}/\mathrm{d}t)/\hat{P} = \rho + \delta + a\sigma_h, \tag{8.146}$$

$$\hat{Q}/\hat{K} = (\partial \hat{Q}/\partial \hat{K})/\beta = (\rho + \delta + a\sigma_h)/\beta, \tag{8.147}$$

$$\hat{s} = \frac{(\mathrm{d}\hat{K}/\mathrm{d}t)/\hat{K} + \delta}{\hat{Q}/\hat{K}} = \frac{\beta(\lambda + \delta)}{\rho + \delta + a\sigma_h}. \tag{8.148}$$

This golden age path is also the optimal (K, P) path provided that the condition $\hat{s} < \beta$, i.e.

$$\rho + a\sigma_h > \lambda, \tag{8.149}$$

is met.

The Hamiltonian formalism has been adopted by the economists from theoretical physics. It is an excellent tool for dealing with problems of optimation met in applications of physics to technology and, no doubt, in applications of economics to the planning of firms and, possibly, of centralized economies. However, to connect the Hamiltonian formalism with fundamental problems of stability implies certain difficulties.

It is well known that the conservative ('autonomous') Hamiltonian systems in physics are dynamically unstable, that is, to use our terminology, either steerable from outside or disintegrating. An example of the former case is, say, the Keplerian motion of planets, while an example of the latter case is the Keplerian motion of projectiles on hyperbolic paths. The notorious instability of physical Hamiltonian systems is also reflected in the problems of 'Hamiltonian economics'.

Let us study a general formulation of Hamiltonian economics suggested by Cass and Shell (1976a, 1976b). We have again the Hamiltonian equations of motion with a perturbation term, but written for per capital variables:

$$\mathrm{d}k_i/\mathrm{d}t = \partial H/\partial p_i, \quad \mathrm{d}p_i/\mathrm{d}t = -\partial H/\partial k_i + \rho p_i; \quad i = 1, 2, \dots, n; \tag{8.150}$$

with the Hamiltonian function

$$H = H(k(t), p(t); v(t)). \tag{8.151}$$

Here k is the n-vector standing for n different kinds of capital goods represented in terms of the ratios of the type of (8.137) (in order to get a fixed point in the state-space). The n-vector p gives the n prices associated with these capital goods, but interpreted in different contexts in different ways. The parameter vector v too is in the general case time-dependent, and it may be given different interpretations and different numbers of real-number components in connection with different problems.

For a constant parameter v and for $\rho = 0$ we are back with a conservative Hamiltonian system, such as we meet in classical physics. What is conserved is the value of H, which is a constant of motion: $H = E$, where E is a real number. In physical applications H stands for the total energy of the system. It is equal to the sum of the kinetic energy $T(p)$ and the potential energy

$V(k)$ of the system. T is a quadratic expression of the momenta p, while $V(k)$ depends on the system investigated. The lowest value E_0 of energy gives a fixed point (k_0, p_0), around which a one-parameter family of closed (hyper)surfaces is grouped, standing for different values of E. Each of them is invariant, and is composed of the satellite trajectories around the centre-point (k_0, p_0). Such a system, of course, is steerable from outside, at least in some neighbourhood of (k_0, p_0). Depending on the parameters of the physical system, it may be disintegrating farther away from that point.

This description of the typical behaviour of autonomous Hamiltonian systems in physics brings us to the discussion of their economic counterparts. Thus, let again v be a constant, and $\rho = 0$, in the matrix

$$\left[\frac{\partial(\dot{k}, \dot{p})}{\partial(k, p)} \right]_0 = \begin{bmatrix} \dfrac{\partial^2 H}{\partial p_i \partial k_j} & \dfrac{\partial^2 H}{\partial p_i \partial p_j} \\ -\dfrac{\partial^2 H}{\partial k_i \partial k_j} & -\dfrac{\partial^2 H}{\partial k_i \partial p_j} + \rho p_j \end{bmatrix} = B, \qquad (8.152)$$

where the derivatives are taken at a fixed point (k_0, p_0). If it has one or more pairs of mutually conjugated purely imaginary characteristic roots, we have again the satellite trajectories around the centre-point (k_0, p_0), at least in some sector of a neighbourhood of this point. If it has no such roots, then for every root μ with a positive real part, $\text{Re}(\mu) > 0$, there is also the root $-\mu$ with the negative real part $-\text{Re}(\mu)$. This makes of (k_0, p_0) in this case a saddle point, having a combined attraction-repulsion field around itself.

In a two-dimensional system ($n = 1$), the domain of stability, i.e. the region of attraction, of the saddle point is a straight line through this point. In a $2n$-dimensional system with $n > 1$ the region of attraction of a saddle point is restricted into an n-dimensional sector of an environment of this point. But in every case, in the domain of stability of a saddle point *every point is a boundary point*: the repulsion field of the saddle point is present in every δ-neighbourhood of this point. Repeating the corresponding linear analysis for the case $\rho > 0$ and reporting the results for the linear system $\dot{x} = Bx$, we can sum up as follows (cf. Kurz, 1968):

Theorem 8.4.1. A linear continuous and conservative Hamiltonian system is dynamically unstable, either steerable from outside (the centre-point case) or disintegrating (the saddle-point case), all over the Euclidean state-space.

Theorem 8.4.2. A linear perturbed Hamiltonian system (8.150) is dynamically unstable, viz. disintegrating, all over the Euclidean state-space, with either a saddle point or an unstable focus or node.

It follows that a nonlinear system of the form (8.150), whether conservative or not, can be stable only in such a special saddle-point case, where *all* the trajectories of the repulsion field near the fixed point (k_0, p_0) make a loop and finally converge to this point when $t \to \infty$. Now we have had examples of loops in nonlinear dynamical systems before (see Section 3.2, Fig. 9; or Section 3.8, Fig. 17). But in these cases the loops were complete, and converged to the fixed point both for $t \to +\infty$ and $t \to -\infty$, which made the fixed point a multiple singular point, in the neighbourhood of which the system is not linearizable. Now we have a simple singular point.

Here we have a loop that should converge to the chosen point (k_0, p_0) when $t \to \infty$, but not for $t \to -\infty$. This latter end of the loop is not known to us, and certainly not to the present Hamiltonian economics. Improbable as such a behaviour of trajectories may seem at first sight, it happens, not of course in the general case, but in cases with a sufficiently reasonable economic interpretation to hold on to such a 'saddle-point property' (Cass and Shell, 1976a) in Hamiltonian economics.

It has been shown (e.g. Cass and Shell, 1976b) that for a concave objective function (e.g. the one in (8.138) for $a > 0$) the Hamiltonian function is concave in k and convex in p. Hence the $n \times n$ matrices $\partial^2 H / \partial p_i \partial p_j$ and $-\partial^2 H / \partial k_i \partial k_j$ are positive semidefinite. Furthermore, if the matrix

$$\begin{bmatrix} \dfrac{\partial^2 H}{\partial p_i \partial p_j} & \rho I/2 \\ \rho I/2 & -\dfrac{\partial^2 H}{\partial k_i \partial k_j} \end{bmatrix} = M \tag{8.153}$$

is positive definite, then all bounded trajectories of a nonlinear system (8.150) converge to a fixed point (Brock and Sheinkman, 1976). Thus the system in this case is globally stable, viz. self-regulating.

It is obvious that the restriction to self-regulating systems and their 'saddle-point properties' has come into Hamiltonian economics in a rather trivial way, viz. from the unnecessary restriction to per capita variables. If, instead of k and p, we use the total capital (vector) K and the corresponding imputed price (vector) P, our Hamiltonian systems, as far as they are dynamically stable, will be self-steering. In fact we did so in connection with the two-dimensional Solow–Swan system (8.138)–(8.144). Indeed this system (and its golden age path) is dynamically stable on the condition of self-steering given in Section 8.3, viz. that $\delta/\lambda > (1 - \beta)/\beta$, where λ is the final rate of growth of both the capital K and the output Q.

In the paper mentioned (Lucas, 1988) Robert Lucas extended neoclassical growth theory to the case where the formation of human capital, i.e. of knowledge and skills, is taken into account along with the physical capital K. He did this by introducing into the Solovian model (8.138)–(8.144) the human capital h obeying the following equation of motion:

$$dh/dt = h_0(1 - u)h. \tag{8.154}$$

Here h_0 is a positive constant, and $1 - u$ stands for the share of the total labour time devoted to the accumulation of human capital (i.e. learning and research), u being the share devoted to production. This implies that the labour force L in (8.138)–(8.144) has to be replaced by the expression uhL, while the Hamiltonian function is

$$H(K, h, P_1, P_2; c, u) = \frac{L}{1-a}[c^{1-a} - 1] + P_1\dot{K} + P_2\dot{h}, \qquad (8.155)$$

with the corresponding four Hamiltonian equations of motion, two conditions of maximization (with respect to c and u), and two transversality conditions.

The completed Lucasian four-dimensional system can be immediately solved for the case where s and u are constants, since in this case the equation of motion of capital again is a Bernoulli equation. The result of this simple calculation is that the condition of self-steering retains its form given in Section 8.3, where now

$$\lambda = g + \left(\frac{1 - \beta + \varepsilon}{1 - \beta}\right)\lambda_h, \quad \lambda_h = h_0(1 - u), \qquad (8.156)$$

λ being again the common final rate of growth of the capital K and the output Q, while λ_h is the rate of growth of the human capital h and $1 - \beta + \varepsilon$ is the exponential weight of human capital in the production function. In this way our condition of self-steering is vindicated against the 'saddle-point school'.

Part 3

Complex Applications: The Self-steering and Self-regulation of Human Societies as Wholes

(Non-measurable Systems)

Note: By *complex applications* of fundamental dynamics we mean applications, where causal recursion cannot be explicitly constructed because of the objective complexity (see Chapter 10) of the system. Such a dynamical system is, accordingly, *'non-measurable'*: the variables that are most essential to its behaviour cannot be empirically directly measured, only mathematically discussed and indirectly estimated.

CHAPTER 9

The Concept of a Self-steering Actor

9.1. Human Acts as Tools of Interaction between Consciousness and the World of Objects

The world of man involves both a subjective aspect, viz. his own consciousness, and the world of objects outside it. In the previous parts we have dealt only with this 'world of objects'. To cope with the new element, human consciousness, we have to add to our theoretical arsenal a new tool, viz. *mathematical action theory*.

We shall consider human action as an interaction between a *subject*, i.e. a conscious actor, and an *object*, i.e. some part of the real world at which the *act* of the actor is directed. The actor may be a human individual, or it may be a collective of human individuals, a collective actor. Such an actor may comprise any number of human individuals, it may be the total population of a human society, or even the whole of mankind, at a given point of time.

In the case of an individual human actor the existence of consciousness is immediately clear. No thing is closer and more immediately known to a human being than his or her own consciousness. In the case of a collective actor it is useful to speak of 'collective consciousness' in connection with collective decisions. But it cannot mean more than a convention referring to features that the individual minds seem to have in common at the moment of making those collective decisions.

The possession of consciousness makes an actor a subject. Thus, we have labelled consciousness as something subjective. What is objective, then? By objective reality, or the real world, we mean the collection of objects with which the actor, in principle at least, may be in interaction by means of acts, and which are outside the consciousness of this actor.

In the last analysis, the distinction introduced here between subject and object, and subjective and objective, only refers to the fact that a human being is as aware of the influences of the external world upon his own affairs as of the impulses he himself is able to give to the external world. These impulses are initiated in his own consciousness by his own subjective

163

intentions, or his own will, and conveyed to the world by means of his acts. Indeed, the point of view of a human being involves both subjective and objective aspects, and this seems to be a fundamental attribute of man.

If a man had a complete knowledge of the world we would possibly have no need for a distinction between consciousness and the real world, between subjective and objective. But then we would not exist as the human beings we are: in that case we would be very different kinds of beings. The distinction between subject and object, and subjective and objective, in the above sense seems to be necessary in any adequate theory of human action.

Human acts, accordingly, can be conceived of as tools of interaction between consciousness and the world of objects. Human action is a two-way traffic. On the one hand, as a consequence of a certain kind of acts we obtain knowledge about the real world. We process this information and compose our cognitive beliefs of what the world is like. The initial information is 'empirical knowledge' acquired by acts of observation, of our own or of someone that we trust to tell faithfully of his observations. Our cognitive beliefs are mainly based on generalizations, and thus are hypothetical.

But on the other hand, man is an active being who chooses his acts in order to push events in the real world in certain directions he wants, and away from some directions that he finds displeasing. Or, as put by philosophers, he wants to realize his 'values' in the world of action, and not only in his imagination. The rules of choice of acts are called the 'norms' of one's behaviour, and we can conceive of them as depending on the values as well as on the cognitive beliefs that the actor has composed on the basis of the knowledge he has obtained from the world. Thus, we get at the kind of visual illustration of the 'categories of action' shown in Fig. 68.

What has been sketched out above is a kind of average philosophical answer to the question about how our conscious decisions to act in a definite way are formed. Of course, no-one knows very much about this subject, and that is exactly why it is just a matter of philosophical opinion. Actually, here we do not need to know anything about the formation of one's behaviour but that one's acts reflect one's subjective intentions, and are determined, in a way we do not know in detail, by one's values as well as by one's cognitive beliefs. We shall go farther than that in Part 5, where a mechanism of the formation of human consciousness will be discussed. But so far we do not need it.

What we shall discuss in this chapter concerns the time-honoured problem of free will. The concept of a self-steering actor is an answer to that problem and, besides, a necessary starting point for what we will call the qualitative theory of social development.

To construct the concept of the self-steering actor we need, first of all, a mathematical theory of acts. Another thing we need is a link between the theory of acts and that of dynamical systems. We shall proceed to do these things in this order.

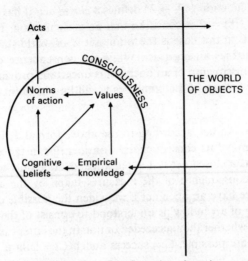

FIG. 68. A schematic illustration of the formation of acts in a conscious actor: human action as a subject–object interaction.

9.2. Mathematical Representation of Acts

Let a complete state-description of the world be given in terms of the set X of the *thinkable total states of the world*. The world is at any moment t in a total state xt that belongs as an element to the set X. Then every *thinkable state of affairs A* can be represented in a unique way by a subset of X, viz. by states where A is true. We can denote this subset of X by the same symbol, A, as the thinkable states of affairs, and the subsets of X determine each other in a unique way. We assume a two-valued logic, so that the complementary set $\mathscr{C}(A) = X \backslash A$ stands for the opposite state of affairs. Let the set of all subsets of X, i.e. the set of all thinkable states of affairs, be denoted by 2^X.

Definitions. Any function $f: X \to 2^X$ is called a *thinkable act*. The state of affairs $f(x) = C$ is called the *consequence* of the act f made in the world state x, while the state of affairs

$$A_C = \{x; f(x) = C\} \tag{9.1}$$

is the *premiss* under which the act f gives the consequence C.

The premisses, accordingly, are the classes of equivalence of the function f, and together give a partition of the set X of all world states:

$$X = + \{A_C; C \in R_f\}. \tag{9.2}$$

Here R_f is the set of values, or range, of the function f. Thus,

$$A_C \cap A_{C'} = \varnothing \text{ (an empty set) for } C \neq C', \tag{9.3}$$

while $C \cap C' \neq \varnothing$ is possible. R_f is a subset of 2^X.

A constant function $f: X \rightarrow 2^X$ defines a *closed act*. It has only one possible consequence, $f(x) = C = $ const., so that R_f is a singleton. The premise of the consequence C in this case is the whole set X of world states. If R_f is not a singleton, f defines an *open act*. While the consequence of a closed act is certain, the consequence of an open act is uncertain, provided that the actor does not know which of the premises is valid at the point of time he makes the act.

Example 1. An act, according to the above formal definition, is entirely defined in terms of its consequences. Consider, for instance, the opening of a window. It is a closed act if it is regarded as done only when the window is open as a consequence of the measures taken by the actor in order to open it. But we have an open act if we widen the possible consequences, i.e. if the 'opening of a window' is understood to consist of the measures taken, irrespective of whether they succeeded or not. In the latter case two alternative consequences are possible, $C = $ success and $\mathscr{C}(C) = $ failure.

Example 2. To stress the significance of the consequences in the above definition of act, let us discuss another simple example. The killing of a dragon, of course, is a closed act if it is defined by the successful consequence, i.e. by a constant function $f(x) = C$ for all $x \in X$, so that $R_f = \{C\}$ is a singleton. If $R_f = \{C, \mathscr{C}(C)\}$, the killing of the dragon is an open, i.e. uncertain, act that may succeed (consequence C) or fail (consequence $\mathscr{C}(C)$). But we get a more exciting story if, say, $R_f = \{C_1, C_2, C_3, C_4\}$, where

$C_1 = $ the dragon is killed and the hero gets the princess,
$C_2 = $ the dragon is killed and the hero gets the princess and half the kingdom,
$C_3 = $ the act fails and the dragon kills the hero,
$C_4 = $ the act fails but the hero manages to escape.

Each different set R_f defines a different act, in view of the above formal definition.

Note 1. The definition of an act given above has been deliberately formulated so that it is broad enough to include all the kinds of acts that ever have been done, or imagined in literature. It is true of most of the acts that even the same act may have different consequences, if made in different situations, depending on the premises of that act. But even this is true only of open acts. Some acts do not permit any alternatives, no matter what the situation. We have called them closed acts. Simple examples are past deeds defined *ex post* on the basis of their consequences (Example 1), paying no attention to the situations in which they were made. This is an act as defined by a bad historian. But even actual acts leaving no alternatives can be imagined, and sometimes done.

Note 2. The passage of time has been stopped twice in the above definition of an act. First we have the world of premises: this is the world at the moment when the deed is done. Time is made to stand still for the second time at the moment when the consequence of the act appears (Fig. 69). Indeed, we proceeded above as if there would be only one set of alternative consequences of which one is realized, depending on which of the premises is true. What happens in the real world is that every act triggers a chain of consequences that follow each other in time, and are causally determined by the 'first' consequence of the act. It is the first, 'immediate consequence' that is intended in the above definition. Of course, what is the immediate consequence remains a convention in so far as we can choose the point of time when we record the immediate consequence of an act. This is also exactly what happens in connection with causal recursion: we can choose the 'unit of time' deliberately, but this possibility of choice in no way limited the scope of the analysis based on causal recursion, as we have seen. We shall see (Section 9.3) that this is also true of the concept of act: indeed the level of generality of the concept of act, as defined above, proves to be exactly the same as that of causal recursion.

Note 3. As for the mathematical representation of the other categories of action mentioned in Fig. 68, we do not need them here, as we shall not be concerned with processes of consciousness (e.g. with the problem of how a decision to act in a certain way takes place in our mind). But of course the 'values' of an actor correspond to an order of preference defined for the consequences of acts, i.e. in the set 2^X of thinkable states of affairs. The 'norms' of an actor define an order of preference of his thinkable acts, i.e. in the set $X \times 2^X$. The 'cognitive beliefs' again define another ordering of states of affairs in the set 2^X, on the basis of their credibility, or subjective probability. Such concepts as we know, are actually applied in the theory of subjective probability. The only difference here is that we have assumed a complete state-description of the world as the basis of action theory.

Note 4. Normally, in a logical discussion, one does not start with a complete state-description of the world. Hence, the meaning of a state of

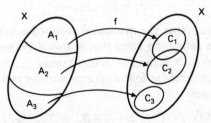

FIG. 69. The mathematical representation of an act as a function mapping the world of premises to that of consequences.

affairs cannot be deduced from its extension, i.e. from the set of objects to which it applies. This is different, of course, when we have a (hypothetical) complete state-description of the world as our theoretical starting point. Now the set of world states where a given state of affairs is true determines its meaning in a unique way: two states of affairs that are true in the same thinkable total states of the world must be considered identical (in a two-valued logic).

9.3. The Theorem on Dual Causality

Every human act has its premises and consequences in the world of objects. But the latter is composed of dynamical systems, in fact it is itself a dynamical system defined in a complete state-description of the world. Or so we are inclined to think, at least when we are speaking of 'universal causality' governing the world. We now make the working hypothesis that universal causality is valid, i.e. that the world is a dynamical system. It follows that human beings, too, are dynamical systems, as well as being actors. Thus, there must be a connecting link between the theory of actors and that of dynamical systems. We shall suppose that the connecting link is the following.

Representation Postulate. Every act f_s made by a dynamical system S can be represented as a function from the state-space X_s of this system to the set 2^X of thinkable states of affairs,

$$f_s: X_s \to 2^X. \tag{9.4}$$

Equation (9.4) does not violate the general definition of act as a function $f: X \to 2^X$, since the state x_s of the dynamical system S can be represented as a function

$$x_s: X \to X_s. \tag{9.5}$$

It follows from Eqs (9.4) and (9.5) that

$$f_s \circ x_s: X \to 2^X, \tag{9.6}$$

which redefines f_s as a thinkable act in the sense of our general definition. If γ is a definite state of the system S, then the class of equivalence of x_s,

$$A_\gamma = \{x; x_s(x) = \gamma\}, \tag{9.7}$$

is the state of affairs that 'the dynamical system S is in the state γ' (Fig. 70).

Our postulate defines an actor that is also a dynamical system. Let us now consider N such actors, or *actor-systems* $S_1, S_2, \ldots, S_N$. Let them simultaneously perform the following acts with the indicated consequences and their premises:

$$f_1(x_1) = C_{y_1} \quad \text{for } x_1 \in B_{y_1} \subset X_1, \quad y_1 \in Y_1,$$
$$f_2(x_2) = C_{y_2} \quad \text{for } x_2 \in B_{y_2} \subset X_2, \quad y_2 \in Y_2,$$

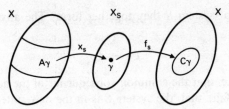

FIG. 70. The functional connection between the premises A_γ and the consequences C_γ of an act f_s made by a dynamical system S in the respective system state γ.

$$f_N(x_N) = C_{y_N} \quad \text{for } x_N \in B_{y_N} \subset X_N, \quad y_N \in Y_N. \tag{9.8}$$

Here the $y_1, y_2, \ldots, y_N$ are indices of the consequences (and of the associated premises), and the $Y_1, Y_2, \ldots, Y_N$ the respective index sets.

The common consequence of these simultaneous acts obviously is the intersection (Fig. 71)

$$C_{y_s} = C_{y_1} \cap C_{y_2} \cap \ldots \cap C_{y_N}, \quad \text{with} \tag{9.9}$$

$$y_s = (y_1, y_2, \ldots, y_N). \tag{9.10}$$

Thus, the simultaneous acts $f_1, f_2, \ldots, f_N$ together generate, in the total system S, the act

$$f_s \colon X_1 \times X_2 \times \cdots \times X_n \to 2^X, \quad \text{where}$$

$$f_s(x_1, x_2, \ldots, X_N) = f_1(x_1) \cap f_2(x_2) \cap \ldots \cap f_N(x_N). \tag{9.11}$$

Let us now put the crucial question: On which condition do the simultaneous acts of our actors together generate a transition of state, x_s

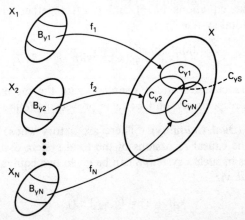

FIG. 71. The co-effect of N simultaneous acts performed by N different actors.

$\rightarrow x_s'$, in the total system S they together form? The answer, of course, is that

$$C_{y_s} = A_{x_s'} \tag{9.12}$$

has to be valid, i.e. that the common consequence of the acts must be equal to the state of affairs that 'the system S is in the new state x_s''.

Under the condition (9.12) we get:

$$g_s \circ f_s : X_s \rightarrow X_s, \quad (g_s \circ f_s)(x_s) = x_s'. \tag{9.13}$$

Here

$$f_s : X_s \rightarrow 2^X, \quad f_s(x_s) = A_{x_s'}, \tag{9.14}$$

$$g_s : 2^X \rightarrow X_s, \quad g_s(A_\gamma) = \gamma \quad \text{for all } \gamma \in X_s. \tag{9.15a}$$

Let the function T_i, defined by

$$T_i : X_i \rightarrow Y_i, \quad T_i(x_i) = y_i \quad \text{for } x_i \in B_{y_i} \subset X_i, \tag{9.15b}$$

be called the *mode of action* of the actor S_i. The modes of action of the actors $S_1, S_2, \ldots, S_N$ define these actors as causal elements. The mode of action of the total system of these causal elements is given by

$$T_s(x_s) = (T_1(x_1), T_2(x_2), \ldots, T_N(x_N)) = y_s. \tag{9.16}$$

It follows from the condition (9.12) that the new total state x_s' is a function of the index y_s. Let this function be denoted by S:

$$x_s' = S(y_s). \tag{9.17}$$

The function S obviously is the *coupling function* of the causal elements $S_1, S_2, \ldots, S_N$. The equations (9.16) and (9.17) together give the causal recursion of the total system S:

$$x_s \rightarrow x_s' = S(T_s(x_s)) = \varphi_s(x_s), \quad \text{with } \varphi_s = S \circ T_s. \tag{9.18}$$

It suffices to associate the states x_s and x_s' with the successive points of time, t and $t + 1$, respectively, and we have proved the following theorem.

Theorem 9.3.1 (Dual Causality). There are actors whose simultaneous acts generate the causal recursions of the total system S they form. The state transitions in such a system S can be given two equivalent forms, an *actor-theoretical*, viz.

$$x_s(t + 1) = (g_s \circ f_s)(x_s t), \tag{9.19}$$

and a *system-theoretical*,

$$x_s(t + 1) = (S \circ T_s)(x_s t). \tag{9.20}$$

In this theorem, the number of actors may be any positive number, even one.

9.4. Self-steering Actors

It follows from the theorem just proved that causal recursion in the total actor-system S concerned obeys the following commutative diagram:

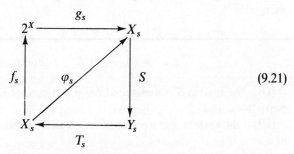

$$\tag{9.21}$$

Depending on which of the branches of this diagram we consider to be the primary, formative one, we can give the system two different interpretations:

Deterministic interpretation. The causal recursion $\varphi_s = S \circ T_s$ of the dynamical system S determines the acts f_s of the actor S: Reactive Paradigm.

Indeterministic interpretation. The acts f_s of the actor S generate the causal recursion φ_s of the dynamical system S: Active Paradigm.

In the former case speaking of an 'actor' has only a metaphorical significance: this kind of actor has no free will, but its acts are determined by the causal structure of the dynamical system it forms. In the latter case the concept of actor has a deeper significance: this kind of actor is capable of creating causal recursion by its own acts, and accordingly by its own subjective intentions, if the acts in turn emerge from such intentions (cf. Section 9.1).

We can show that the latter, indeterministic, interpretation is possible only for self-steering actor-systems, i.e. for actors S such that the dynamical system S is self-steering. For this purpose consider the set of past states of S at the 'present moment' t:

$$X_t = \{xt'; t' < t\}. \tag{9.22}$$

What we know of the causal recursion of the system S at the moment t is confined to the restriction of the function φ_s to this set:

$$\varphi_s | X_t. \tag{9.23}$$

All prediction, of course, is based on causal recursions (whether probabilistic or strict). Therefore, prediction implies that at least one of the following three assumptions is valid for the system concerned. Either the system itself reproduces its past states, as happens, for instance, in celestial mechanics. Or the system can be repeatedly put back to the same initial state x, and the same trajectory through x can be run again and again, as happens in a classical experiment. Or the system can be replaced, for an experiment, by another, identical system.

Suppose that none of these assumptions hold good, but instead

(I)   the dynamical system S has an *expanding, unbounded state-development*, i.e.

$$X_t \subset X_{t'} \subset X_{t''}, \quad X_t \neq X_{t'} \neq X_{t''}, \quad \text{for any } t < t' < t'', \qquad (9.24)$$

$$X_{R^+} = \bigcup_t \{X_t; t \in R^+\} \text{ is unbounded}; \qquad (9.25)$$

(II)   the system S is *unmanipulable*, i.e. it cannot be returned to any of its past states; and

(III)   the system S is *irreplaceable*, i.e. there is no second system identical to it.

Obviously such a dynamical system is unpredictable as a matter of principle, not only in practical terms. The property (I) is tantamount to the assertion that the same state never returns in the system, not even in the approximate sense as in non-periodically pulsating self-regulating systems, with their quasi-closed trajectories. In the latter systems (9.24) is true but not (9.25), and the past states make a kind of ghostly reappearance, as the positive half-trajectories xR^+ visit every δ-neighbourhood $S(x, \delta)$ of x an infinite number of times.

The condition (9.25) excludes even the ghostly reappearance of states, and leaves self-steering systems, along with the systems that are steerable from outside and conceivable as degeneracies of self-steering systems, as the only kinds of goal-directed systems having the property (I). But only for self-steering actor-systems can we apply the property (I) to justify their indeterministic interpretation: only for such actor-systems can we assume that the extension of the domain of definition of the causal recursion (Fig. 72)

$$\varphi_s | X_t \rightarrow \varphi_s | X_{t+1}, \qquad (9.26)$$

is created by the acts f_s of this actor-system.

To make use of the possibility of indeterministic interpretation permitted by the property (I) for self-steering actor-systems we define:

Definition.   A self-steering actor-system S, for which the indeterministic interpretation of Theorem 9.3.1 holds good, and which is unmanipulable and irreplaceable, will be called *self-steering actor*.

Some important conclusions follow immediately from this definition and Theorem 9.3.1:

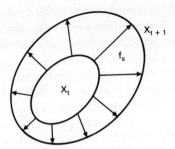

FIG. 72. A schematic illustration of expanding state-development in a self-steering actor-system, indicating the effect of its own act f_s on the expansion of the set X_t of materialized states.

Corollary 1. The total behaviour of self-steering actors is unpredictable and indeterministic, also in a probabilistic sense.

Corollary 2. Self-steering actors are capable of *ultra-self-organization*, i.e. of generating not only changes of values of parameters in their causal recursions, but also causal recursions themselves. No other dynamical system has this property.

Corollary 3. Conscious actors, capable of making decisions based on their own will, can be represented by self-steering actors, and only by them.

Human actors, the only kind of conscious actors with a free will that we know of, obviously are, in terms of the present dynamical theory, self-steering actors. They are the only case of self-steering actors we know. (This is not to deny some forms of rudimentary self-steering in higher animals as well.)

Self-steering actors are the only case in which the indeterministic interpretation of Theorem 9.3.1 obtains. For all other dynamical systems only the deterministic interpretation is possible. Thus, all other actor-systems reduce to deterministic causal systems. (In the case of the 'chaotic systems' with quasi-closed trajectories this may need some discussion: these systems, though practically unpredictable because of technical difficulties, are nevertheless, as a matter of principle, deterministic.)

In fact, as far as the non-self-steering 'actor-systems' are concerned, the combination of actor and system theories performed in this chapter did not bring anything new to their theory: they remained the ordinary dynamical systems that they are in the theory of dynamical systems. Thus, we could drop the word 'actor' entirely and speak only of dynamical systems, unless some new elements would appear that justify the use of this word. In the following, we shall indeed meet such new elements, and introduce a second actor-concept, viz. that of 'self-regulating actor'.

Note (A Dynamical Solution to the Problem of Free Will). The concept of a self-steering actor obviously offers a third way to the solution of the problem of free will. This solution avoids the handicaps of the two trivial solutions called *finalism* and *determinism*. We need not accept finalism, where one of the fundamentals, viz. universal causality, is sacrificed, and human action is reduced exhaustively to the subjective world of ideas. Neither need we accept determinism, which sacrifices the other fundamental, free will, and makes of 'free will' a euphemism for what is really a strictly lawful causal process. In the concept of a self-steering actor, universal causality has been retained simultaneously with the existence of genuinely free will.

Not even the latest excuse for finalism, sometimes called *neofinalism* (e.g. von Wright, 1971) is needed. In this variety of finalism, which has been popular lately, only some aspects of human action are interpreted as inaccessible to causal explanation, to be only 'understood', not 'explained'. But even in this respect the theory of the self-steering actor offers a different interpretation of man: causal recursions are *created* by some acts of self-steering actors, but *afterwards*, once they exist, the resulting causal structures can be investigated, and even the acts in question given *ex post* causal explanations. (In fact this is what scientific history and most of social science are about.)

CHAPTER 10

Self-regulating Actors and Actor-hierarchies

10.1 Subjective and Objective Complexity

Let two players, called Disturber D and Regulator R, be engaged in a 'game of survival', where the *arsenal* of D consists of a finite number $n(D)$ of moves, and that of R of the finite number $n(R)$ of moves. The game is played in rounds, each of which consists of a move α_j of D followed by a move β_k of R, the result being a 'state of the world' a_{jk}. Regulator survives the round, if a_{jk} belongs to the *survival set E*, which is a subset of the set of all world states,

$$X = \{a_{jk} \, j = 1, 2, \ldots, n(D); \, k = 1, 2, \ldots, n(R)\}. \tag{10.1}$$

Let Y be the trajectory of the world states representing a match played, with the understanding that Disturber has tried every move at least once. For the sake of simplicity, let us assume that the same element is not repeated in the same column of the table of world states (Fig. 73). Let us also assume that Regulator tries to keep the number of states visited, i.e. the number $n(Y)$ of states included in Y, as small as possible, and that this is his only aim (ignoring survival for a moment).

Let Disturber start with α_1, to which Regulator responds with β_1. To the

		R			
		β_1	β_2	············	$\beta_{n(R)}$
	α_1	a_{11}	a_{12}	············	$a_{1n(R)}$
	α_2	a_{21}	a_{22}	············	$a_{2n(R)}$
D	$\vdots$	$\vdots$	$\vdots$		
	α_j	a_{j1}	a_{j2}	············	$a_{jn(R)}$
	$\vdots$	$\vdots$	$\vdots$		
	$\alpha_{n(D)}$	$a_{n(D)1}$	$a_{n(D)2}$	············	$a_{n(D)n(R)}$

FIG. 73. The game of survival between Disturber D and Regulator R, showing the arsenals of the two players, and the results a_{jk} of their moves.

second move, α_2, of D, Regulator must change his move too, if he is not to increase the number of states visited. Let him play the move β_2. Continuing in the same way we observe that in his pursuit of minimizing $n(Y)$ Regulator has to play all his moves at least once, provided that $n(D) \geqslant n(R)$. After having played each of them once, $n(R)$ different world states have been visited. But if $n(D) > n(R)$, the game may go on, and Regulator has to play each of his moves at least as many times over as the number $n(D)/n(R)$ indicates. It follows that $n(Y)$ cannot be smaller than this ratio, i.e.

$$n(Y) \geqslant n(D)/n(R). \tag{10.2}$$

This holds good for any game of survival of this type. By defining the *variety* of a finite set A by

$$H(A) = \log n(A), \tag{10.3}$$

we can rewrite (10.2) as

$$H(Y) \geqslant H(D) - H(R). \tag{10.4}$$

A necessary condition of survival of R, of course, is

$$n(Y) \leqslant n(E), \quad \text{or} \quad H(Y) \leqslant H(E). \tag{10.5}$$

This kind of example was indeed used by W. Ross Ashby himself (Ashby, 1970, pp. 202–206) by way of a didactic introduction to his law of requisite variety, of which (10.4) is a very special case. Though, of course, formally correct, these examples lack an important element that first makes sense of this law, namely the presupposed objective complexity of the world concerned. Worse still, if we move in our mind to more complex games, such as chess, we come into the realm of a wrong kind of complexity, viz. a subjective one. In chess, and in other complex games, the rules are always simple, and the complexity follows merely from the number of different possibilities being too great for a player to keep them simultaneously in his mind. This is subjective complexity.

Objective complexity is there if the exact rules of the real game of survival played in the world are themselves complex, for instance too complex to be written down during the life-span of man; or so complex that the existing amount of paper in the world does not suffice for these rules to be written down. In such a situation the players do not, and cannot, know the exact rules, according to which the game of survival is being played. This is the situation implied by Ashby's law, and we had better start again from square one, and approach this law through another route.

10.2. Survival through Self-regulation: Ashby's Law

From the point of view of cybernetic theory, the survival of a dynamical system implies the ability of the system to eliminate the effects of external disturbances on the motion of state of the system, partially at least (cf. a system steerable from outside), and thus to prevent the disintegration of the system. But if the system in question is a living being, and survival means keeping alive, we have a narrower concept of survival. Such a being is, from a dynamical point of view, a self-regulating system having the ability of self-organization. It is facing an environment that has a certain arsenal of disturbances ready to use against it. The system confronts the challenges by various acts of self-organization, i.e. by changing the values of parameters in its causal recursion. If it succeeds in restricting the effects of disturbances on its state not only bounded, but bounded in a certain domain E of state-space, it survives the threats posed. The *region of survival E* expresses the physiological and anatomical conditions of survival for this particular living being.

Thinking of a living being, or a community of living beings, in such a situation we define:

Definition. A self-regulating and self-organizing dynamical system S, with the causal recursion

$$\varphi(x, \lambda): X \times \Lambda \to X, \tag{10.6}$$

where x is the total state and λ the parameter vector, is called a *self-regulating actor* having the *arsenal of regulatory acts* Λ, if a region of survival $E \subset X$ is indicated in the state-space X.

This kind of 'actor' may or may not have consciousness. A self-regulating actor is just a representation of whatever living organism, or set of living organisms, in terms of dynamical theory.

Let us now place a self-regulating actor in an environment where its survival is challenged by various threats called 'disturbances'. It may happen that the interplay of these disturbances and the regulatory acts, to which the actor resorts to keep itself alive, is extremely complex. For an external observer who wants to describe and explain this struggle for existence, the simplest approach to cope with the complexity it displays is to let each environment be represented by the probability distribution that the appearance of various disturbances in that particular environment seems to obey. Thus, he associates a probability $p(d)$ to each disturbance d in a given environment. Similarly, he associates a probability $p(\lambda)$ to each regulatory act λ performed by the actor.

Following this train of thought he can classify the different environments according to their *variety of disturbances* $H(D)$, and the different self-regulating actors according to their *variety of regulatory acts* $H(R)$, defined as the entropies

$$H(D) = \sum_{d \in D} p(d) \log \frac{1}{p(d)}, \tag{10.7}$$

$$H(R) = \sum_{\lambda \in \Lambda} p(\lambda) \log \frac{1}{p(\lambda)}. \tag{10.8}$$

Here a finite set of disturbances, D, and a finite set, Λ, of regulatory acts have been assumed, for the sake of simplicity. But the concept of entropy, as we know, applies to any kind of set, whether finite or infinite, and to qualitative as well as quantitative variables.

Entropy is the most general measure of variety. Besides, it gives the most general measure

$$H(A) - H_B(A) = H(B) - H_A(B) \tag{10.9}$$

of the mutual dependence of two variables A and B. Here the conditional entropy, for finite sets A and B given by

$$H_B(A) = \sum_j p(B_j) \sum_i p(A_i/B_j) \log \frac{1}{p(A_i/B_j)}$$
$$= \sum_j p(B_j) H(A/B_j), \tag{10.10}$$

is used. Of the general properties of the concept of entropy we need here, in addition to (10.9), only the following:

$$H(A) \geqslant H_B(A) \geqslant 0, \tag{10.11}$$

which holds good for any sets or variables A and B.

We can now discuss Ashby's general *law of requisite variety*. It pertains to a regulatory process

$$D \to R \to Y \leftrightarrow X, \tag{10.12}$$

where a complex self-regulating actor, in possession of the variety $H(R)$ of regulatory acts, tries to prevent the external disturbances, having the variety $H(D)$, from pushing its state off the region of survival, $E \subset X$. The variable Y stands for the result of regulation, and $H(Y)$ is the part of the variety of disturbances that has nevertheless passed through the regulatory shelter of the actor. There is a 1–1 mapping between the values of Y and the states $x \in X$, so that a necessary condition of survival of the actor is

$$H(Y) \leqslant H(E). \tag{10.13}$$

The purpose of regulatory acts, accordingly, is to make the variety $H(Y)$ as small as possible.

Ashby's law gives a limit to this possibility. It is based on the following assumption (Ashby, 1970, pp. 207–208).

Ashby's Regulation Postulate. A passive regulator R does not reduce the variety of disturbances, i.e.

$$H_R(Y) \geqslant H_R(D) - K, \tag{10.14}$$

where K is a constant reduction due to a possible structural shelter.

Here the passive regulator, as is evident from (10.14), means one that reacts with the same act whatever the disturber D does. By applying the identity (10.9) to $H_R(D)$ and using (10.11) the condition (10.14) gives the desired law (illustrated in Fig. 74):

$$H(Y) \geqslant H(D) - H(R) + H_D(R) - K. \tag{10.15}$$

The expression on the right-hand side indicates the minimum variety $H(Y)$ that can be achieved by regulation.

Ashby's law states that the threat imposed by the environment to the survival of a complex self-regulating actor, i.e. a living being, or a community of such beings, can be reduced

(i) by making the variety of available regulatory acts, $H(R)$, as large as possible;

(ii) by a structural shield, K, such as, for example, the buildings and clothes of man, or the shell of a tortoise, or the skin of animals; and

(iii) by reducing the *uncertainty of regulation*, $H_D(R)$, as much as possible. If $H_D(R)$ approaches zero, R becomes a function of D,

$$H_D(R) \to 0 \Rightarrow R = f(D), \tag{10.16}$$

and there is only one regulatory response to each disturbance d: uncertainty has entirely disappeared.

Point (i) above had in fact been the starting point of Ashby's theoretical speculation. He came to that by observing the functioning of an electro-magnetic apparatus, called a *homeostat* (see e.g. Ashby, 1972, or Ashby, 1970), which he had constructed to imitate the self-organizing self-regulation that is supposed to take place (cf. Section 2.4) in a living organism. The total state of the homeostat could be indicated by the position of four needles, and every time one or more of them came to the border of the region of survival a new electric coupling scheme (selected at random) was automatically switched on, and changed, until a coupling was found that sent the needles back into the allowed domain.

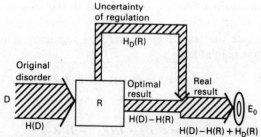

FIG. 74. The reduction of entropy (disorder) in a simple process of regulation, as studied by Ashby (in Ashby, 1970, originally printed in 1956).

Ashby observed that the homeostat was always able to return to its state of equilibrium (all needles at their zero position), after having been disturbed, provided that the number of different alternative coupling schemes, i.e. regulatory responses of the apparatus, was large enough. Hence, he formulated a preliminary idea of what was later mathematically specified and put into the form of his law of requisite variety: 'only variety in R (regulation) can force down the variety due to D (external disturbance)' (Ashby, 1970, p. 207). The mathematical discussion indicated that, in addition to the variety of regulatory acts, measured in terms of the entropy $H(R)$, the uncertainty of regulation, $H_D(R)$, had to be taken into account (Ashby, 1970, p. 208).

It follows from the existence of the uncertainty of regulation that we have to distinguish between the *optimal regulatory ability* $H(R)$ and the *effective regulatory ability*

$$H_{\text{eff}}(R) = H(R) - H_D(R) \qquad (10.17)$$

of our actor.

Since the increase of entropy means growing disorder, and decreasing entropy means an increase of order, Ashby's law can also be interpreted to mean that

order is created in the regulatory processes maintaining life.

As the negentropy law concerns complex self-regulating actors in interaction with their environment, and thus open systems, it is not in contradiction with the physical *entropy law* stating that entropy (disorder) grows in closed dynamical systems. But, of course, it weakens the philosophical argument for a 'cold death', based on the entropy law: if life plays an essential role in the universe, a structured universe may survive after all.

10.3. The Improvement of Self-regulation by Actor-Hierarchies

Suppose that we have a complex self-regulating actor in possession of a regulator, R_1, which is insufficient for its purpose, which is to bring the variety of external disturbances, $H(D)$, down to the level necessary for the survival of the actor. Suppose next that we have another regulator, R_2, that is able to regulate the result, Y_1, of the first regulator. Then the optimal result of the first regulator,

$$H(Y_1) = H(D) - H(R_1) \qquad (10.18)$$

can be further improved to

$$H(Y_2) = H(D) - H(R_1) - H(R_2). \qquad (10.19)$$

If even this reduction of the variety of disturbances is not enough, we can add further regulators, until, say, a regulator of order m, R_m. The optimal reductions of $H(D)$ achieved will then be

$$H(Y_m) = H(D) - H(R_1) - H(R_2) - \cdots - H(R_m).\qquad(10.20)$$

Thus, the sequence of m regulators together make up a total regulator R whose optimal regulatory ability is the sum of those of its parts:

$$H(R) = H(R_1) + H(R_2) + \cdots + H(R_m).\qquad(10.21)$$

In the general case, the optimal result is not achieved because of the uncertainty of regulation, which reduces the effective regulatory ability (10.17). If nothing is done to it there will be, in our sequence of m regulators $R_1, R_2, \ldots, R_m$ a total uncertainty that amounts to (see Fig. 75)

$$H_D^0(R) = H_D(R_1) + H_{Y_1}(R_2) + \cdots + H_{Y_{m-1}}(R_m).\qquad(10.22)$$

But this can be reduced by arranging for each regulator R_k a *controller*, or *governor*, G_k, that specializes in decreasing the uncertainty of the regulatory action of R_k. In other words, G_k tries to co-ordinate the regulatory acts of R_k so as to fit better to the appearing disturbances Y_{k-1} (with $Y_0 = D$). If this succeeds, the uncertainty of R_k will be reduced by some positive amount $H(G_k)$ to

$$H_{Y_{k-1}}(R_k) - H(G_k).\qquad(10.23)$$

The total effect of the m first-order controllers will be that the uncertainty of the total regulator R will be reduced by the amount

$$H(G^1) = H(G_1) + H(G_2) + \cdots + H(G_m),\qquad(10.24)$$

the remaining uncertainty being

$$H_D^0(R) - H(G^1).\qquad(10.25)$$

Hence, we can go on and build up a *hierarchy of control* (cf. Fig. 75) as well, until we reach the top controller G^s, the total wisdom of all the governors being

$$H(G) = H(G^1) + H(G^2) + \cdots + H(G^s).\qquad(10.26)$$

The remaining ignorance about regulation will be

$$H_D(R) = H_D^0(R) - H(G).\qquad(10.27)$$

Neglecting the trivial constant K, the result of regulation obtained in the total *hierarchy of regulation and control* (Fig. 75) constructed above will be

$$H(Y) \geqslant H(D) - H_{\text{eff}}(R),\qquad(10.28)$$

with the effective regulatory ability

$$H_{\text{eff}}(R) = H(R) - H_D^0(R) + H(G).\qquad(10.29)$$

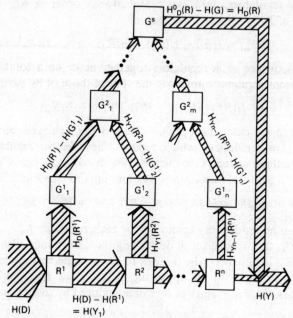

FIG. 75.   The larger reduction of entropy (disorder) achieved in a hierarchy of regulation and control due to self-regulating actor hierarchy, as investigated by Aulin, 1982.

Depending on the magnitude of the *control entropy* $H(G)$ the effective regulatory ability achieved by hierarchy lies between a minimum and a maximum as follows:

$$H(R) - H_D^0(R) \leqslant H_{\text{eff}}(R) \leqslant H(R). \qquad (10.30)$$

Thus, we have shown that by arranging a number of complex self-regulating actors into a hierarchy of regulation and control, where some actors concentrate on regulation and others on control, the effective regulatory ability of this community of actors can be greatly improved. Let us call this kind of community an *actor-hierarchy*. It is itself, of course, a complex self-regulating actor, a collective one.

An obvious result involved in the above construction could be stated as follows:

The weaker the average regulatory ability and the larger the average uncertainty of available regulators, the more *requisite hierarchy* is needed in an actor-hierarchy for the achievement of the same effective regulation.

This is a rudimentary form of the law of requisite hierarchy, to which we shall soon give a more exact formulation.

CHAPTER 11

The Law of Requisite Hierarchy

11.1. Human Society as a Complex Self-regulating and Self-organizing Actor-Hierarchy

In human society the population is maintained by social production. The machinery of regulation, R, in human society thus comprises the production and distribution of goods and services. The threats against human life that are resisted by the production and distribution of goods and services consist of hunger, thirst, coldness, heat, diseases, storms, beasts, hostile human communities, etc. Together they are the source of 'disturbances' D. The result of regulation, Y, is composed of the remaining hunger, thirst, coldness, heat diseases, etc., that could not be eliminated by the produced and distributed goods and services. The resulting process of regulation in society is illustrated by Fig. 76.

There are many kinds of units of production or distribution in human society. Households produce manpower and consume goods and services. Various institutions of production, such as factories, farms, and firms of different kinds, produce goods and consume labour force and the goods called raw materials and capital goods (machines, buildings, etc.). The service institutions, like hospitals, schools, and various offices, produce services and consume manpower and goods. There are also special institutions, such as shops and the many intermediary commercial companies, that distribute goods and consume manpower and some goods. Here a terminology referring to a modern industrial society has been used. But for at least some of these units there are functional counterparts even in primitive human communities. All these units of production or distribution, say $r_1, r_2, \ldots, r_n$, together make up an input–output net, which can be arranged into the successive steps of production and distribution,

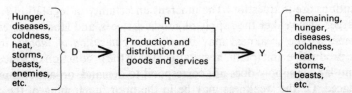

FIG. 76. The variables of disturbance (D), regulation (R), and the outcome of regulation (Y) in the regulatory process (or production) of human society.

$$R_1 \to R_2 \to \cdots \to R_m. \qquad (11.1)$$

Each of the units of production or distribution normally has some sort of control unit, called 'management', or 'administration', or 'government', or simply 'leadership'. These units, say $g_1, g_2, \ldots, g_n$, make up, when scattered in their proper places in the sequence (11.1), and combined into the units $G_1, G_2, \ldots, G_m$, the first step in the control hierarchy of human society. These controllers are those closest to the actual 'doers' in production or distribution. Depending on how we choose the units $r_1, r_2, \ldots, r_n$, they may represent various layers of management, beginning with the foremen of small groups to the management of large companies. What exactly the r- and g-units chosen are is not important here.

The controllers $G_1, G_2, \ldots, G_m$ in the first tier of the hierarchy are, in turn, controlled and co-ordinated by higher units of control, and these again by even higher ones. We can continue our inspection of social control hierarchy up to the uppermost leadership, G_s, of the society in question, whatever it or he or she may be. Thus, all in all, human society is an actor-hierarchy composed of complex self-regulating actors, in the last analysis of individual human beings. It is itself a complex self-regulating actor-hierarchy.

It remains to give an interpretation, in terms of production, distribution, and management, of the entropies $H(R)$, $H_D^0(R)$, and $H(G)$ that govern the behaviour of human societies *as wholes*. The following formulations cannot be wide of the mark.

(i)   The optimal regulatory ability $H(R)$ of human society is a compound measure of the technological level and occupational skills in society, being the larger the higher the technological level and the better the skills. In other words, $H(R)$ is a measure of the *technical level of 'productive forces'* (labour + means of production) in society. Here we mean the productive forces actually in use, whether they are owned or loaned.

(ii)   The uncertainty of regulation inherent in productive forces, $H_D^0(D)$ is, of course, a measure of how well or badly the produced and distributed goods, services, and manpower *hit their targets*, i.e. satisfy the needs for them in society. For consumer goods and services the needs of the human population are concerned. For capital goods, unfinished products, raw materials, and services 'consumed' in production the needs of productive institutions are in question. The inherent uncertainty of regulation $H_D^0(R)$ is the larger, the weaker the supply of goods, services, and labour corresponds to demand. The weakness may be in the *organization of labour*, inside a single productive unit (plant, hospital, etc.) or in the production or educational system: labour supply does not correspond to demand, or labour inputs are misplaced. Or the weakness may be in the poor functioning of the *market system*: the produced goods and services have no demand.

(iii) The control entropy, $H(G)$, of a given society is a measure of how much the poor organization of labour and the poor functioning of the market system in society have been compensated for by a hierarchy of control. Since the total control entropy is a grand total of the control entropies of all single controllers,

$$H(G) = \sum_{i,j} H(G_j^i), \tag{11.2}$$

over the different tiers and individuals of the hierarchy, we can consider it as a measure of the *largeness of hierarchy* in society: in a great number of single controllers G_j^i the individual differences between them in 'managerial skills' can be assumed to cancel out.

11.2. The Law of Requisite Hierarchy for Undeveloped Economic Systems

Here, an undeveloped (and thus undeveloping) economic system means undevelopment of both production and distribution. More exactly, the situation is defined by constancy of both the optimal regulatory ability $H(R)$ and of the inherent uncertainty of regulation $H_D^0(R)$ of the economic system of a given society, the former on a low level, the latter on a high level:

$$H(R) = a, \quad H_D^0(R) = b, \tag{11.3}$$

where a and b are constants.

It follows that the effective regulatory ability

$$H_{\text{eff}}(R) = H(R) - H_D^0(R) + H(G) = a - b + H(G) \tag{11.4}$$

of the economic system in question can grow only by increasing the hierarchy of society, measured by the control entropy $H(G)$. The maximum utilizable value of $H(G)$ is, of course, the initial total amount of uncertainty in regulation, $H_D^0(R)$: you cannot compensate, by hierarchy, for more uncertainty than there is in the first place.

There is also a minimum acceptable hierarchy obtained from the condition of survival of the population: to satisfy (10.13) with the result of regulation obeying (10.28) we must have

$$H_{\text{min}}(Y) = H(D) - H_{\text{eff}}(R) \leqslant H(E). \tag{11.5}$$

This now gives, in view of (11.4),

$$H(G) \geqslant [H(D) - H(E)] - (a - b). \tag{11.6}$$

Thus, we have the following limits:

$$H_{\text{max}}(G) = b, \quad H_{\text{min}}(G) = [H(D) - H(E)] - (a - b). \tag{11.7}$$

Hence we get:

In a society of an undeveloped economic system, the effective regulatory ability of the economic system can be increased, and thus the ability of population to survive improved, only by increasing the control $H(G)$ and thus social hierarchy from the minimum necessary for the survival of population, $H_{min}(G)$, towards the largest utilizable hierarchy corresponding to the value $H_{max}(G)$ of control entropy (the *law of requisite hierarchy for undeveloped economic systems*).

Since the pursuit of greater security is a very strong motive in all kinds of human societies, the law of requisite hierarchy predicts a trend towards growing social hierarchy in all societies of an undeveloped economic system (Fig. 77).

Example 1. Economic development has rarely in human history entirely stopped. But it was closest to a stop, when compared with later times, during the era of *archaic societies*, which started as small primitive communities and ended up with the formation of the great 'temple states' of the ancient Middle East and Orient. This was the historical birth of the state, and the state in question was a theocratic state with priests in command. It was also the purest example of a 'class society', with a strict social hierarchy, at the bottom of which were slaves who did most of the physical work. The development of archaic societies is a striking example of growing hierarchy in societies of undeveloped economic systems. It started in the region stretching from the Mediterranean to the Indian Ocean, but similar developments towards ever-greater hierarchy took place later, as we know, in other parts of Asia, in Africa, and in America. The best-known temple states were those in Babylonia (from about 4000 BC), in Egypt (from 3000 BC), in India, in Persia (c. 500 BC), and later still on the American continent (the states of the Incas and the Aztecs, c. AD 1000).

Example 2. A similar development towards greater internal hierarchy seems to have taken place in many African societies, when the African states gained independence in this century, after the colonial era. When left on their own, with relatively undeveloped economic systems, usually an increased

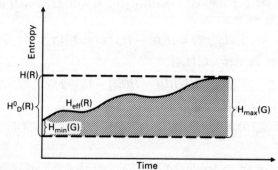

FIG. 77. The law of requisite hierarchy for a stagnated economy: the effective regulatory ability can be improved only by increasing hierarchy.

concentration of power on a small local elite followed, often leading to dictatorships. The tendency to a greater concentration of power, and thus to a greater internal hierarchy, has also been discernible in the poor countries of Asia and South America. This theme, of course, like the problem of the reasons for Third-World grievances in general, makes good kindling for ideological controversies. But it is interesting to note that the law of requisite hierarchy suggests a direct connection between the tendency to greater internal hierarchy and an undeveloped economic system.

11.3. The Law of Requisite Hierarchy for Developing Economic Systems

'Developing economic system' is used here to mean development of both $H(R)$, i.e. technology and occupational skills, and of $H_D^0(R)$, i.e. the markets and the organization of labour, so that $H(R)$ increases and $H_D^0(R)$ decreases;

$$\mathrm{d}H(R)/\mathrm{d}t > 0, \quad \mathrm{d}H_D^0(R)/\mathrm{d}t < 0. \tag{11.8}$$

Thus, the curve of the effective regulatory ability $H_{\mathrm{eff}}(R)$ runs between ever narrower rising limits (see Fig. 78), and accordingly rises itself, on average. At the same time the need of the control entropy

$$H(G) = H_{\mathrm{eff}}(R) - [H(R) - H_D^0(R)] \tag{11.9}$$

and thus of hierarchy diminishes, and approaches zero on the long run. In other words:

In a society with a developing economic system the effective regulatory ability of the economic system increases on average, improving the means

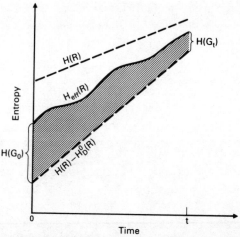

FIG. 78. The law of requisite hierarchy for a developing economy: the effective regulatory ability is improving simultaneously with a reduction in hierarchy.

of survival of the population, at the same time as the need for hierarchy decreases (the *law of requisite hierarchy for developing economic systems*).

As a consequence there is, in a society with a developing economic system, a situation favourable to the dismantling or relaxation of the internal social hierarchy *without jeopardizing the growth of security of the population*. The emergence of such a situation for the first time in human history surely has been a remarkable turning point. Thus, the law of requisite hierarchy supports the 'Comte–Pareto hypothesis', stated by Auguste Comte at the beginning of nineteenth century and completed by Vilfredo Pareto towards the end of it, to the effect that the history of human societies can be divided into two great developmental periods that are qualitatively different from each other.

Example. When did the great turn of history really take place? This again, of course, is a good opening to an endless ideological dispute. Here, however, it is only the connection between the beginning of economic development and the turn from increasing to decreasing social hierarchy, predicted by the law of requisite hierarchy, that we want to illustrate by a historical example. A striking example is the birth of the Western type of societies in Western Europe upon the ruins of the disintegrated Roman empire. The era of archaic societies came to an end everywhere with a more or less violent disintegration of the latest archaic empires based on slavery. But only in Western Europe did this death produce a new type of society with a more relaxed hierarchy and a briskly developing economy, first in the Mediterranean city states, and later on a larger scale elsewhere in Europe.

CHAPTER 12

The Cybernetic Concepts of Social Development and Underdevelopment

12.1. Human Society as a Collective of Self-steering Actors

Human society, as a collective actor composed of human individuals, is itself a self-steering actor. But its self-steering comes entirely from the self-steering of its individual members: the greater the self-steering of its individuals, the greater the self-steering of society. Subjectively, corresponding to the objective concept of self-steering is the notion of freedom. In the following pages we shall consider *human freedom* as a subjective expression for self-steering, i.e. human freedom is self-steering as experienced by the human individual. Or, just as well the other way round, self-steering is the objective, as a matter of principle but not in practice measurable, expression of human freedom.

How to measure self-steering? A workable operational rule for the measurement of self-steering is impossible. What is sufficient for theoretical purposes is the principle of its measurability. This principle can be given the following mathematical formulation.

As a self-steering actor a given human society S is simultaneously a dynamical system. The dynamical system S has a state-space X, of the mathematical properties of which we first have to state the most general one, viz. that it is a topological space. Let a Boolean algebra F of subsets of X be given, that is: a set F of subsets of X, closed under the formation of differences, countable unions, and countable intersections, such that $X \subset F$. Then the combination (X, F) is a measurable space. Let a measure v be given in this space, i.e. a function

$$v: F \to R^+, \quad \text{such that } v(\varnothing) = 0;$$

and

$$v\left(\bigcup_{n=1}^{\infty} A_n \right) = \bigcup_{n=1}^{\infty} v(A_n) \text{ for any } A_1, A_2, \ldots \in F;$$

such that

$$A_j \cap A_k = \varnothing \text{ for all } j \neq k. \tag{12.1}$$

If xt is the total state of the dynamical system S at the moment t, we shall use the following notations:

D_{xt} = the domain of self-steering of the society S at the point of time t; $\hspace{3cm}$ (12.2)

$v(D_{xt})$ = the degree of self-steering in the society S at the moment t. $\hspace{1cm}$ (12.3)

With these notations we have formally defined a measure of self-steering in the society S, viz. its v-measure (12.3). Of course it is not an operational definition, but a formulation that is tantamount to a statement of existence, and as such a statement it will be used in the following.

What exactly is stated to exist by this formulation? What is the human freedom measured, as a matter of principle, by the degree of self-steering in human society? Since self-steering processes can be assumed to be representations of the total intellectual processes in human consciousness, and of their mutual interactions in human society, the core of the human freedom here intended is intellectual freedom, i.e. the freedom of thought and expression in the given society S. But as the self-steering intellectual processes in the human mind are the initiators of self-determined human actions in general, the meaning of human self-steering extends to the freedom of action in the most general sense.

No-one can hope to measure, in any reliable operational terms, what is the degree of intellectual liberty, and thus of the freedom of action, in a given society at a given point of time. But instead of pursuing such a hopeless aim of direct measurement, we can choose an indirect approach that is more promising, theoretically at least.

12.2. The Limits of Toleration of Self-Steering in Human Society

What is the relation between the two roles of human society as a self-steering actor, on the one hand, and as a complex self-regulating actor-hierarchy on the other?

We know that the former role is related to the nature of man as an intellectual being: self-steering processes stand for the total intellectual processes occurring in human consciousness, and for their mutual interactions in human society. A self-regulating actor-hierarchy, on the other hand, is a means of survival for the population of human society. However, these two aspects of human existence are, from a dynamical point of view, two aspects of the *same* dynamical system S. Thus, they are necessarily interrelated. We shall suppose that when expressed on the most general level their mutual relationship is, accurately enough for our present analysis, given by the following rule (Aulin, 1982, 1986a):

The Rule of Inverse Proportionality. The more hierarchy there is, measured by the control entropy $H(G)$, in a given human society, the less self-steering there is:

$$v(D) \sim 1/H(G). \tag{12.4}$$

This is not an independent postulate but a rule of thumb, whose validity as a reasonable first-order approximation follows from the concept of a cybernetic system as a feedback system. The more demanding cybernetic properties, of which the highest is self-steering, imply more feedback circuits in the coupling structure of the system in question. But the more hierarchy there is in the coupling structure, the less feedback can there be. At the extreme end of a perfectly hierarchical coupling we have a tree graph, with no feedback circuit at all. Thus, as a general rule, hierarchy reduces self-steering. No doubt the rule (12.4) is not accurate, but it can be supposed to be accurate enough for a first approach of the problem of self-steering in human societies.

It follows from the rule (12.4) that we have the following

Corollary. The developmental degree of economy in human society determines an *upper and lower limit of toleration of self-steering* in society. These limits cannot be exceeded without engendering serious consequences to the society in question.

Indeed, from the existence of the upper and lower limits of the hierarchy that can be used to reduce the uncertainty of regulation, viz.

$$H_{max}(G) = H_D^0(R) \tag{12.5}$$

and

$$H_{min}(G) = [H(E) - H(D)] - [H(R) - H_D^0(R)], \tag{12.6}$$

it follows, by the rule (12.4), that there are corresponding limits of self-steering:

$$v_{min}(D) \sim 1/H_{max}(G), \tag{12.7}$$

$$v_{max}(D) \sim 1/H_{min}(G). \tag{12.8}$$

Since $H(E)$ and $H(D)$ can be thought of as given, the former defining the survival of the population, the latter defining the environment in which the society in question lives, the limits of self-steering are determined by the variables $H(R)$ and $H_D^0(R)$, i.e. by the developmental level of economy in this society.

What are the 'serious consequences' mentioned in the Corollary? To answer this question we have to think of the consequences that follow, according to theory, from crossing the limits of hierarchy, $H_{min}(G)$ and $H_{max}(G)$. Crossing the former limit means that the existing control hierarchy of our society becomes less than is necessary for the survival of the population:

$$H(G) < H_{min}(G). \tag{12.9}$$

It follows that the population starts dying out. By Eq. (12.8) this implies that when the actual self-steering $v(D_{xt})$ of our society at the point of time t exceeds the upper limit of tolerance, i.e.

$$v(D_{xt}) > v_{max}(D_t), \tag{12.10}$$

the *survival of population is threatened* (see Fig. 79).

The maximum control entropy $H_{max}(G)$ indicates the largest amount of hierarchy that can be used to reduce the inherent uncertainty of regulation, $H_D^0(R)$, in a self-regulating actor-hierarchy, such as human society. Obviously, there can be more hierarchy in society, but the excess hierarchy cannot be utilized. This is why we have to interpret the 'too low' self-steering,

$$v(D_{xt}) < v_{min}(D_t), \tag{12.11}$$

as implying the existence of *unnecessary hierarchy* in society (Fig. 79).

Note. From the law of requisite hierarchy for a developing economy it follows that (cf. Fig. 78)

$$H_{min}(G) \rightarrow H_{max}(G) \rightarrow 0. \tag{12.12}$$

This does not imply that the corresponding limits of self-steering should approach each other. This is easily seen from a simple example. Consider, for instance, the function

$$f(x, y) = 1/x - 1/y, \text{ with } x = H_{min}(G), y = H_{max}(G). \tag{12.13}$$

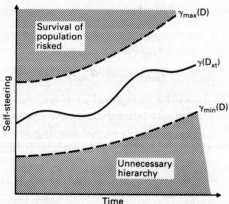

FIG. 79.   The limits of toleration of self-steering in human society, and the Scylla and Charybdis looming behind them.

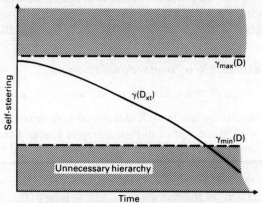

FIG. 80.   The inevitable decrease, in the long run, of human freedom (self-steering) under the conditions of underdevelopment.

Let us put $y = kx$, where k is a positive constant that indicates the direction in which we want to approach the origin. In the now-relevant directions we have $k > 1$, i.e. $H_{max}(G) > H_{min}(G)$, and we get

$$f(x, y) = (1 - 1/k)/x \to +\infty \quad \text{with } x \to 0. \tag{12.14}$$

We can see that it is more likely that the difference $v_{max}(D_t) - v_{min}(D_t)$ increases rather than decreases with time in a society with a developing economic system.

12.3.   The Concept of Underdevelopment

(1) The defining characteristics

This is the case of the law of requisite hierarchy for undeveloped economic systems. In other words, both the optimal regulatory ability $H(R)$ and the uncertainty of regulation $H_D^0(R)$ of the economic system are constants (or nearly so). From the law of requisite hierarchy and from the rule of inverse proportionality we then have the following consequences that define the *type of underdevelopment*, or the *type of archaic development*.

(I)   Security can be created only by hierarchy: the effective regulatory ability of the economic system can be increased only by increasing hierarchy in society.

(II)   There is a conflict between security and individual freedom: with increasing social hierarchy the freedom of human individuals, or self-steering in society, decreases (Fig. 80).

It follows that the type of underdevelopment leads to a continuing growth of hierarchy, and thus an incessant decrease of individual freedom in society. The purest example of this type comes from the archaic societies, as was noted previously, but present-day examples can be found in the Third World.

(2) The vicious circle of underdevelopment

The successive steps of this circle can be characterized as follows.

(i) The economic system is underdeveloped and thus undeveloping: the technological level of the means of production is low and the occupational skills of manpower poor (i.e. the optimal regulatory ability $H(R)$ of the economic system, is weak), at the same time as the organization of labour is bad and the market system poorly developed (i.e. the inherent uncertainty of regulation, $H_D^0(R)$, of the economic system is large). This is the phase of *economic stagnation.*

(ii) Hierarchy keeps growing and power concentrates steadily, leading to a dictatorship and, as an extreme development, to a totalitarianism of the archaic type. This is the phase of *political stagnation.*

(iii) Self-steering, i.e. human freedom, decreases in society, as a consequence of which intellectual and economic initiative shrinks, and is replaced by general apathy. This is the most extreme phase of a long-lasting underdevelopment: *cultural stagnation.*

(iv) Back to square one: economy cannot develop.

It is economic stagnation that starts the circle of underdevelopment. But in the course of time, cultural stagnation is likely to become the worst obstacle to the beginning of economic and social development.

(3) The collapse of economy and the disintegration of society

Unless the vicious circle is broken, the degree of self-steering in society sooner or later crosses the lower limit of acceptable self-steering (Fig. 80), and unnecessary hierarchy emerges. At this point of archaic development the possibility of maintaining production and distribution with the help of increased control of society has been exhausted. As a consequence the economy collapses, and society disintegrates into smaller parts, where the vicious circle is easier to break because there is less internal hierarchy, and thus less internal suppressive power. This is what happened with the great archaic empires in the end. A similar development in some African or other Third-World societies of our age cannot, of course, be excluded. But the situation in the present time is in an important respect better than it was in the archaic era: now a group of briskly developing economies exists, and the drive of these economies may draw even the poorest societies off their vicious circle, provided that cultural stagnation can be broken.

12.4. The Concept of Social Development

(1) The defining characteristics

We can now start with the law of requisite hierarchy for developing economic systems. Thus, we have a situation where the optimal regulatory ability, $H(R)$ of the economic system grows and the inherent uncertainty of regulation, $H_D^0(R)$, decreases with time. From the previously mentioned law and from the rule of inverse proportionality we deduce the *type of developing society* with the following characteristics.

(I)   Security is created by economic development, not by hierarchy: the technology and skills are improving ($H(R)$ grows), just like the organization and allocation of manpower and the sensitivity of markets ($H_D^0(R)$ diminishes), thus the effective regulatory ability of the economic system increases.

(II)   The conflict between collective security and individual freedom has disappeared: social hierarchy can be dismantled, and its relaxation contributes *both* to an improved effective regulatory ability of the economic system *and* to increasing self-steering, i.e. human freedom, in society (Figs 78 and 79).

Without doubt the best example of this type of developing society is given by the history of the so-called Western societies, where a simultaneous progress of economy, political democracy, and intellectual liberty was shown to be possible. But the cybernetic argument proposed here makes it likely to be a lawful phenomenon, not just a historical accident. We can again formulate it in terms of a 'circle'.

(2) The virtuous circle of social development

The successive steps of this causal circle are as follows.

(i)   The economic system is developing, as indicated in point (I) above. The start on this orbit implies an *economic breakthrough from underdevelopment*.

(ii)   Hierarchy can be reduced without jeopardizing the security of the population. If implemented, it means increasing *political democracy*.

(iii)   Self-steering increases in society, and the greater individual freedom experienced leads to a growth of intellectual and economic initiativeness. This is the highest manifestation of favourable social development, viz. that of *intellectual liberty*.

(iv)   Back to square one: the economy blooms.

Lately, many Third-World societies, the so-called 'newly industrialized countries', have accomplished the economic breakthrough from under-development, thus proving that favourable social development is not a monopoly of a certain geographic area. Once a society is on the orbit of development, its further favourable development, according to the above argument, also makes up a self-maintaining process that goes on unbroken, unless destabilized by outside forces (or perhaps by the internal process called 'habituation' by the physiologists?). If it continues unimpeded, a necessary consequence is an incessant *growth of the middle classes* in society: every step in the reduction of hierarchy adds to their growth.

CHAPTER 13

The Governability of Human Society

13.1. The Crises of Governability as Crises of Self-steering

If the variety $H(E)$ of the set of states in which the population of the society S concerned survives, and the arsenal of 'disturbances' of environment, $H(D)$, can be considered as given, the limits of toleration of self-steering in that society are determined by the developmental level of its economic system, expressed by $H(R)$ and $H_D^0(R)$. We have, by (12.5)–(12.8):

$$\left.\begin{array}{c} H(R) \\ H_D^0(R) \end{array}\right\} \Rightarrow \left.\begin{array}{c} H_{\max}(G) \\ H_{\min}(G) \end{array}\right\} \Rightarrow \left.\begin{array}{c} v_{\min}(D) \\ v_{\max}(D) \end{array}\right\}. \tag{13.1}$$

Thus, it is the developmental level of economy that is represented by the limits of self-steering: with a favourable development of the economic system these limits rise; when economy flops they too go down. The dependence of the limits $v_{\min}(D_t)$ and $v_{\max}(D_t)$ on time indicates the ups and downs of the economy in the society in question. (Warning: this is not the same as the fluctuations of gross domestic product! More than that is involved, viz. the development of technology and occupational skills, the sensitivity of markets, and the organization of work.)

The actual self-steering that has been reached in society at the point of time t is expressed by the measure $v(D_{xt})$, where xt stands for the total state of society at that moment. The degree of actual self-steering depends on political and intellectual activities, or on the lack of them, in society. Various political forces inside the society may advance or thwart the development of actual self-steering, by relaxing hierarchy and bureaucracy or increasing them, respectively. And intellectual liberty can be expanded by courageous public challenges of petrified attitudes. Or it can be diminished by the lack of civil courage in population.

The governability of a society depends both on its economic development and on the force and direction of political and intellectual activities in that society. In short, we shall consider a society *governable* as long as the actual self-steering in that society keeps between the limits of toleration:

$$v_{\min}(D_t) < v(D_{xt}) < v_{\max}(D_t). \tag{13.2}$$

It follows that society may become ungovernable in two ways: either the upper or lower limit of toleration of self-steering is exceeded. Let us discuss them separately.

Case 1: The upper limit of tolerance has been exceeded

This may happen for two entirely different reasons: either by a sudden burst of self-steering over the normally developing upper limit, or by a sudden slump of the upper limit below the quietly developing self-steering. Besides these extreme cases, of course, all the less dramatic encounters of the two lines are possible, where one of the curves slowly approaches the other, or they both approach each other. In these cases both of the mentioned reasons are operative. Let us study each of them separately, in its extreme manifestation.

Case 1(a)

Too many liberties taken by a group of individuals in society, or too much hierarchy is dismantled by excessive social reforms. As a consequence, there is a burst of self-steering that brings the curve of $v(D_{xt})$ over the limit $v_{max}(D_t)$, as illustrated in Fig. 81a. A case in point is terrorism of some individuals that disrupts law and order, and thus threatens the survival of parts of the population. Modern times do not lack examples. We can think of the present situation in Northern Ireland or Lebanon, for instance. But a full-scale social revolution also leads, for some period, to a similar breakdown of law and

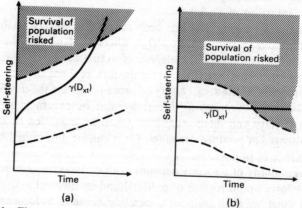

FIG. 81. The crises of governability due to excessive self-steering: (a) over-measured reforms in a developing economy, and (b) a sudden economic or environmental catastrophe.

order, where the survival of some part of the population is threatened. From revolutionary frenzy it is only a small step to excessive social reform generally, such as attempting to dismantle the traditional hierarchy far too quickly and taking away too much at a time. Again law and order may be shaken, and the survival of the population threatened.

In all the above situations, society is ungovernable as long as the situation lasts; following the observation that legal society has become unable to protect them from the terrorists, or from revolutionary fury, or against general chaos, people take law and order into their own hands, trying to survive as best they can.

Case 1(b)

A sudden unfavourable environmental change, or a sudden collapse of the economic system. If the environment, in which the human population of a given society lives, suddenly becomes dangerous in a new and unexpected way, this means a rapid increase in the arsenal of disturbances, $H(D)$, as a consequence of which the upper limit of tolerance of self-steering,

$$v_{max}(D_t) \sim 1/H_{min}(G) = 1/[H(D) - H(E) - H(R) + H_0(R)] \qquad (13.3)$$

suddenly goes down. If the fall is great enough, even the normal course of the curve of actual self-steering, $v(D_{xt})$, brings it over the upper limit (see Fig. 81b), since what was normal before is not normal any more, in the changed environmental conditions. Again the survival of the population is threatened.

A sudden worsening of the environment must have been a frequent experience for small primitive communities that could unexpectedly face a new hostile tribe or new dangerous beasts. Later on, nomadic communities especially were exposed to this kind of peril. An epidemic disease is a still later environmental threat: as shown by Le Roy Ladurie, the 'microbial unification of the world' first took place after the great explorers had opened the routes for seafaring and trade all over the world (Le Roy Ladurie, 1973a).

A collapse of the economic system means a sudden drop of the optimal regulatory ability $H(R)$ together with a quick increase of the uncertainty of regulation, $H_D^0(R)$. Hence, by Eq. (13.3), the upper limit of tolerance of self-steering again comes down in the way illustrated in Fig. 81b, crossing the curve of actual self-steering. This is another case in which the survival or the population is threatened, and law and order may be lost. This rarely occurs in modern industrialized society. The 'great depression' of 1929–1933 in the United States, for instance, was very far from this class of catastrophe. No; to meet genuine examples of the collapse of an economic system we have to go farther back in history.

An example, in which a sudden deterioration of environment was combined with the collapse of an economic system, is the disaster in Europe that followed the arrival, in 1348, of the plague. Brought to Europe by seamen coming to the port of Marseilles from the Middle East, the plague soon spread all over Europe and reduced populations, for instance in England and in France, to half of their previous magnitudes. As a consequence, agricultural production also dropped accordingly, and an economic catastrophe was a reality. In France, for example, production did not reach the pre-1348 level again until the middle of the fifteenth century, one hundred years later (Le Roy Ladurie, 1973b).

Just as previously, in Case 1(a), the society in question is ungovernable for so long as the situation illustrated by Fig. 81b lasts. Legal society loses the control of events, and people are left to survive on their own.

Case 2: The lower limit of tolerance has been passed

This too can happen for two different reasons, corresponding to the two extreme cases that (a) the curve of actual self-steering goes down and passes the lower limit of tolerance, or that (b) the lower limit starts to rise quickly and crosses over the curve of actual self-steering. Again all combinations of these two reasons are possible. Let us study, however, each of the reasons in their 'purest' extreme manifestation.

Case 2(a)

A continual concentration of power and increase of hierarchy in society. With the increase of actual hierarchy, actual self-steering, according to the rule of inverse proportionality, (12.4), decreases. If the trend continues, it passes the lower limit of tolerance and enters the domain where *unnecessary hierarchy* appears in society (Fig. 82a). This time the survival of the population

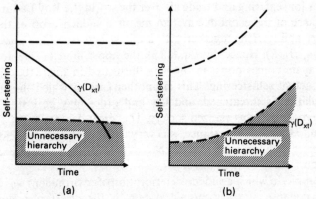

FIG. 82. The crises of governability due to excessive hierarchy: (a) a stagnating, underdeveloped economy with the characteristic suppression of human freedom, and (b) a sudden start of economic growth leaving social hierarchy untouched.

is not threatened – such a thing happens only with too little hierarchy. But too much hierarchy is an evil too, because it means suppressing people, depriving them of individual freedom, and reducing their initiative, both in intellectual and economic areas. In the worst case it leads to cultural and economic stagnation, along with absolute power and totalitarianism – in other words, to underdevelopment. This was the case with the ancient archaic societies, which disintegrated in the end, showing that the kind of situation illustrated by Fig. 82a can make society permanently ungovernable, the only outlet being through disintegration into smaller parts. Thus Case 2 is worse than Case 1, where ungovernability could only be temporary. As surmised before, some Third-World societies may be suffering from this worse kind of ungovernability today.

Case 2(b)

The beginning of 'economic growth' in a society where hierarchy is not correspondingly dismantled. There is a historical situation where this case is more likely to appear than in any other times, viz. just when a traditionally immobile or only slowly developing economy for the first time starts the brisk development characteristic of modern society. In such a situation both limits of tolerance of self-steering, since

$$v_{min}(D) \sim 1/H_{max}(G) = 1/H_D^0(R)$$

and

$$v_{max}(D) \sim 1/H_{min}(G) = 1/[H(D) - H(E) - H(R) + H_D^0(R)], \qquad (13.4)$$

start to rise quickly, after having been constant or nearly so until then. It follows that if the traditional hierarchy keeps lingering on, the curve of actual self-steering automatically crosses the lower limit of tolerance sooner or later (Fig. 82b). This means that unnecessary hierarchy is again present, even if actual hierarchy is constant, since what was tolerable before is not tolerable any longer.

If the economic system continues its lively development, while ancient hierarchy does not move, we have what can really be called a *revolutionary situation*, meaning that if a full social revolution is ever to happen in the society in question, then this is the most likely point of time for it. Indeed, both of the most typical social revolutions, the French and the Russian, took place in such situations. The French economy, after having been developing rather slowly, took a great leap forward during the eighteenth century, while the autocracy that Richelieu had helped to create did not budge. In Russia, after a thousand years of underdeveloped economy, a boom started towards the end of the nineteenth century, producing in the 1890s the largest annual rates of growth that the gross national product in Russia, or later in the Soviet Union, could ever show. This, together with the immobile regime of the tsar, created the conditions favourable for revolution.

These facts are no news to anyone who has studied economic history. Only confident people, whose hopes for better times are already raised by recent economic development, are ready to take part in revolutions. But Marx was entirely wrong in supposing that full-scale social revolutions would be necessary tools of historical development. They are neither necessary, nor desirable, but simply symptoms of serious political failure leading to a loss of governability of society. Their damaging effects are discernible long after the event in the unstable development of society. This is a subject to which we shall soon return for a closer inspection. Before that, a glance at governable social development and at its conditions is in order.

13.2. The Freedom of Action of Two Opposite Political Forces as the Condition of Governability of Developing Society

If the governability of society and its development means keeping the society between the limits of tolerance of self-steering, thus preventing both unnecessary (i.e. non-utilizable) hierarchy and the threats to the survival of the population, then two political forces must exist and have a free access to power in society:

(a) A political force that prevents actual self-steering from exceeding the upper limit of tolerance that would destroy law and order, and impose threats to the survival of the population (Fig. 83a). This means placing the maintenance of law and order in the first place among political priorities. Let this be called a *conservative political force*.

(b) A political force that is able to prevent actual self-steering from passing the lower limit of tolerance, and thus save the society from the

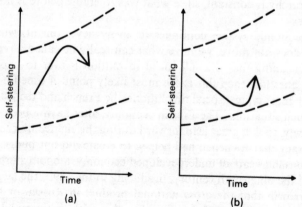

FIG. 83. The workings of the two opposite political forces in a favourably developing society: (a) conservatism, and (b) radicalism.

formation of unnecessary hierarchy (Fig. 83b). This puts the social reforms that are prone to promote human emancipation and human freedoms uppermost in the list of political priorities. Let this be called a *radical political force*.

It is conceivable, of course, that responsible political leaders switch their attitudes from promoting emancipation to law and order, and vice versa, as the situation – that of Fig. 83a or that of Fig. 83b – may demand. If the whole population could be equally well informed of each topical situation it would be conceivable that radical and conservative forces could unite in one and the same political movement. But in the kind of mass societies that we are living in modern times, with their enormous populations, this is less probable. Therefore, it is likely that conservative and radical forces materialize in two different political movements, if they are to materialize at all. Hence we get:

The Principle of Political Democracy. The crises of governability of society can be best avoided, if a radical and a conservative political movement, opposing each other, can freely act in society, and, as necessary, come quickly to power according to the demands of social situation.

Example. It seems that this principle explains the successes of Anglo-American two-party parliamentarianism (in England and the U.S.A.), i.e. a political system where only two major parties exist, and the mandate of government is automatically given to the winner of general elections. In continental European many-party systems, the change of governmental policies is often the object of lengthy negotiations between a number of parties, and a U-turn (needed from the situation of Fig. 83a to that of Fig. 83b, or vice versa) is possible only after a crushing victory over the opposition. In a one-party system, of course, a total change of governmental policies is next to impossible.

Note. In a favourably developing society, i.e. in a society where governability is never lost and the economic system is improving all the time, social hierarchy must be progressively dismantled with advancing economic development. This obviously is a direct consequence from the continual rise of the limits of tolerance of self-steering and from the demand that actual self-steering has to remain between these limits. But this does not mean that there would be a political panacea called 'moderate radicalism' that would help overcome all situations. Neither does it mean that the content of radicalism and conservatism would be always the same, and recognizable, for instance, from the names of political parties. This is because every new social situation has its own *particular historical content*, along with the formal content defined above, to what is radical or conservative in that situation. What was radical earlier may soon become conservative, as even the former radical attitudes, like all attitudes of people, are prone to stiffen to rigid prejudices. Or, to take an example, the policies called 'progressive' or 'liberal' were no doubt radical during the period of building up the welfare state, but they soon created hierarchies and bureaucracies of their own, to be dismantled by an entirely new type of radicalism. Thus, each situation creates its own radicalism that in its policies may even bear a closer resemblance to past conservatism than to past radicalism, which is the immediate target of its attacks. So goes the merry-go-round of human history, unexpectedly, unpredictably, according to the self-steering scheme.

13.3. Full-scale Social Revolution as a Loss of Governability of Society

Let us again consider a society that has failed to dismantle its traditional social hierarchy when the economic system for the first time settled on the orbit of development. Thus, we have the situation illustrated by Fig. 82b. What comes next?

If a radical enough 'revolutionary political force' exists to implement a full-scale social revolution, society is pushed through four successive phases (Fig. 84):

(1)   The lingering revolutionary situation: not much alteration in prevailing social or political conditions.

(2)   The heroic period of revolution: a sudden burst of self-steering, accomplished by a popular mass attack of traditional institutions of power, overthrows the government and sets up some provisional revolutionary authority.

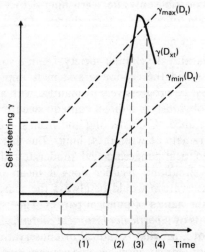

FIG. 84.   The characteristic phases of a full-scale social revolution: (1) the revolutionary situation, (2) revolution, (3) the terror and anarchy period, and (4) the ensuing build-up of compensative hierarchy.

(3)   The period of anarchy: actual self-steering exceeds the upper limit of tolerance, as a consequence of which law and order gives way to terror and anarchy, during which the survival of some parts of the population is threatened.

(4)   The restoration of power and privilege by building up *compensative hierarchy*: a new hierarchy and a new 'revolutionary' leadership stabilizes its power, sometimes no less suppressive than that of ancient regime, and the society returns to normal conditions.

The establishment of compensative hierarchy is necessary in order to restore the lost governability of society. The four successive phases given above are not a theoretical prediction, but simply a description in terms of the concept of self-steering of what seems to be the normal course of a full-scale social revolution, such as the French or Russian.

Example.   In the French Revolution beginning in 1789 the four successive phases are easy to identify:

(1)   In the eighteenth century the French economy began the first great 'getting-into-motion', after hundreds of years of 'immobile history' (Le Roy Ladurie, 1974), while autocracy remained as rigid as before.

(2)   The heroic period can be placed in the time between the start of revolution until 8 August 1792, when the Jacobins grabbed power.

(3)   The regime of terror followed, and came to an end with the execution of Robespierre on 26 July 1794.

(4)   Then the building of a compensative hierarchy began, first by the 'directorial government', later by Napoleon. However, the after-effects of the great revolution were long visible in the unstable development of French society through the nineteenth century, with the successive alternations of restored kingdom and new republics, until the final victory of parliamentarianism with the establishment of the Third Republic in 1870.

It is not difficult to identify the corresponding periods in the Russian Revolution and its aftermath: the heroic Leninist phase, the period of anarchy in the time of 'war communism', the restoration of hierarchy during the Stalinist era, and various attempts at 'liberalization' later on.

Methodological Note on Possible Future Extensions of Current Theory of Mathematical Dynamics

The approach to the theory of human action taken in this book has been based on the concept of the goal-seeking system in the strict sense in which this concept was introduced in Section 2.3. Let us now consider the mutual relation, or rather similarity, between goal-seeking and dissipative systems, and discuss the possible significance of the latter in the shaping of purposeful behaviour.

(1) Let a dynamical system be goal-seeking in a state x. In view of the definitions of self-steering and self-regulation, which are the two alternative forms of goal-seeking behaviour, this means (see Section 2.3) that the trajectory xR^+ attracts all the trajectories in a neighbourhood of xR^+, which thus asymptotically approach it with $t \to +\infty$. This implies, according to the theorem of orbital convergence (Theorem 4.6.1), that the volume element $\Delta V_{dF}(xt)$ of the space $V_{dF}(xt)$ orthogonal to the trajectory xR^+ at the point xt shrinks to zero with $t \to +\infty$. Thus, a goal-seeking system eliminates, in a neighbourhood of the trajectory xR^+, all the dimensions of state-space that are orthogonal to xR^+ at the moving point xt of this trajectory.

The similarity to a dissipative system lies in the fact that the latter also eliminates one or several dimensions of state-space. Let us again consider the trajectory xR^+, assuming only that the system is dissipative all along it. Then we know, by the definitions given in Section 1.4, that the volume element $\Delta V(xt)$ of the total state-space shrinks to zero with $t \to +\infty$. In other words, at least one of the dimensions of state-space is eliminated in the course of a *dissipative process* that we have just described.

But the process of elimination quite generally, even as a dissipative process, is very important in most human activities. The elimination of 'inessential' degrees of freedom is a preliminary phase in all purposive action, before reaching the final decision to pursue a definite goal. Thus, we can assume that dissipative processes are to become important as preparatory steps toward proper goal-directed behaviour in human action, i.e. toward self-steering, both in social action and in the acts of an individual. Even in purposive animal behaviour dissipative processes may have a similar role.

(2) Another possible extension of mathematical dynamics, as introduced in Parts 1–3, may be due to a general *mathematical theory of self-organization*, once such a theory is at our disposal. At present we do not have such a theory – only some fragments closely connected with self-organization, of which the best-known are the theories of Feigenbaum and Hopf bifurcations.

Due to our lack of knowledge of the mechanisms of self-organization, we were compelled to take recourse, in Chapter 10, to a stochastic discussion of self-organization in terms of entropy. While the cybenetic negentropy laws so obtained, i.e. the laws of requisite variety and requisite hierarchy, may retain their significance in an overall discussion of human societies as wholes, a more detailed picture of the mathematical mechanisms of self-organization would give us important details and better tools for the assessment of human development.

The first elements for a mathematical theory of self-organization of human societies seem to be included in the recently developed self-organizing dynamic input–output model (Aulin-Ahmavaara, 1987), briefly discussed at the end of Section 8.2 of this book. Whether along the lines indicated in that pioneering model, or in some other way, a more detailed understanding of the mathematical mechanisms of self-organization may make up the next essential step in our understanding of social processes (just as the physico-chemical mechanisms of self-organization have helped the understanding of biochemical processes, cf. Nicolis and Prigogine, 1977).

What follows here, however, is a historical (Part 4) and anthropological (Part 5) illustration of the fundamental dynamical concepts, those of self-steering and self-regulation especially, of the existing dynamical theory.

Part 4

Historical Illustrations of Self-steering in Different Types of Human Societies

Prologue

David Hilbert once had a student of mathematics who stopped his studies to become a poet. Hilbert remarked: 'I never thought he had enough imagination to be a mathematician'.

It takes a lot of imagination to understand the significance of mathematical concepts in a new context. These illustrations are meant to help to get that imagination on the move.

CHAPTER 14

The Birth of the Western-style Society of Relaxed Hierarchy

14.1. The Origin of East–West Differences: Economic Development in Western and Eastern Europe since the Twelfth Century

In view of the qualitative theory of social development presented in Part 3 of this book the fundamental difference between Western and Eastern societies today is not one of 'capitalism' versus 'socialism', but lies deeper down. While Western Europe reached its period of development, with the establishment of feudal states and, in particular, a little later, with the birth of lively commercial centres in the Mediterranean and other city communes, nothing comparable occurred in Eastern Europe until almost a thousand years later. The major difference between Western and Eastern societies is, therefore between a society having a long tradition of relaxed hierarchy on the one hand, and a society with a thousand years of strict hierarchy on the other. This reflects a difference in economic development.

We can start to study the East–West difference by examining the development of agricultural productive forces in Europe since the twelfth century. According to the law of requisite hierarchy, a necessary condition of relaxed social hierarchy is the improvement of the regulatory ability of the economic system. This condition was for many centuries fulfilled only in the western parts of Europe.

There is a measure of the technological level attained by agricultural productive forces called *yield ratio*. If one sown seed produced, on average, a harvest of five, we have a yield ratio of 1:5. The growth of this index is a good enough indicator of the progress of farming methods, both tools and skills included. Statistics are available in Europe over a surprisingly long interval of time, beginning as long ago as AD 810. The numbers for West Europe and Russia are as follows*:

Century	Western Europe	Russia
1100–1300	1:3 to 1:4	1:3
1300–1500	1:5	1:3
1500–1600	1:6	1:3
1600–1700	1:7	1:3
1700–1800	1:10	1:3

*The numbers are based on B. H. Slicher van Bath, 'Yield ratios 810–1820', in *Afdeling Agrarische Geschiedenis, Bijdragen*, No 10, Wageningen (1963).

The numbers are averages. There was a variation over countries and years in Western Europe, and there was an especially large fluctuation over years and areas in Russia, where deviations could be as large as 1:2 to 1:5 from one year to the next. Only the leading countries in Western Europe have been counted, excluding, for instance, the Scandinavian countries, where progress was slower because of climate: 1:6 was not reached there until the eighteenth century. Let it also be emphasized that yield ratios tell only of productivity, and not, of course, of the magnitudes of the total yield. The latter depends on the size of the cultivated area as well as the yield ratio. The population can thus grow and yet retain a constant standard of living, provided that the total area cultivated enlarges, even though yield ratio remains the same. This is what happened in Russia, where there was no shortage of land to limit the size of the cultivated area. So the Russian population increased from the estimated 5–12 million in the sixteenth century to 17–18 million in the mid-eighteenth, to 68 million in 1850, to 124 million in 1897, and to 170 million in 1914. But the growth of population was entirely supported by a drastic growth in cultivated area, as a consequence of politics of expansion, not by an increase in yield ratio, which did not permanently take place before the nineteenth century.

There is an interesting limit in yield ratios that is worth mentioning. It is generally agreed that a yield ratio of 1:5 is necessary for supplying food for any significant city population that does not participate in agriculture. We can see from the above numbers that yield ratios started to grow in Western Europe from the twelfth century, and reached the limit 1:5 during the fourteenth and fifteenth. All these centuries were the formative years of Western European trade and wealth, first noticeable in the Mediterranean city states. During these centuries a remarkable growth of city populations, which were engaged in trade and handicraft, occurred in Western Europe.

In Russia and Eastern Europe the situation was different. Cities also developed there, but for the most part the city population had to engage themselves in agriculture as well as in other subsistence activities. Russian tradesmen, for instance, lived in particular areas of the town that were assigned and hired to them by the tsar, and most of them had to farm as well. On the other side of the fence the rural population in Russia used to occupy themselves with handicrafts, which further reduced the differences between urban and rural people. Over most of the centuries discussed, the proportion of the Russian population that was virtually living as rural workers was over 99% (the initially tjaglo-obliged part of the population*).

But our interest in yield ratios, from the point of view of the law of requisite hierarchy, comes from the fact that they are fairly good indicators of the regulatory ability of agricultural productive forces. Thus, the main

---

* See Richard Pipes, *Russia under the Old Régime*, p. 98. Weidenfeld & Nicolson, London (1974).

tale these numbers tell is that in Western Europe there was a continual improvement in the regulatory ability of these principal productive forces of the time ever since the twelfth century, with a slight speeding up during the sixteenth and seventeenth, and a greater leap, finally, during the eighteenth. If this is graphically represented in terms of the index of self-steering, we get a rise in the curves marking the lower and upper limits of self-steering (see Fig. 85): both of these limits rise from the twelfth century, first slower and then more rapidly, until in the eighteenth century they begin to show steeper growth. But this is the case only so far as Western Europe is concerned. For Russia we would have an entirely different picture, where both limits of self-steering keep constant and indicate a persistent economic stagnation through all these centuries until the nineteenth.

14.2. Western-European Feudalism and its Relaxed Hierarchy of Power

The pronounced difference between Western and Eastern Europe in economic progress between AD 1100 and AD 1800 is a factor that makes a useful starting point for our developmental analysis of European societies. If we ask for reasons of this difference, we can hardly overlook the collapse of the Roman empire and its power in Western Europe about a thousand years before the corresponding end of slave empires in the East. It was not until 1480 that the grand duke of Moscow, Ivan III, refused to pay the tribute to the khan, and thus ended the power of the Mongol empire in Russia.

After the decline during the fifth and sixth centuries AD of the Roman empire in Western Europe, no 'leading force' existed that had been able to

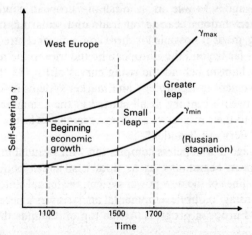

FIG. 85. The average development in Western Europe of the regulatory ability of the economy, as indicated by the rising limits of self-steering. The picture is based on statistics collected by Slicher van der Bath (1963).

claim a similar kind of power to that the Roman emperors had wielded. The Pope, after all, had no armies, at least not on a permanent basis. Although the Catholic Church retained its spiritual might, it did not have military power of its own, but had to rely upon the conversion to Christianity of those German warlords who one by one now conquered Rome. It was the first time in the history of mankind that ideological and military power parted. And it was this split of power that can be considered decisive for the whole development that followed in Western Europe.

The new organization of power in Western Europe, the feudal state, stood only for a military, and not a religious leadership. As much as many kings desired it, the Pope would not, and could not, without giving up the very foundation of the Catholic faith, surrender any part of his religious authority to earthly leadership. But then no king could claim a position comparable to the absolute power of the Pharaohs, or even the Roman emperors, as his power could no longer be given a religious backing by deriving it from God. It was the Pope, not the king, who represented God on earth. As this split between divine and earthly power became established in Western Europe, so began the relaxation of power and hierarchy, which thus created the prerequisites for the individual initiative that was soon to manifest itself in the flourishing life of Italian cities, the engines of a new age.

That it was the Italian cities, Venice, Genoa, Florence, Milan and later, especially, the free ports of Western Europe, Lisbon, Amsterdam, London, Hamburg, where the European way of life was born, is significant. Small was beautiful then too, as it is always with the creative individual that a new idea is born and action begins. Small, active groups around such individuals, working in a relaxed atmosphere of lively, small, independent communities – this was the formula of discovery, much before the age of Silicon Valley, in the Greek city states as well as in Western-European commercial centres. Technological inventions bearing on trade and seafaring, from astronomy to shipbuilding, made it possible for these commercial centres to spread their market systems far beyond the European boundaries to the rest of the world. This is the mechanism behind the rising curves of Fig. 85 that indicate the general improvement in technology and market systems in Western Europe ever since the twelfth century. It all started in the commercial centres, and hence radiated to their respective economic regions, as has been so brilliantly expounded by Fernand Braudel.*

But why engage in a hopeless competition with Braudel in the description of these European beginnings, as he does it so much better? Let me just indicate the outlines of the new power system, the feudal state, whose relaxed hierarchy according to our developmental analysis was a necessary condition of all Western-European progress. At the top of it stands the king (see Fig.

---

* Fernand Braudel, *La Méditerranée et le monde méditerranéen à l'époque de Philippe II*, 2nd edition, Colin, Paris (1966) first printing in 1947), and *Civilisation matérielle, économie et capitalisme, XVe–XVIIIe siècle*, Vols 1–3, Colin, Paris, (1967–1979).

86). But the feudal king as a power-holder was much less than a Pharaoh or other theocratic emperor in the ancient slave empires. Lacking a religious background his power was far from being absolute. He was mainly the highest warlord in his state. This is understandable, as the situation where kings were made, after the collapse of Roman power, was a chaos of criss-crossing armies with their various warlords, fighting each other for power. A king's position accordingly originated as that of a winning warlord. But soldiers had to be drawn from some fixed population, and agriculture was still the main source of wealth and subsistence. These two factors connected the king and other feudal lords to landed property and a settled population cultivating that land. Add the religious hierarchy that was necessary, even within the feudal state, and you have collected the pieces from which feudal order was composed.

In that order the main strings of power stretched from king through an echelon of various vassals down to ordinary people, the bulk of whom were serfs (see Fig. 86). For reasons to be explained the diagram in Fig. 86 refers

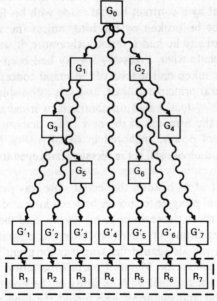

FIG. 86.   The hierarchy of power in a typical Central-European feudal state (eleventh century).

	G_0	= king
Highest nobility	G_1, G_2	= throne vassals: princes, dukes, highest counts, arch-bishops, cardinals, bishops
	G_3, G_4	= vassals of throne vassals: lower counts, Freiherren
Lower nobility	G_5, G_6	= lords of castle, land-owning knights
	$G'_1 - G'_7$	= knights in the service of other nobles
	$G_1 - G_7$	= serfs and tenants

to the conditions of continental Western-European feudalism, such as existed mainly in France and Germany by the eleventh century. The people next to the king were the king's vassals, also called throne vassals or barons, some of them being the high officials of the Church and the rest being princes, dukes and highest counts, all of whom together were sometimes convened by the king to sessions of the king's council. This was an organ where ecclesiastical and military power met, under the leadership of the highest military chief, the king.

It was up to the king to decide when to wage a war, and against whom. But through the king's council he consulted his next vassals. The secular throne vassals were also his military commanders. Though, originally, all landed property in a feudal state belonged to the king, a great part of it was from the beginning given as a fief to the throne vassals, both the clerical and secular ones having their share. And the throne vassals in turn had vassals, and the latter had lower vassals, down the whole echelon of the feudal state, through counts and *Freiherren* (in Germany) to lords of a single castle (see Fig. 86).

All this landed aristocracy practically owned their fiefs, since the fief was granted to a vassal by a contract he had made with his feudal lord, and the contract could not be broken by the lord, unless the vassal violated the pledge of *feudal loyalty* he had given. Furthermore, it was not for the king or the lords to decide when a vassal's loyalty had been violated, but such decision had to be taken on the level of the vassal concerned, by his equals. This was the general principle at least, and it is astonishing to find out how well such a power system, based on contracts, worked in feudal times. Or shall we say that the firmness of the new order demonstrated how deeply the real relations of power, anchored in social reality, had changed since archaic times? That, obviously, is the message conveyed to us by the strength of feudal loyalty.

From the outset of its history, the feudal state was programmed to be a stage for a continual struggle for power between king and aristocracy. Every feeble king had to make concessions to his vassals, and the latter would keep all the concessions firmly in their grip later on: a well-known collective uprising of British vassals against the king was the event in 1215 that led to the document known as the Magna Charta Libertatum. On that occasion the armies of vassals, together with an army of the London city commune, besieged King John and his army, and forced the king to the concessions stated in that remarkable declaration of rights. Things never went as far as that in continental feudalism, which was to remain much more hierarchy-bound, and was also to suffer the consequences, as we shall see later.

But even in the feudal state on the Western-European continent, which was the standard, and oldest, model of feudalism, it was usual that some of the strongest vassals could leave their fiefs as a heritage to their descendants. This, of course, much consolidated the position of the nobles in regard to the king, and was prone to reduce the effectiveness of any attempts on the king's part to establish absolute power. Though some kings in later feudal ages, like Louis XIV (1643–1715) in France, and even after the feudal ages in England Henry VIII (1509–1547), could gather quite a power in their hands, they still never equalled a Pharaoh or other emperors of the archaic theocracies. The time of absolute power had passed, as far as Western Europe was concerned.

The main duties of vassals consisted of their military service to their lords. But in normal times the service was limited. It could consist, for instance, for the owner of a single castle, of 40 or 60 days annually. War was a different thing. In wartime each vassal had to take his own army, drawn from the population that inhabited his land, to join the king to defend the country, or, more often than not, to conquer lands of other kings. Apart from their military duties the vassals had to collect and pay taxes to the king. But in normal times – if peaceful times can be called normal during those ages full of military escapades – much of the king's power was actually delegated to the vassals. It was the lord of the castle, or in more serious cases some higher vassal, not the king, who usually settled quarrels and dispensed justice. In France from the thirteenth century, for the worst strifes, mainly those between vassals themselves, there was the organ called the Parliament of Paris, which despite its name was a purely juridical institution.

The net effect of all that protection of vassals' rights was a considerable reduction in the power of the king. This indeed must be regarded as the starting point of Western democracy: later on we shall see how the control of power by means of a system of contracts enlarged to cover most uses of power in Western society. But the relaxation of hierarchy that started in the feudal state extended to lower classes as well. A serf was not a slave. He had an independent economy and household, which most slaves in an archaic society could not even dream of. He did not live in his lord's house, except if he happened to be one of the service personnel of the house. He lived in a feudal village clustered around a castle. The labour tax that he owed to his lord was limited by a contract, and he and his family did not belong, body and soul, to the feudal lord like a slave and his family had belonged to their master.

Yet continental feudal society was a typical two-class society, just like the archaic slave empires had been, only with lesser hierarchy within and between classes. A serf was not permitted to move from his village, or to choose his feudal lord. The first-night's right of the latter with his villeins' daughters, for instance, reminds us that there were personal humiliations and violations of the right of self-determination, even in the most private domains of life, that were considered more or less normal in a feudal society. All this despite the common Christian belief in the equality of men before God.

But Christianity and its egalitarianism did have great significance as a guarantee of the general awe felt in front of contracts, which had been enforced on oath before God. When Pipin Petit, in AD 751, wanted to depose the last and weak Meroving king, and to take the crown, he asked the Pope to settle who was to be king – he who had the power, or he who had the name? The Pope, of course, gave the desired response, after which the Franks could appeal to the 'will of Heaven', and enthrone the new king. That the Pope was invoked in such an urgent question concerning the state indicates the power of the faith. The religious hierarchy that interlaced the feudal state was indeed an important element of feudalism. It provided the ideological backbone that supported the feudal system.

Even in continental feudalism there were people who did not actually belong to either of the two clear-cut social classes, who were neither nobles nor serfs. Mainly, they were town-dwellers engaged in handicraft and trade, not forgetting lower clerics, who certainly could not be counted among the members of the upper class, but still were important to them as a means of reaching, ideologically, the lower echelons of society. Since people living in cities, especially those in major ports, were the first to encounter new technological discoveries or improvements in the market system, they were well placed to have a good share of the increased material welfare.

On the other hand, as the feudal power, being based on landed property and military escapades, had little interest in city life, the cities were permitted liberties that had been unthinkable elsewhere. The main ports were even allowed a far-reaching autonomy in all local issues. Hence, much of the intellectual life outside the king's court was centred in cities. But in continental feudalism the city bourgeoisie never gained control, nor even any notable position, in matters of state. The state remained firmly in the hands of the king and the nobles.

14.3. The Germ of Western Democracy: the Local Power of the Middle Class in Medieval England

Things were different in English feudalism. In England the city bourgeoisie, together with the free but noble land-owners, gained some control in matters of state as well. In medieval England state power and national politics were not solely matters for the aristocracy, as they were on the continent at the same time. But then the whole story of feudalism was different on the other side of the Channel.

As far as continental feudalism goes, it is impossible to name any single year in which it began. A feudal system had already started to take shape in the kingdom of the Franks during the Meroving kings (481–751), but became more pronounced only during the age of Charlemagne (768–814). At the same time it also spread, with the Frankish empire, which by then comprised France, Germany, Northern Italy and what is now called the Benelux countries, to the major part of continental Europe. In England, feudalism was established much later. As we know, it was brought there from the outside by the duke of Normandy called William who, in 1066, invaded England and set up the Norman power, the hallmark of the English feudal system.

In English feudalism the prediction of our theory (Section 12.4) on the appearance and later expansion of a middle class was realized for the first time in the history of mankind. But why was English feudalism to be so different from the continental model? Why could not William and his men simply impose on the British the same social order that obtained on the continent at the same time? To suggest any detailed answers to questions like this from a general theory of social development would mean stretching the theory too far. On the other hand, neither is the professional historian, who has the privilege of knowing the details, necessarily better placed in settling such general issues. So let me try a rational guess.

I would like to point out two factors acting in succession. First, when the Normans conquered England they were a tiny French-speaking minority, albeit one in control of the state they created. Their small number and the linguistic barrier effectively impeded their mastering of the local administration. On the local level the circumstances compelled them to leave the reins, from the very outset, partially in Anglo-Saxon hands. They bought them by allowing these 'free-holders' to keep their land, thus creating a class of land-owning people who were not nobles – an unheard of situation in any other feudal system. The Norman gentry who at first controlled local affairs soon merged with them and the rest of the Anglo-Saxon population.

Secondly, the Anglo-Saxon social order that was left to prevail in local matters was largely an old German tribal system with its rather undeveloped hierarchy, much less influenced by the Roman power than were the tribal systems in Germany or in Gaul. Britain, after all, had been a distant province that the Roman hierarchy could never subdue to the same extent that had been possible on the continent. Furthermore, the Anglo-Saxons themselves were newcomers to Britain, and met there only the remnants of a weakening Roman power. All this adds up to the fact that the Anglo-Saxon social order, left to dominate the local administration in feudal England, was much less hierarchical, and permitted much more space for local initiative, than any administration known in continental feudalism.

So the Anglo-Saxon local administration in feudal England became the germ out of which political democracy in the Western sense was to evolve, and gradually to extend to matters of state as well. Not coincidentally, similar forms of political democracy, though on a smller scale, were simultaneously or even earlier developing in Scandinavia and in Iceland – both of them places that were even more remote to the power centres of the Western-European and Eastern-European slave empires, but still close enough to the Mediterranean civilization to feel the pull of the new technology and seafaring.

What did the germ of Western democracy in medieval England look like?

The main thing, of course, was the very existence of autonomous local government in the countryside, and not only in the cities. The old division of the country into shires was retained, and in each shire the sheriff (from the word *shiregerefa*), appointed by the king, supervised both the most important tasks of local administration; viz. justice and the collection of taxes. Once a month in each shire there was a meeting, with the sheriff in the chair, in which every Anglo-Saxon freeholder of the shire – as well as the Norman gentry – had the right to participate. In these meetings decisions were made upon such issues of local administration as general law and order, taking care of the roads, etc.

More important still was the institution of Justices of the Peace and juries for the dispensation of justice in criminal as well as in civil cases. Here again, local freeholders, together with the gentry had the decisive say. The juries were composed of free laymen who had to bring all the locally known aspects of a case to the knowledge of the judge. The latter was initially either the sheriff or a Justice of the Peace, but since the reign of Henry II (1154–1189) mainly an itinerant judge sent by the king to hear cases. The possibility offered by juries for people to stand up for themselves, and for their neighbours, in front of the king's judges, tended to revive interest in individual initiative on the local level. Add to this the invigorating influence of autonomy and lively trade in great cities, London in particular, and you have mentioned all the factors that made for the birth of the English middle class – the first in the world. The seed out of which the middle class was to grow was a compound consisting of Anglo-Saxon freeholders, the Norman gentry (the lower nobles living in the countryside), and the city bourgeoisie.

It is interesting to follow how this compound of various elements began to develop, and gradually gained a share of political power too. The position of a Norman king in England as the leader of a minority speaking a foreign language was precarious from the very beginning of the Norman reign. The times being what they were, the king was incessantly at war with his continental neighbours, and it was constantly necessary to resort to feudal levy, both in the form of heavy taxes and of taking men as soldiers for the army. It only took an arbitrary and murderous king to upset the vassals, and the Anglo-Saxon population was ready to rally under their leadership against the king. Such a king was John, called Lackland, Henry II's son who governed England (1199–1216) after his elder brother Richard. A conspiracy led by the archbishop of Canterbury, Stephen Langton, formed an army and encircled the deserted king, and on 15 June 1215, on an island in the Thames opposite the now famous meadow of Runnymede, compelled him to sign the remarkable document known as Magna Charta.

Or was it so remarkable? There are notable historians, among them Winston Churchill,* who are inclined to consider it overemphasized. Just a

---

* Winston S. Churchill, *A History of the English-speaking Peoples*, Vol. 1 (1956).

paper, they say, with a lot of beautiful words on it, but not of much consequence in *Realpolitik*. I think that Churchill, for instance, with his characteristic admiration for strong just men – like himself – as the true makers of history, did overlook the great significance that the tradition of political contracts, begun by Magna Charta, was to have in English society. Not that they were fully respected but they were respected enough to gain, with time, an importance never met before in the history of mankind.

Therefore, it is worth while listing some points of the contents of Magna Charta, even though many of them were to be severely violated during the following centuries. One of the most important things declared in this document was, as it proved, the fairly permanent shift of power from the king to the king's council insofar that henceforth the king had to have the council's consent to the collection of taxes that were necessary for raising and equipping an army. Considering the fact that waging a war was a feudal king's principal right and occupation this was no small concession. The fact that the council was able to hold firmly on to this handle of power made the king's council the germ of the English parliament, or rather the House of Lords. In England, too, the king's council was composed mostly of throne vassals, called peers, and of the highest clerics.

The city of London and other town communes that had taken part in the rebellion also got important rights with Magna Charta. But in addition to the reshuffle of power the document also contained some curious, never before heard-of declarations of general human rights. Among them were the right of a free man to move freely in the country, and the right, proclaimed in the famous clause 42 of Magna Charta, of a free man to be tried according to the law of the country as interpreted by his equals (a reference to the system of juries). The mentioned clause can indeed be taken as a first attempt at a separation of juridical from executive power. It implied, and even stated explicitly, that a free man could no more be detained, or declared an outlaw, or deported from the country, nor could his property be confiscated, by the king or his men, arbitrarily, without the verdict of a lawful jury. It is astonishing to see how much, despite the violence of those times, clauses like this could gain influence in English society. It was the starting point for *Anglo-Saxon human rights* to be used as tools in the control of power for the benefit of the common people.

Henceforth, in principle at least, the king had to ask for the consent of the highest nobility whenever he wanted men and arms for a war. And at any time when king and nobility were at odds, both sides tended to seek the support of the lower gentry and other free people who mattered in local administration. This implied further concessions in favour of the latter. In 1252 the king's council was composed of 74 members of gentry, two from each of the 37 counties. In 1261 some representatives of the greatest cities were enrolled. In 1291 this was extended to include two representatives from every city. In 1301 Edward I made further concessions in the central issue of power by stating that the king could not plead 'urgent necessity' as a reason for imposing taxation without the consent of the council. This, even Churchill admits, meant that 'a long stride had been taken toward the dependence of the Crown upon Parliamentary grants' (Churchill, 1956, p. 233).

In 1341, finally, the members of the gentry and the cities in the king's council were qualified to form their own House of Commons, as distinct from the House of Lords. This was not only the birth of English parliament with its two chambers, but it also meant the establishment of the English middle class as a separate entity partaking of the government of the state through the House of Commons, its own organ of political power (cf. Fig. 87) and the birth of a genuine *three-class society*.

At first the House of Lords held almost all the power in parliament. But

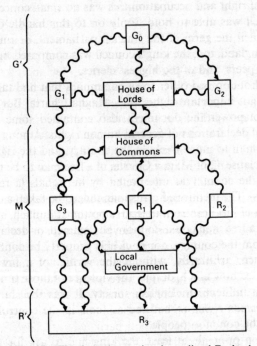

FIG. 87. The three-class society in medieval England.

Governing class G′ = G − R	G_0 = king G_1 = secular peers; G_2 = highest clergy
Middle class M = G ∩ R	G_3 = gentry, lower clergy; R_1 = Anglo-Saxon freeholders R_2 = merchants, craftsmen, other city bourgeoisie
Class of serfs R′ = R − G	R_3 = serfs, tenants working in agriculture

the continual bargaining over taxes and armies was prone to shift power gradually in favour of the House of Commons. By the way, what was the power wielded by the medieval English parliament? Speaking in modern terms it was purely legislative by nature. The king, not parliament, was the highest organ of the state. The king held executive power until 1689, that is until the creation of modern parliamentarism, in which a government that is responsible to parliament is an essential part. Even Magna Charta had contained a reference to such an arrangement, but it remained a dead letter for over four centuries. Medieval parliament was in fact, and until the age of Edward I in name as well, the 'King's parliament'. The king could convene parliament whenever he liked, but usually because of wars he was compelled to do that once or even twice a year, for some weeks at a time.

Both chambers had the right of initiative in legislative matters. After a parliamentary discussion the king undersigned the motion, if carried. The final text of an Act was at first compiled in the House of Lords, but from the end of the fifteenth century the House of Commons wrote the text, whose words could not be altered. The freedom of speech in political issues was, as a matter of course, confined within the walls of parliament. The House of Lords had this liberty in its own right, as the successor to the former king's council. The House of Commons formally asked the king for the right of free speech each time that parliament congregated. The right was then solemnly granted by the king. Furthermore, the king was allowed to know, or to pretend to know, of parliamentary discussions only what was formally reported to him by the Chancellor of the House of Lords and his counterpart in the House of Commons, the Speaker.

It is surprising that all these gentle habits were preserved, and to a certain extent even respected, through the violent war games of the fourteenth and fifteenth centuries. Just to remind the reader that, despite her more developed system of social contracts, even medieval England was no exception amidst an extremely violent and cruel age, we may quote Churchill on how Edward II met his final fate in 1327: 'His screams as his bowels were burnt out by red-hot irons passed into his body were heard outside the prison walls, and awoke grim echoes which were long unstilled' (Churchill, 1956, p. 251).

Neither was the development of the medieval parliament without its ups and downs. A serious setback upon parliamentary power was inflicted by the parliament that was called in 1397, of which Churchill has this to say:

Never has there been such a Parliament. With ardour pushed to suicidal lengths, it suspended almost every constitutional right and privilege gained in the preceding century.... All that had been won by the nation through the crimes of John and the degeneracy of Edward II, all that had been conceded or established by the two great Edwards [I and III: the latter reigned fifty years, 1327–1377], was relinquished. And the Parliament, having done its work with this destructive thoroughness, ended by consigning its unfinished business to the care of a committee of eighteen persons. As soon as Parliament had dispersed Richard [II] had the record altered by inserting words that greatly enlarged the scope of the committee's work (Churchill, 1956, p. 303).

But better times were nigh. During the Lancastrian rule (1399–1461) medieval parliament blossomed as never before. The general interest in its work was reflected in the epithets given to the parliaments that convened in the period: 'the Good Parliament', 'the Mad Parliament', 'the Merciless Parliament'. During this time parliament began to follow the expenditure of taxes, and to receive accounts from the officers of the state for that purpose. This was an important step taken towards parliamentary control of the executive power, which was not to be fully realized until much later, in 1689.

CHAPTER 15

Underdevelopment as Illustrated by Russian Social History from the Ninth to the Nineteenth Centuries

15.1. The Backwardness of the Russian Economy and the Defeat of the Norman Power

If the difference between English and Western-European continental feudalism was only one of degree, Russian feudalism was in its own genre entirely different from both of them. If we consult the qualitative theory of social development presented in Part 3, we find an obvious reason in the fact that the regulatory ability of agricultural productive forces in Russia, measured in terms of yield ratios, did not, on average, improve at all during all the centuries beginning with the ninth and going on till the nineteenth. During these thousand years the economic circumstances in Russia thus corresponded to those of primary underdevelopment in archaic societies, contrary to the favourable conditions of social progress that prevailed in Western Europe.

Asking about the possible reasons for the backwardness of the Russian economy over the centuries of Russian feudalism brings us first to the particular geographical factors that moulded the habits of cultivation in Russia. Huge areas of land and forest were available and, accordingly, unlike that in Western Europe, Russian agriculture was for a long time to remain based on the burning of forest, exploiting the land for tilling for some years, and then moving on when land was exhausted.

In this way you can even increase the cultivated area to fit the growth of population, but you are unlikely to change your methods of agriculture. This at least is what happened in Russia, whose population during those thousand years increased enormously, while the productivity of tilled land remained at the same low level, bringing forth a minimal surplus. Or, in words of the eminent historian, Richard Pipes:

The trouble with Russian agriculture was not that it could not feed its cultivators but that it never could produce a significant surplus. The productivity lag of Russia behind western Europe widened with each century. By the end of the nineteenth century, when good German farms regularly obtained in excess of one ton of cereals from an acre of land, Russian farms could barely manage to reach six hundred pounds. In Russia of the late nineteenth century, an acre of land under wheat yielded only a seventh of the English crop, and less than a half of the

French, Prussian or Austrian. Russian agricultural productivity, whether calculated in grain yields or yields per acre, was by then the lowest in Europe. . . . Scandinavia, despite its northern location, attained already by the eighteenth century yield ratios of 1: 6, while the Baltic provinces of the Russian empire, where the land was in the hands of German barons, in the first half of the nineteenth century had yields from 1: 4.3 to 1: 5.1, that is, of a kind which made it possible to begin accumulating a surplus (*ibid.*, pp. 8–9).

What could have kicked the stagnant technological development of Russia into life, and thus caused productivity to grow, would have been the appearance of lively markets. But whence could they have turned up? In Russia there was none of the vivacious intellectual and commercial intercourse that flavoured and enlivened Mediterranean life. The Russian peasant stood on his vast and flat land, facing the high sky, immovable and stern, making the passions grow but paralysing the activities of the mind.

For a fleeting moment in the early history of Russia there was a possibility that could have pushed the Russian mind into moving in step with the Western-European one. It was the time in the ninth century when the Normans came down the Volga and Dnieper, raided Constantinople, and opened a trading route through Russia. Trade between Western Europe and the Near East went that way as long as the Islamic empire closed the Mediterranean to the Christians. But with the crusades the route was made available again, and with the conquest by the Christians of Constantinople in 1204 most traffic along the eastern route was over. So the enlivening influence of the Western-European type of markets upon Old Russia, or Rus as it is sometimes called, remained short-lived. Rus, the Norman state in the south of Russia with Kiev as its capital, quite otherwise than the Norman state in feudal England, never became a seed of future social progress. In the thirteenth century it was invaded and conquered by the Mongols.

Yet the Norman influence lingered on in the north of Russia, in the city states of Novgorod and Pskov, through the Mongol era until the fifteenth century. The northern city states were the eastern ends of trade, over the Baltic Sea, with the Hanseatic League of which they became active members. This important connection with Western Europe kept them in contact with Western trends, and actually the city states in northern Russia developed societies that with their autonomy and city bourgeoisie greatly resembled their opposite numbers in Western Europe. Witness Pipes:

Politically, Novgorod began to detach itself from the other Kievan principalities around the middle of the twelfth century. Even at the height of Kievan statehood, it had enjoyed a somewhat privileged position, possibly because it was the senior of the Norman cities and because proximity to Scandinavia enabled it somewhat better to resist Slavization. The system of government evolved in Novgorod resembled in all essentials that familiar from the history of medieval city states of western Europe. The bulk of the wealth was in the hands not of princes but powerful merchant and landowning families. The task of expanding the territories of the principality, elsewhere assumed by the princes, was in Novgorod carried out by business entrepreneurs and peasants. Because they played a secondary role in the growth of Novgorod's wealth and territory, its princes enjoyed relatively little power. Their main task was to dispense justice and command the city-state's armed forces. All other political functions were concentrated in the *veche* which after 1200 became the locus of Novgorod's sovereignty. The veche elected the prince and laid down the rules which he was obliged to follow. The oldest of such contractual charters dates from 1265 (*ibid.*, pp. 36–37).

[By mutual agreement, Pskov separated itself from Novgrod in 1348 and formed an independent city state.]

The thriving existence of city states in northwestern Russia came to an end when Ivan III, Great Prince of Moscow, defeated Novgorod's armies in 1471, and annexed the city in 1477. After that, the chronology of Novgorod reports of massive land expropriations and, in 1484 and again in 1489, massacres combined with deportations of its leading citizens to inland Russia. In 1494 the Hansa depot in Novgorod was shut down. The story of Novgorod ends here. What remained was a minor Russian town among many others equally suppressed by the Moscow tsar. One last item in its history tells of new massacres of its inhabitants when in 1570 Novgorod was razed on orders of Ivan IV.

In this way the extermination of the remnants of Norman régime in Russia, begun by the Mongols, was completed by the Muscovite tsars. The influence of Western-European freer markets and relaxed hierarchy upon Russian society was so uprooted, and Russian feudalism ushered to its Eastern course, closer to the social order in ancient archaic societies than to the new European model. As indicated by the constant average yield ratios, the regulatory ability of the economic system did not develop. This implies, by the law of requisite hierarchy, a continual trend to greater hierarchy, and thus to decreasing self-steering as the only means of improving the effective regulatory ability of the economic system. Thus, Russian feudalism preserved the archaic model of underdevelopment in Eastern Europe well into the nineteenth century, without the possibilities of development that feudal order had given to Western Europe.

15.2. The Continual Growth of Centralized Power in Russia

It remains to have a closer look at some characteristic details of Russian feudalism. At first, in view of the statistics on yield ratios we have, as was noted by Pipes, to reject what has been the orthodox Soviet interpretation of the Russian peasant's life before the October Revolution. It was by no means the continual deprivation claimed by the Marxist view. In fact, the orthodox version has already been criticized in 1959 by a Soviet historian:

We confront a paradox. A scholar investigates the condition of peasants in the period of early feudalism. Their condition is already so bad it cannot deteriorate further. They are perishing completely. But then, later on, they turn out to be worse off yet; in the fifteenth century still worse, in the sixteenth, seventeenth, eighteenth, nineteenth centuries, all the time worse and worse. And so it goes until the Great October Socialist Revolution. It has already been pointed out, very correctly... that the living standard of peasants is elastic and capable of shrinking. Still it cannot shrink ad infinitum. How did they survive?*

* A. L. Shapiro, *Ezhegodnik po agramoi istroii Vostochnoi Evropy*, Akademia Nauk Estonskoi SSR, Tallin, 1959, p. 221 (as quoted by Richard Pipes, *op. cit.*, p. 8).

The answer, obtainable from the available statistics concerning productivity is that they survived very well, but they only survived: productivity, as measured by yield ratios or by the outcome per acre, did not increase, thus the technological level and the regulatory ability of productive forces did not improve, thus the standard of living of Russian peasants did not rise. But this of course is different from living in a state of continual deprivation. Or, as stated by Pipes:

Like the rest of Europe, Russia averaged in the Middle Ages ratios of 1:3, but, unlike the west, it did not experience any *improvement* in yield ratios during the centuries that followed. In the nineteenth century, Russian yields remained substantially the same as they had been in the fifteenth, declining in bad years to 1:2, going up in good ones to 1:4 and even 1:5, but averaging over the centuries 1:3 (slightly below this figure in the north and slightly above it in the south). Such a ratio generally sufficed to support life.... Recent computations of the incomes of Novgorod peasants in the fifteenth century, and of the Belorussian-Lithuanian peasants in the sixteenth (both inhabitants of northern regions with the inferior, podzol, soil) do indeed indicate that these groups had managed to feed themselves quite adequately (*ibid.*, p. 8).

In reality what was deteriorating during all the centuries of Russian feudalism was not the material standard of living of the peasants – which remained constant on average – but their social status and human rights. This is also a prediction of the law of requisite hierarchy for the Russian period of underdevelopment. Under those conditions there is a continual trend towards increasing social hierarchy, and this trend became true in theocratic slave empires in the archaic ages, and again in Russian feudalism over all its many centuries. At the bottom of the social echelon it means increasing oppression – that is, arbitrary treatment and deprivation of rights. This was also the fate of Russian peasants in feudal times. While the position of their Western-European counterparts was improving all the time, that of Russian peasants deteriorated.

But the effects of increasing social hierarchy are not only felt on the lowest social strata. They touch other social layers too. Here again the archaic model of Russian feudalism made Eastern development part company with development in the West. While the king's power in Western-European feudalism was ever more delegated to the nobility, in England even to the bourgeois middle classes, nothing of the kind happened in Russia. Contrary to such a democratic trend, Russian feudalism developed, just like the ancient theocracies in the Near East, toward the concentration of all-pervading power in the hands of one man depicted, in the central myths of society, as the representative of God. In Russia he was called the tsar. In the process, what had existed of feudal nobility in Russia was gradually reduced to a status of officers of the state in the service of the tsar, entirely dependent on him. If there had not been the Russian peasant, whose position developed to that of a serf rather than to that of an ordinary slave, one could indeed say that Russian feudalism was step by step exterminated by the tsars, and replaced by a totalitarianism similar to that in ancient slave empires in the Near East. (And really, in Stalin's time, at the climax of Russian totalitarianism, even slavery, in the form of forced-labour camps, was back again.)

According to the hierarchy law, increasing social hierarchy, and thus a concentration of power, is a lawful phenomenon under the conditions of underdevelopment, such as obtained in feudal Russia. But we can illuminate the workings of that law in this case by following the step-wise crumbling of Russian feudalism. To do that we must go back to its beginnings long before the times of the tsars. In fact, we must even return to the era prior to the Mongolian conquest of Russia – a conquest that Marx thought had brought the 'Asian mode of production' to Russia. Contrary to the assumption of Marx, the peculiar, Russian type of feudalism with its seed of self-destruction already existed before the arrival of the Mongols.

The seed alluded to was the non-contractual character of Russian feudalism from its very beginnings. In words of Richard Pipes:

Vassalage represented the personal side of western feudalism (as conditional land tenure represented the material). It was a contractual relationship by virtue of which the lord pledged sustenance and protection, and the vassal reciprocated with promises of loyalty and service. This mutual obligation, formalized in the ceremony of commendation, was taken very seriously by the parties concerned and society at large.... What do we find in Russia? Of vassalage, in its proper sense, nothing. The Russian landowning class, the boyars, were expected to bear arms but they were not required to do so on behalf of any particular prince...Although some historians claim that the relations between princes and boyars were regulated by a contract, the fact that not a single document of this kind has come down to us from Russian (as distinct from Lithuanian) territories raises grave doubts whether they ever existed. There is no evidence in medieval Russia of mutual obligations binding prince and his servitor, and, therefore, also nothing resembling legal and moral 'rights' of subjects, and little need for law and courts (*ibid.*, pp. 50–51).

The point is undoubtedly important. As a matter of fact the contractual vassalage in the Western-European feudal state was the origin of law whose obligations were binding on all citizens, including the holders of power. In Russian feudalism this important element of law was lacking. Pipes also stresses the importance of the unwritten, moral obligations implied by a feudal relation in the Western sense but lacking in the East:

the feudal contract, beside its legal aspect had also a moral side: in addition to their specific obligations, the lord and vassal pledged one another good faith. This good faith, imponderable as it may be, is an important source of the western notion of citizenship. Countries which had no vassalage or where vassalage entailed only a one-sided commitment of the weak toward strong, have experienced great difficulties inculcating in their officials and citizens that sense of common interest from which western states have always drawn much inner strength (*ibid.*, p. 50).

The weakness of the Russian feudal system – if system it was – made it easy for the Mongols to impose their absolute power principle on conquered Russia. The Russian nobles had only to transfer their loyalty to the new Mongol lords, binding as this loyalty already was on the weaker partner only, thus involving the possibility of the absolute loyalty required by the new power-holders. The Mongol conquest of Russia occurred in 1237–1241, and the western branch of the Mongol empire called the Golden Horde, with Sarai on the Lower Volga as its capital, was founded about 1243, and was to last for more than two centuries as a period during which Russia was governed by an archaic slave empire. It did not collapse until the fifteenth century, one thousand years after the collapse of the Western-European slave empire, Rome. Furthermore, there was the great difference between the Golden Horde and the Roman empire in that the latter was a more modern kind of slave empire that recognized, though did not always observe, the law, while the régime of the Mongol empire was an absolutism of the ancient type.

Thus, there were two main reasons why the Mongolian conquest contributed to the backwardness of Russian feudalism when compared with Western-European feudalism: the time lag so engendered, and the absolute type of régime imposed on the Russians. The time lag, of course, is what it is, but as for the differences between the Romans and Mongols some qualifications must be added. It is much too simplistic to assume, for instance, that absolute power would exclude all refinement in manners, and all intellectual life. It never did quite that. The upper classes in ancient theocracies, in Egypt, Mesopotamia, or Persia, were not without education or intellectual interests, as we know: astronomy could thrive, as well as poetry. The Mongols did not display quite that kind of educational achievement, but an English visitor to the Mongolian court appraised their cultural level as higher than that of the Russians at the same time. Thus, the difference between the Mongol and Roman régimes was not simply one between the rule by barbarians and civilized people.

Yet undoubtedly it was a difference between the levels of civilization achieved by the leading classes in each case. When a Roman emperor, in the period of one-man power in Rome, violated the law, he knew that he was doing so, while a Mongol khan, however brutal his behaviour, never broke the law because there was no law. For a Roman emperor there existed a moral obligation to do otherwise, while a Mongolian ruler could commit cruelties believing that it was his inherited right and duty to do so. This is the crucial difference between *despotism*, such as we meet it in the later epoch of the Roman empire, and *absolutism* – the kind of rule that obtained in the ancient, theocratic slave empires in the Near East as well as in the Golden Horde. The absolute quality of power was the principal inheritance left by Mongolian rule to Russian feudalism. 'Russian life', Pipes writes:

became terribly brutalized, as witnessed by the Mongol or Turco-Tatar derivation of so many Russian words having to do with repression, such as kandaly and kaidaly (chains), nagaika (a kind of whip) and kabala (a form of slavery). The death penalty, unknown to the law codes of Kievan Rus', came in with the Mongols. During these years, the population at large first learned what the state was; that it was arbitrary and violent, that it took what it could lay its hands on and gave nothing in return, and that one had to obey it because it was strong. (*ibid.*, p. 57).

Here, I would like to emphasize that *absolutism* and *totalitarianism* are not identical notions. Totalitarianism is absolutism grown all-pervading. First, when absolute power reaches all sectors of human life it becomes totalitarian in the proper sense of the word. The rule of the Golden Horde, though it was even totalitarian in its own territory, never became that in Russia simply because the interests of the Mongolian rulers were limited there. The Mongols never permanently occupied Russia. What they wanted was a regular tribute in the form of slaves (mainly children) and taxes. They employed Russian princes as collectors of the tribute, and stationed their own men only in cities and towns to watch that all that could be taken was taken. 'A prince confronted with popular dissatisfaction had merely to threaten with calling in the Mongols to secure obedience – a practice that easily grew into habit' (Pipes, *op. cit.*, p. 57).

15.3. The Development from Absolute towards Totalitarian Power

The development from 'merely' absolute towards totalitarian power first began in Russian society (cf. Fig. 88) when the Mongolian empire broke up in the fifteenth century. Now the period of 'partial princes' in Russia was over, and the Great Prince of Moscow began to assume the position of the Mongol khan. The first of them to have this ambition, Ivan III (1462–1505), in 1480 left the tribute to the khan unpaid for the first time. By then he had already defeated and annexed Novgorod, and through the massacres and deportations of the 1480s he cruelly crushed the centre of normal social order in the 'north-west' (the Soviet euphemism for the medieval city states in Russia). In 1485 the principality of Tver, and in 1489 that of Viatka, were annexed to Moscow, and in 1510 Ivan's successor Basil III (1505–1533) arranged the same fate for the city state of Pskov. When Ivan IV 'the Terrible' (1533–1584) came to the throne he already governed a remarkable empire.

Which crown, by the way, was in question? As a matter of fact it was not yet a crown – Ivan IV was the first to be crowned 'tsar' in 1547. This was an

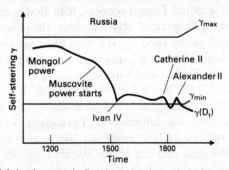

FIG. 88. Social development in Russia under the period of underdevelopment, as shown by the indices of self-steering. After the absolute power of the Mongols started a development toward totalitarianism. (Warning: the picture serves only to illustrate the concepts of the qualitative theory of self-steering. The details in the curve of self-steering are just an educated guess, not based on any measurements! – A.A.)

old title used in Russian initially for the Emperor of Byzantium, and later for the Mongol khan, the first undisputed personal sovereign of Russia. The word itself is short for the Russian spelling of Caesar (tsezar). But from 1547 it was adopted by the Great Prince of Moscow, Ivan IV, to indicate his newly-established sovereignty over the Russian empire. In 1552 this empire was expanded to include the khanate of Kazan and in 1556 that of Astrakhan.

On coming to the crown, Ivan IV first undertook reforms, among others, of local government. But in 1564 he started the notorious measures to which he owns his epithet 'Terrible'. The *oprichina terror* began, to last until 1572. During those years Ivan crushed what was left of the independent position of feudal nobility in Russia, and a lot of the nobles physically as well. Their landed property was taken to the possession of the tsar, who from that time considered himself to be the owner of the whole land, although he did not formally own all of it. Those princes and boyars who survived were taken into the service of the tsar to administer for him the many economic, juridical, military, and other affairs, into which his absolute power now began to penetrate. In return the new officers of the state were given land and peasants, no more in terms of a real 'votchina' but mostly as 'pomestja' that could be taken away by the tsar at any moment. Many former nobles got only salaries.

Here we have to stop for a moment to consider the concept of *votchina*. This ancient Russian term referred to a feudal property, or rather to everything that feudal ownership involved in Russia. As a feudal relation was not based on a contract, then what was its foundation, one may ask. The answer is that a feudal lord in Russia was considered to own both his land and his peasants as an inheritance from his father, and this was what was understood by the word 'votchina'. Such a concept of feudal ownership in fact put the Russian peasant somewhere between a serf and a slave. 'All the people consider themselves to be kholops, that is, slaves of their Prince', a sixteenth-century German traveller* remarked. Some years after the oprichina terror the noted French scholar, Jean Bodin, called the Russian type of government a 'seigneural' régime, where 'the prince has become the lord of property and persons and ... rules over them like the head of a family rules over his slaves', adding that this form of régime still existed in only two European states, Turkey and Russia.†

Here we come to the second most important historical factor that ushered the social development in Russia on its unfortunate path of underdevelopment, viz. the influence of Byzantium, which acted as the mediator of the ancient, archaic modes of power and ownership, such as 'votchina', from the slave empires of the Middle East into Russia. Religion

---

* Sigismund Herberstein, *Rerum Moscoviticarum Commentarii*, Basle (1571), p. 49 (quoted by Pipes, *op. cit.*, p. 85).

† Jean Bodin, *Six livres de la république*, (1576), in English titled *The Six Bookes of Commonweale*, first published in 1606, and reprinted in 1962 in Cambridge, MA, Vol. 2, pp. 197–204.

was a powerful mediator of these traditional ways of life, as was noted by Richard Pipes:

The fact that Russia received its Christianity from Byzantium rather than from the west had the most profound consequences for the entire course of Russia's historic development. Next to the geographic considerations discussed in the opening chapter of this book, it was perhaps the single most critical factor influencing that country's destiny. By accepting the eastern brand of Christianity, Russia separated itself from the mainstream of Christian civilization which, as it happened, flowed westward. After Russia had been converted, Byzantium declined and Rome ascended. The Byzantine Empire soon came under siege by the Turks who kept on cutting off one by one parts of its realm until they finally seized its capital (in 1453). In the sixteenth century, Muscovy was the world's only large kingdom still espousing eastern Christianity. The more it came under the assault of Catholicism and Islam, the more withdrawn and intolerant it grew. Thus, the acceptance of Christianity, instead of drawing it closer to the Christian community, had the effect of isolating Russia from its neighbours (Pipes, *op. cit.*, p. 223).

Thus, the economic and other powers of the lord over a peasant were tightly tied together, not separated from each other as they used to be in Western-European feudalism, where even the relation between a lord and serf was based on a kind of contract stating their mutual give-and-take. When, as a consequence of oprichina, the whole country was made a 'votchina' of the tsar, it implied that his absolute power was now to be extended, within the bounds of his empire, to all aspects of human life, in principle at least. This is why the oprichina can be singled out as a crucial step from absolutism towards totalitarianism in Russia.

By means of oprichina the tsar cleared away the worst potential enemy that there was on his way to totalitarian power, viz. feudal nobility. After that it was a minor operation to crush the emerging city bourgeoisie, and to return it to a status near that of the peasants. A decisive act, of course, had already been committed in the fifteenth century, when Ivan III had defeated the army of Novgorod. Even here Ivan IV completed the work. He let the city of Novgorod be entirely demolished in 1570. Generally speaking the oprichina terror was not merely directed against the nobles. Even the common people, in cities and towns especially, got a share of it:

Individual Moscow streets, small towns, or market places and particularly large vochiny, upon being designated by the tsarist order as part of the oprichina, became the personal property of the tsar, and as such were turned over to special corps of oprichniki. This crew of native and foreign riff-raff were permitted with impunity to abuse or kill the inhabitants of areas under their control and to loot their properties (Pipes, *op. cit.*, p. 95).

Even more far-reaching in preventing the growth of a bourgeois middle class was the distinction between servitors and commoners, imposed on the whole country by the Muscovite rule. The former were in the service of the state, that is, the tsar, while the latter, mainly composed of peasants, traders, shopkeepers and artisans, were the so-called tjaglo-bearers, and formed the bulk of the population in feudal Russia, about 99.7%. The word 'tjaglo' designated the load of taxes and labour which the commoners owed the tsar. While the servitors, also called 'big men' (muzhi or ljudi), served the state, the commoners – called 'little men' (muzhiki) were all more or less in a position of serfs of the state. But serfdom in Russia, even on the part of peasants, was instituted rather late. After all, the initial agriculture in Russia had been based on a nomadic life, as was mentioned before, and peasants had traditionally been tenants rather than serfs. First, between approximately 1550 and 1650, they were turned into serfs, but then they were accompanied by artisans, shopkeepers, and traders as well. The fact that all these 99.7% of people ultimately belonged to the votchina of the tsar made life for them rather harder: in addition to the labour that could be demanded by their landlords they could now be demanded also to help with the construction of whatever the state wanted to build, from fortifications to roads and bridges; they could also be made to feed the armies.

What must have been hard to impose on Russian peasants and tradesmen, in view of their nomadic traditions, was the abrogation of free movement, the next step towards a totalitarian control of people undertaken by the tsar.

Free movement proved more difficult to terminate among commoners than among servitors. A landowner could be discouraged from going into someone else's service...there was always his landed estate or that of his clan to provide collateral. But it was a different matter to try to keep in place farmers or tradesmen who had no title to the land which they tilled, no career status to worry about, and for whom nothing was easier than to disappear without trace in the endless forest.*

Yet free movement of that 99.7% of people could be suppressed by severe measures, and the Code of 1649 meant that excuses or mitigating circumstances for runaway peasants were no longer accepted. Therefore, this year is mostly mentioned as the date on which serfdom proper was installed in Russia. But from the midsixteenth century edicts had been given which forbade traders and artisans to leave their towns, and fixed them as serfs to their *posad* – a particular area in town where they were allowed to ply their trade. Richard Pipes quotes a Soviet historian, S. M. Soloviev, to indicate one of the consequences of such restrictions:

The chase after human beings, after working hands, was carried out throughout the Muscovite state on a vast scale. Hunted were city people who ran away from tiaglo wherever they only could, by concealing themselves, bonding themselves (as slaves to private persons), enrolling in the ranks of lower grade clerks. Hunted were peasants who, burdened with heavy taxes, roamed individually and in droves migrated beyong 'the Rock' (the Urals). Landlords hunted for their peasants who scattered, sought concealment among other landlords, run away to the Ukraine, to the Cossacks.*

How hard were the measures employed to stop free movement of the Russians, not only within the Russian empire but beyond its boundaries too, is shown by the severity of the punishments for unauthorized travelling abroad:

No one was allowed to escape the system. The frontiers of the state were hermetically sealed. Each highway leading abroad was blocked at frontier points by guards who turned back travellers unable to produce the proper authorization, a document called proezzhaia gramota obtainable only by petition to the tsar. A merchant who somehow managed to get away abroad without such papers suffered the confiscation of his belongings; his relatives were subjected to

* Cited by Richard Pipes, *op cit.*, p. 108, from a Soviet work published in 1961.

torture to elicit information on the reasons for his trip, and then exiled to Siberia. The Code of 1649 provided in Chapter VI, Articles 3 and 4, that Russians who had gone abroad without authorization and then, upon their return, were denounced for having done so, had to be questioned as to their motives; if found guilty of treason they were to be executed, but if making money was their purpose, then they were to be beaten with the knout. The principal reason for such draconic measures was fear of losing servitors and income. Experience indicated that Russians familiar with foreign ways did not wish to return home.... It was never forgotten that of the dozen or so young dvoriane whom Boris Godunov [the tsar in 1598–1605] had sent to England, France and Germany to study, not a single one chose to come back. (Pipes, *op. cit.*, pp. 110–111).

But the drastic prevention of free movement, and the fixing of even tradesmen and artisans as part of the tjaglo-bearers to their pieces of land in posads, of course paralysed any development towards the formation of sensitive markets. Besides, the tsar used his enormous authority as the hereditary owner, votchinnik, of the whole country, by simply confiscating the property of any merchant who began to grow richer than the others. This precluded all accumulation of private commercial capital, and thus an effective investment in productive forces, and in technological innovations in particular. The centralized state bureaucracy, which administered the economy of the tsar, had no motives to advance either investments or a market system.

The middle class, and with it a dynamic economy, thus could not develop in feudal Russia, which remained a strictly two-class society. The divide between the classes ran along the same line that separated the tjaglo-bearers, that 99.7% of the population, from the remaining 0.3%, the service class. Although the Romanov dynasty, on its accession to the throne in 1613, wishing to solidify its power had fairly generously distributed land to the boyars and dvorianes, these were just servitors to the tsar, who could at any moment 'take back' their property, if they proved disobedient. Thus, even the fact that in the second half of the seventeenth century 80.3% of the tjaglo households were owned by the nobility or by the church, and only 9.3% outright by the crown, did not mean the existence of private interest in the Russian economy. The service-nobility of feudal Russia was not disposed to invest its landed property in commercial undertakings, for the simple reason that all successful commercial undertakings were in due time confiscated by the tsar.

The fact that there was no genuine middle class in feudal Russia is most clearly illustrated by her city life, or rather by the lack of anything one is accustomed to call urban life according to Western-European standards. In the middle of the seventeenth century, for instance, the whole 60% of the inhabitants of Russian cities were servitors, that is, bureaucrats on the various steps of the tsarist echelon, and only 32% were tjaglo people, including traders and artisans. So the distribution of the city population was very different from that in a Western city. Add to this the fact that tradesmen and artisans were also fixed to the land, and had to do some farming to feed themselves, and you have a picture of life in Russian cities and towns that hardly resembles that in contemporary Western-European cities, the centres of throbbing commercial life.

But just saying that feudal Russia was a two-class society may give the incorrect impression that it was something comparable with the feudal societies on the Western-European continent, which also had a prominently two-class structure. It was not. The power of the Russian tsar, as an inheritance of the Mongol khan, was absolute; the power of the feudal king in Western Europe never was. And ever since Ivan the Terrible the stronger of the Russian tsars strived for an all-pervading, totalitarian power, while the development in Western Europe, and in England in particular was advancing in the opposite direction, towards a relaxation of social hierarchy. Although the bureaucratic machinery of the tsar was far too ineffective to achieve the goal that was first successfully reached by Stalin, with his huge totalitarian state apparatus, some of the hallmarks of a totalitarian state actually appeared rather early in the history of tsarist Russia.

One of them, and indeed the most important, was a massive system of *mutual denunciation* established among the citizens. In fact this practice seems to have been begun by Ivan IV, in the notorious Code of 1649 that made it compulsory to inform the government, that is, the tsar, of any actions in the country that could be detrimental to governmental interests:

The range of such anti-state crimes was broad; included were offences which in the language of modern totalitarian jurisprudence would be called 'economic crimes', such as concealing peasants from census-takers or misinforming the Office of Pomestja about the true extent of one's land-holdings. The Code placed great reliance on denunciation as a means of assuring that the state obtained the proper quantity of service and tjaglo. Several of its articles made denunciation of anti-government 'plots' mandatory under penalty of death. The Code specified that families of 'traitors' (including their minor children) were liable to execution for failure to inform the authorities in time to prevent the crime from being committed (Pipes, *op. cit.*, p. 109).

Furthermore, the kind and nature of a totalitarian justice, with its characteristic *tendency to ignore the difference between intention and deed*, is already apparent in these early pursuits. The same holds true of the application of *collective responsibility* as a means of encouraging the mutual denunciation among the members of the collective:

In the seventeenth century crimes against the state (i.e. against the tsar) came to be known as 'word and deed', that is, either expressed intention or actual commission of acts injurious to the gosudar (the emperor). Anyone who pronounced these dreaded words against another person, caused him to be arrested and subjected to torture; as a rule, the accuser suffered the same fate, because the authorities suspected him of having concealed some information. 'Word and deed' often served to settle personal vendettas.... Denunciation would not have been half as effective a means of control were it not for the collective responsibility inherent in tjaglo. Since the taxes and labour services of anyone who fled his tjaglo community fell on its remaining members (until the next cadaster, at any rate), the government had some assurance that tjaglo payers would attentively watch one another. Shopkeepers and artisans were particularly keen to note and denounce any attempts of their neighbours to conceal income. So the state watched its subjects and the subjects watched one another (Pipes, *op. cit.*, pp. 109–110).

A system of denunciation, of course, not only implies the nonexistence of the freedom of speech, in the usual sense of forbidding public speech, but it excludes the liberty of thought in private as well. The latter variety of the suppression of intellectual liberty is a speciality of totalitarianism. Until 1703 all domestic as well as foreign news in Russia was reserved for the ears and eyes of the tsar and his top officials. After that the news was conveyed by official reports (kuranty) prepared by the Office of Ambassadors.

The more enlightened Russian sovereigns, such as Catherine II (1762–1796) or Alexander II (1855–1881), could not turn the general tide that irreversibly streamed toward a totalitarian society. The new Criminal Code published in 1845 by Nicholas I returned the old idea of the Code of 1649 by decreeing the penalty of death for subversive thought as well as for such a deed. Consideriing its Chapters 3 and 4, about political crimes, one can only agree with Pipes who wrote:

These two sections, covering fifty-four pages, constitute a veritable constitutional charter of an authoritarian regime.... Politics has been declared by a law a monopoly of those in power; the patrimonial spirit, for centuries a nebulous feeling, has here at long last been given flesh in neatly composed chapters, articles and paragraphs. Particularly innovative in these provisions is the failure to distinguish deed from intent – a blurring of degrees of guilt characteristic of modern police states.... One is justified in saying, therefore, that Chapters Three and Four of the Russian Criminal Code of 1845 are to totalitarianism what the Magna Carta is to liberty (*ibid.*, pp. 293–295).

Only the bottomless inefficiency of the tsarist bureaucracy, combined with its lack of enthusiasm, prevented it from setting up a truly totalitarian society – both obstacles were removed by Stalinism.

CHAPTER 16

The Gradual Breakthrough of Modern Western Society in England from the Fourteenth to the Nineteenth Centuries

16.1. The End of English Feudalism and the Rise of Independent Farmers and City Bourgeoisie

During the thirteenth and fourteenth centuries, when it had not even begun in Russia, serfdom was, in the most advanced society of the time — viz. that of England — already coming to an end. Because of the many wars between feudal kings, peasants had a rising price as soldiers, and they were able to make themselves tenants and, especially in England, even to buy land, thus becoming 'yeomen' or 'franklins', joining the middle class.

The process was precipitated by the plague, which arrived in England in 1348 and quickly reduced the total population of four million to half that. The value of the labour force increased, and many peasants were now able to change their labour tax to wages payable in money, and with the continual shortage of manpower their wages grew.

In this situation, as if sent by Heaven – but, as a matter of course, a natural phenomenon related to relaxed social hierarchy – a new religious movement appeared preaching egalitarianism. Its leader was the 'first Puritan' John Wickliffe (1324–1384), a professor of theology at Oxford, who took his lead from the Bible, and turned against the worldly hunger for power that he thought had got a grip on the Catholic Church and on its priests. Together with friends he translated the Bible into English, the language of the people, and with this Old Vulgata in their hands waves of 'poor priests' now surged over the country disseminating their early Protestant message. They found an audience among the newly liberated yeomen in particular, and a new clerical party was born, called the Lollardians, that also demanded land reform from parliament.

One of the Wickliffian 'poor priests' called John Ball became famous with his song

> Whan Adam delfe and Eve span,
> Who was than a gentleman,

to the tune of an ever-since well-known popular melody.

Beginning in 1339 there was the Hundred Years War, which finally ended in 1453, between England and France, increasing in many ways the price of soldiers and the self-esteem of peasants. In 1381 a rebellion of peasants under the leadership of Wat Tyler burst out, and required the abrogation of serfdom altogether. The king (Richard II, 1377–1399) was forced to give charters to peasants that granted their freedom, but these were withdrawn later after the revolt had been crushed and its leaders beheaded. However, such was the social situation in England, where the middle class was emerging and feudalism coming to an end, that the position of peasants improved rather than deteriorated after those events. Things were different in France, where a similar uprising of peasants – the first of the kind in the world – called La Jacquerie was, in 1358, ruthlessly routed, with severe consequences for the defeated peasants.

Indeed, the end of feudalism in England was near. The Norman nobility was badly impoverished, and with the War of the Roses (1455–1485) about the throne, between the red rose of the York family and the white rose of the Lancasters, feudal nobility almost entirely exterminated itself, leaving one million dead – one-third of the total population at that time – with some eighty royal princes in that number. The reign of the York family (1466–1485) was a violent time, but when the war in 1485 ended, and the first Tudor, Henry VII (1485–1509) came to the throne, both feudal lord and serf had become rare animals. The Tudors were to favour the formation of independent peasantry. Later, there was again to be a nobility with landed property, but never again fiefs and serfs. The political hegemony of feudal lords, and herewith feudalism itself, was over in England.

Yet the city bourgeoisie, the new dominant class, was not yet ready for a take-over. An intermediary period between feudalism and bourgeois power followed, during which the kings of England filled up the vacuum of power so formed; sometimes very willingly indeed. It is not without justification that this period in England's history is mostly named according to the reigning kings – the Tudors (1485–1603) and Stuarts (1603–1649 and again 1660–1688), although after Cromwell's time (1653–1658) the kingship was never the same again. But before then the kings of England intermittently – not unlike the French kings a little later – strived for absolute power. The first of them, and the nearest to success, was the second Tudor, Henry VIII (1509–1547).

It is worthwhile to note that attempts at absolute power in England brought into political life similar features as elsewhere – these features are universal. It is very well known that Henry VIII, and to some extent his father before him, set up a network of spies and informers. That was certainly needed, for the king let it be declared in all the churches, in March 1534, that he required, henceforth, the absolute loyalty of all citizens, on the basis of 'Henry VIII being immediately next unto God, the only and supreme head of this Catholic Church of England, and Anne his wife, and Elizabeth daughter and heir to them both, our Princess'. To pronounce malicious words about the king was made high treason. Churchill remarks: 'As the brutality of the reign increased many hundreds were to be hanged, disembowelled, and quartered on these grounds' (*ibid.*, Vol. 2, p. 51).

The father, Henry VII, had gathered, by both legal and illegal means, the enormous wealth of 1.8 million pounds for the fund of the state. The son enriched the crown further by confiscating, as the self-declared head of the Catholic Church of England, the property of the Church, thus destroying the only remaining powerful rival with landed property.

But it would be unjust to remember the first two Tudors only for their greed and hunger for power. The father excelled in legislation, and the son developed local government, especially the system of Justices of the Peace. On his initiative Justices' manuals, which were to run through innumerable editions, were published later in the century.

The Tudors were indeed the architects of an English system of local government [Churchill writes] which lasted almost unchanged until Victorian times. Unpaid local men, fearless and impartial, because they could rely on help from the King, dealt with small matters, sitting in the villages often in twos and threes. Bigger matters such as roads and bridges and sheep-stealing came before quarter sessions in the appropriate town. It was a rough justice that the country gentlemen meted out, and friendship and faction often cut across the interests of both the nation and the Crown. If in the main they carried the directions of the Crown to the people, the Justices could also on occasion, by turning a deaf ear to official advice, express popular resistance to the royal will. What they did in the counties they could also sometimes do in the House of Commons. Even as Tudor rule advanced toward its climax the faithful members of Parliament were not afraid to speak their minds (*ibid.*, Vol. 2, pp. 33–34).

It was not by chance that the Tudors paid so much attention to issues of local government. If the king's pursuit of absolute power was one mark of the period, the peasants' removal from the status of serf or tenant to that of yeoman was another. England became the first country in the world where a class of independent farmers emerged from former serfdom. This not only increased the significance of local government, but also deeply affected the economy of Tudor England. During the sixteenth century the yield obtained in England doubled, as the yeomen who themselves owned the land they were tilling were, of course, more interested in introducing better technology and more efficient working habits than the serfs or tenants had been. At the same time England was really engaging in world trade and, as a consequence of a hectic commercial exchange, the prices of some wares rose sixfold, and wages only twofold. This inflation reduced people to beggary and vagabondage – forms of unemployment at that time. Even many yeomen on their newly acquired small estates were hard up. Who after all was the winner?

The question brings us to the third, and maybe most remarkable, characteristic of the period, which was the rise of the bourgeoisie. London and other English ports had become lively commercial centres and starting points for those daring escapades that took English traders and pirates all over the world, and back again with their ships full of wares from distant countries. Even inland England developed an article for which there seemed to be an increasing demand in world trade: wool. Ever more land was turned over to pasture, and the process was precipitated by the dissolution of the monasteries, as a consequence of the measures taken by Henry VIII against the Church.

In some counties as much as one-third of the arable land was turned over to grass, and the people looked in anger upon the new nobility, fat with sacrilegious spoil, but greedy still (Churchill, *op. cit.*, Vol. 2, p. 71).

The wealth gathered from wool and other trade began to accumulate in the hands of the city bourgeoisie and the new landowners, and was transformed into capital invested in all the promising economic pursuits offered by broadening commercial perspectives. In the glorious age of the last Tudor, Elizabeth I (1558–1603), England, after having defeated the Armada of Spain in 1588, became a sea power on a world scale, and now trade really started to prosper. England's first colony in North America, Virginia, was founded in 1584, and in 1599 the East India Company was constituted. Sea routes were opened both westward and eastward. Britannia was not yet the only one to rule the waves, but she was one of them.

16.2. The Short-lived Threat of Full Revolution: Cromwell's Time and a Temporary Drop of Self-steering (Human Freedom) in English Society

Even in England the newly started economic growth, combined with lingering power hierarchy, in the form of the sovereign's autocratic pursuits, produced something of a revolutionary effect, if not quite a proper revolutionary situation (cf. Fig. 89). The Protestant movement, which in England had started in the fourteenth century, was not only a revolt against the clerical power of the Catholic Church, but had, in England especially, secular, egalitarian ends as well. The religious tradition of Wickliffe and the Lollardians was going strong among the new classes of yeomen and bourgeoisie, and emphasized not only the necessary Christian purity of everyone's personal life, but the dignity of every human being. No-one was to be feared but God. These Puritans already had the majority in parliament in the Elizabethan era, but they did not turn against the queen. However, the establishment in 1562 of the new Church of England, though it was not Catholic but Episcopal, soon raised the wrath of the Puritans who regarded it, and the High Churchmen leading it, as one further attempt at priestly power.

Out of these elements an open political conflict, with a deep religious undertone, developed during the age of the Stuarts. The Anglican Church, under the leadership of its arrogant archbishop Laud, indeed proved to be more priestly than the Catholic Church itself, and rallied tightly around the

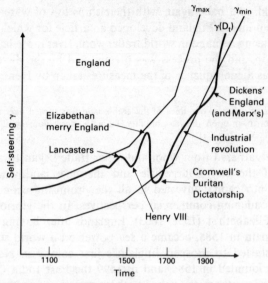

FIG. 89. The history of England as told by the indices of self-steering. Note Dickens' England, misunderstood by Marx as a revolutionary period: despite class hierarchy and the relative poverty of many development was not stagnating but rushing on: social progress was only temporarily lagging behind the stormy economic growth. (Warning: the picture serves only to illustrate the concepts of the qualitative theory of self-steering. The details in the curve of self-steering are just educated guesses, not based on any measurements!–A.A.)

royal autocracy, which it declared to have been confirmed by God. During the Personal Rule of Charles I, an enlightened autocrat, Parliament was not convened for eleven years, but when it at last was, in April 1641, it carried the Triennal Bill that provided for the summoning of parliament at least once in every three years. Furthermore, it made itself in practice undismissable, and indeed this 'Long Parliament' sat from 1640 until 1653, when it was displaced by Oliver Cromwell's nearly totalitarian régime. This time, and the period up to 1689, was the formative era of modern parliamentarism, and needs closer examination.

The Puritans, with their leader Pym, were the prevailing force in parliament, while High Churchmen dominated court. So, when parliament, in a new bill called 'Grand Remonstrance' challenged the whole executive power of the king, the religious grouping of the impending conflict was clear. The parliamentary party became known as the Roundheads, and those who rallied around the king the Cavaliers. In social terms the parties of the ensuing civil war are harder to define. 'The "new classes" of merchants and manufacturers and the substantial tenant-farmers in some counties were claiming a share of political power, which had hitherto been almost monopolised by the aristocracy and the hereditary landlords', Churchill writes. But he adds:

Yet when the alignment of the parties on the outbreak of the Civil War is surveyed, no simple divisions are to be found. Brother fought against brother, father against son.... The greater part of the nobility gradually rallied to the Royalist cause; the tradesmen and merchants generally inclined to the Parliament; but a substantial section of the aristocracy were behind Pym, and many boroughs were devotedly Royalist. The gentry and yeomen in the counties were deeply divided. Those nearer London generally inclined to Parliament, while the North and West remained largely Royalist. Both sides fought in the name of the King, and both upheld the Parliamentary institution... in Rake's compact phrase, 'One party desired Parliament not without the King, and the other the King not without Parliament' (*ibid.*, Vol. 2, pp. 185–186).

Such a dispute is not a Marxist class war, in which one class wants to annihilate another, but rather a controversy between two opposite *politics* in a developed society, quite in line with our theory of social development (Fig. 83). In this conflict the Puritans stood for what must be considered a radical cause, as they, initially at any rate, wanted to dismantle the hierarchy of power in favour of the lower classes. The Royalists clearly opposed this, and supported the existing hierarchy, thus representing a conservative force. But since parliamentarism in the modern sense was not yet there, the dispute, which essentially was about who should have how much control of executive power, had to be settled in the feudal way, by arms.

And to arms one took. In the English civil war (1642–1648), unique in history, best characterized as one between two common parliamentary parties, neither of them extreme, the Roundheads and the Cavaliers, the fortune of war was now on this side and now on that. But the Puritans triumphed in the end. In the words of Churchill,

the middle class, being more solid for Parliament, had beaten the aristocracy and gentry, who were divided. The new money-power of the City had beaten the old loyalties. The townsfolk had mastered the countryside. What would some day be the 'Chapel' had beaten the Church (*ibid.*, Vol. 2, p. 208).

The victory of the Puritans in the civil war was as little of 'historical necessity' as that of the Royalists had been. But in the end, the way in which it happened was – as it always is – even more significant than the name of the victorious side. One of the warlords of the parliamentary troops, later their supreme chief, obtained an enormous personal power, not only over his own 'Ironsides' but gradually over the whole parliamentary party of the conflict. This man, Oliver Cromwell, was, after the execution of the king, Charles I, in 1649, something of the first totalitarian dictator of the modern type in the world. His régime, usually taken to include the years 1653 to his death in 1658, because he was acclaimed 'Lord-Protector' in 1653, indeed comes closer to outright totalitarianism than anything that either preceded or followed it in the history of the Anglo-Saxon nations. Yet it was not quite totalitarianism comparable to that of Hitler or Stalin, or Castro or Mengistu, not to speak of the totalitarian régimes such as those of Idi Amin or Pol Pot. We can entirely agree with Churchill, who lists the points of difference as follows:

Nevertheless the dictatorship of Cromwell differed in many ways from modern patterns. Although the Press was gagged and the Royalists ill-used, although judges were intimidated and local privileges curtailed, there was always an effective vocal opposition, led by convinced Republicans. There was no attempt to make a party around the personality of the Dictator, still less to make a party state.... Few people were put to death for political crimes, and no one was cast into indefinite bondage without trial (*ibid.*, Vol. 2, p. 250)

But it is worth noting that even in England, as anywhere else, political fanaticism of a religious type, whether in pursuit of Christian purity or socialism, inevitably led to a concentration of an immense power in the hands of a single man, as soon as the fanatical faction obtained power. And even though not all the features of modern totalitarianism were present, a great deal of them were. 'The maintenance of all privilege and authority in their own hands at home and a policy of aggression and conquest abroad absorbed the main energies of Cromwell and his Council', Churchill remarks (*ibid.*, Vol. 2, p. 247). On the ideological front, the observance of Puritan ethics was controlled by similar means as a 'counter-revolutionary action' in more secular types of totalitarianism:

Soldiers were sent round London on Christmas Day before dinner-time to enter private houses without warrants and seize meat cooking in all kitchens and ovens. Everywhere was prying and spying.... All over the country, the May-poles were hewn down, lest old village dances around them should lead to immorality or at least to levity... sumptuary laws sought to remove all ornaments from male and female attire (Churchill, *op. cit.*, Vol. 2, p. 248).

Hypocrisy was an inevitable consequence and, in promoting one's career, speeches garnished with Old Testament phrases were useful, just as some centuries later Marxist and 'peace' phraseology is in some other parts of the world. But it soon transpired that England, with her battered but still vital democratic institutions, was something apart. The quasi-totalitarian Cromwellian régime in England became one of the shortest-lived of all approaches to totalitarianism.

To the mass of the nation, [Churchill writes, *ibid.*, Vol. 2, p. 149] the rule of Cromwell manifested itself in the form of numberless and miserable petty tyrannies, and thus became hated as no Government has ever been hated in England before or since.

16.3. The Rapid Increase of Self-steering: The Emergence of Two-party Parliamentarism, Industrial Revolution, and the Anglo-Saxon Freedom of Speech

On the death of Cromwell in 1658, the Puritan ideology had exhausted much of its popular appeal, and the parliamentary tide brought the opposition to power. The Cavalier parliament (1660–1678), the longest in the history of England, saw to it that no military concentration of power was possible, that the Church of England was now 'by law established' and the only one to be taught in schools and universities (1662), and that the parliamentary control of finance was set up, thus bringing the responsibility of the cabinet to parliament closer. So, after all, the Puritan revolution proved to be rather a violent excess in a parliamentary movement, where the two opposite political forces were each in turn able to govern the country. Indeed the Cavalier and Roundhead traditions led to the two political groupings that, until 1914, were to dominate the British political scene, viz. the Tories and Whigs respectively. This is the first historical example of the conservative/radical political division that, according to the hierarchy law, must exist in any favourably developing society based on technological progress (cf. Fig. 83). The words 'Tory' and 'Whig' were used as political labels for the first time during 1679 and 1680, and henceforth politics in England were exercised in secular rather than religious terms.

After the formative period, say 1640–1680, of the opposite parliamentary forces in England, and after the last Catholic king of England, James II Stuart (1685–1688) had made his hopeless attempt to bring the English governmental system down to the two-class model of continental feudalism, the establishment in February 1689 of modern parliamentarism in England was almost just a formal final act. Some more or less melodramatic histrionics – like Prince William's bloodless war – were involved in those three months of 'revolution' between November 1688 and February 1689, as always in connection with political happenings of some consequence, but on the whole it all went rather smoothly.

Ever since 1689, accordingly, Britain has had a parliamentary system in the modern sense, in which the highest organ of executive power, the cabinet of ministers, is responsible to parliament. It is not by chance that by that time the two opposite political forces, according to the hierarchy law characteristic to the developmental phase of human society, had taken shape in England, and were ready to compete for power on peaceful terms. The alternation of power between the radical Whigs and the conservative Tories, in conformity with the general picture depicted in Fig. 83, launched the country into an unforeseen economic growth sometimes called primary capital formation, sometimes *industrial revolution* (see Fig. 89).

Quite in accordance with the hierarchy law, the relaxation of the old hierarchy of power, involved in the establishment of the new parliamentary system, triggered a tremendous amount of activity – both economic and intellectual – among the social strata that were ready to take part in social affairs but had been thwarted in it by the old system. The accumulation of capital that had started with the wool trade back in the sixteenth century was pushed on by new means. The merchants and land-owners that now dominated the House of Commons, the stronghold of power, used appropriate legislative measures to enforce yeomen to sell their land on terms favourable to the new owners – those very same merchants and land-owners. The capital obtained was invested in trade and industrial enterprises, which since about the mid-eighteenth century were greatly encouraged by many technological innovations: the spinning-jenny, steam machines, and others.

This part of English history is very well known, not least because of Marx's efforts to represent it as a general model of social development in terms of class conflicts. And we must do him justice in so far that a new leading class, big bourgeoisie, was indeed born as if 'from the womb' of the preceding society, and that this class in turn was to suppress and exploit the class of industrial workers, simultaneously formed by poor town people and by the yeomen who had been compelled to sell their land and move into cities, now working for the bourgeoisie in their factories.

But first the advantages of the new situation. The period of primary capital formation in England, from the bourgeois takeover in 1689 until about the mid-nineteenth century, made the country number one in the world, both in wealth and power. Britain became the first industrial society. The average standard of living improved greatly, and the population more than doubled during the eighteenth century. Indeed, the population of England and Wales, taken together, increased from 5,134,516 in 1700 to 6,039,684 in 1750 and to over 13 million in 1800. Unlike the case in Russia, where the growth of the population was due to the expansion of the cultivated area, that growth in Britain must be ascribed to improved technology and markets, that is, in cybernetic terms, to the increased regulatory ability and decreased uncertainty of the economic system.

The achievements of the period were not confined to economic success. One has to point out that the Habeas Corpus Act in 1679, passed just on the threshold of the bourgeois era, reinforced the right of every citizen, if sued for something, to have a public trial before a jury, and not to be arbitrarily arrested. The rise to power of the bourgeoisie in 1689 was accompanied by the Declaration of Rights, which once again confirmed the human rights claimed in Magna Charta. The censorship of books, pamphlets, and journals prior to their publication was abolished in 1679, but still this did not bear on parliamentary speeches. The ban on reporting them in public was lifted first in 1763, in connection with the famous Wilkes affair, when John Wilkes, the editor of the journal named *North Briton*, deliberately violated the law, and was temporarily detained for it, until a popular protest forced Parliament to change the law. As a consequence the free press in the Western sense, entitled to criticize power-holders without being punished for it, was born. Between 1782 and 1789 *The Times*, *The Morning Chronicle*, and *The Morning Post* started their careers as prominent representatives of the 'third state power'.

In this way an important, qualitatively new stage was reached in human emancipation. Freedom of speech in its modern Western sense, was launched on its stony path. Even today this step is still awaited in the greater part of the world outside Western Europe and North America; that is, outside that Atlantic family of societies, where the Anglo-Saxons still play the leading role, both economically and intellectually. Well, these have been the advantages of 'capitalism'.

Now to the losses brought about by it, and emphasized by Marx – and depicted so vividly by Dickens. Most yeomen lost their land and were transformed into industrial workers living in towns and in the slums of great cities. What made their fate heavier than it had otherwise been was the simultaneous loss of political rights. The right to vote in Britain was traditionally associated with landed property. In 1654, for instance, the limit was set at the ownership of an estate with a capital value of two hundred pounds. Yeomen who had sold their estate were at the same time disqualified from voting in general elections. This is why they, in their new positions as industrial workers, became free game for exploiters: they were bereft of their political rights, and had no means to defend themselves against bourgeois legislation, politically, and against the exploiters, economically. So here Marx was right: a social class in control of the state, rightly identified as the bourgeoisie, *was* exploiting another class or social stratum, viz. the former yeomen turned into industrial workers, who had lost their political rights because of their economic position. And Dickens was right in displaying in his novels those startling scenes where miserable, decent people were shown to be in the claws of greedy usurers. By the way, Dickens still is the major source of popular information about capitalism used in the Soviet Union.

But Marx was badly mistaken about the cause and remedy of the situation that British workers found themselves in. The culprit was not the Anglo-Saxon political system and parliamentarism in its entirety, but a hole in it, exploited by the bourgeoisie: the traditional link between suffrage and landed property in England. After all, parliamentarism in England was a much older system than was industrial work: hence the hole in legislation. And the remedy was not the annihilation of the bourgeois class, and its replacement by the working class in control of the state, through a revolution. This end was never followed by the majority of English workers, to the obvious disappointment of Marx. What the workers' movement in Britain really was after was the acquirement for workers of full political rights *within* the existing parliamentary system, and an economic status allowing a decent life as a citizen equal to others. Dickens, accused by Marxists for a toothless moralism, was closer to the truth than Marx: it *was* a moral issue, a case of raising people to rectify what was wrong in the existing law – not one of abolishing that law.

Early milestones of the workers' movement were the People's Charter in 1831, demanding general suffrage and secret general elections, and the establishment in 1884 of the Fabian Society. If, as far as general suffrage goes, the British political system was slow to catch up with the democratic achievements of the French Revolution, the English trade union movement showed the way to other peoples. In the end, after the First World War, the Labour Party, which had been built up as the political arm of the unions, was ready to take on the role of radical political force in parliament. Instead of launching a revolution it became an established part of the Anglo-Saxon political system, thus attesting to its strength.

Like every earlier power system, the European trade union movement finally developed its own bureaucracy and hierarchy. In due course it evoked a new type of radicalism to democratize the union movement and to reduce its centralized power. This is the normal course of social history: what was once radical sooner or later becomes an obstacle to further development, and is replaced at the forefront of social development by new radical forces.

But the forefront of the economic and social progress of the Anglo-Saxon social system, and of the world, had by the end of the nineteenth century already moved over the ocean, to the United States of America. On 4 July 1776 the American colonies had declared their independence from Britain, and soon, after the victorious fight for independence, the Congress was divided into the Senate and the House of Representatives thus following the British model of a two-chamber parliament. As in Britain, two opposite political parties were formed, thus extending the Anglo-Saxon political system to the emerging new world power. Two-party parliamentarism has ever since proved its superiority to other political systems by way of the enormous growth of individual freedom (self-steering) and wealth in the United States of America, which in the twentieth century became the engine of economic and intellectual development in the world, and has thus been envied and resented by others like few superpowers before.

CHAPTER 17

The Full-scale Social Revolutions and Their After-effects on the European Continent

We have seen how the Anglo-Saxon social system was capable of social development, where power hierarchy was relaxed in pace with the simultaneous improvements in the economic system, due to technological progress and resilient markets. Social progress as a whole had its ups and downs, but the growth of self-steering never soared into the unrealistic heights where the unity or survival of the nation is endangered, nor did it plunge permanently into the area of unnecessary hierarchy (see Fig. 89).

On the continent of Europe things went an entirely different way. Because of the more rigid hierarchical structure of its feudal social systems, some of these societies were led into the abyss of full-scale social revolution, with its detrimental after-effects. The severity of these effects depended on the depth and rigidity of the preceding hierarchy of power, as predicted by the qualitative theory of social development introduced in Part 3.

17.1. Delayed Social Development in France and the Instability Following the Great Revolution

Feudal society in France lasted three hundred years longer than in England, until the end of the eighteenth century, and even longer in Germany. Besides, continental feudalism retained its rigid two-class structure until the violent end of it, being incapable of developing a clear-cut middle class as long as society remained feudal.

Harder feudalism, harder serfdom. The rebellion of French peasants in 1358, known as La Jacquerie, never had the favourable consequences that its English counterpart, Wat Tyler's uprising in 1381, had on the social position of peasants in England. And the Huguenots, as the Protestants in France were called, never had the same possibilities in their fight against the Catholic Church as their Puritan brothers across the Channel. Nevertheless, in about the mid sixteenth century, a quarter of the French population is estimated to have been Huguenots, among them a great number of enlightened gentry and bourgeoisie. But the religious wars that started off in 1562 and ended with the Edict of Nantes in 1598 made it quite plain for the Huguenots that, even though they were at last given the right to pursue their religion, they would never succeed in persuading the majority of the French people into converting to their faith. Finally, in 1685, after ruthless persecutions by Louis XIV of the two million Huguenots still in the country, even the Edict of Nantes was abrogated, as having by now become 'needless'. Indeed, by that time, most of the Huguenots had either joined the Catholic Church or fled over to the Netherlands, England, or Switzerland to escape the persecutors.

We know that in the sixteenth century, during the Huguenot wars, France as well as Western Europe generally had already crossed over the limit of 1:5 yield ratio, after which a city population that is, bourgeoisie, engaged in trade and administration rather than agriculture, becomes possible. But the bourgeoisie in France, for reasons given above, was too weak to produce a successful Protestant movement comparable to that of the Puritans in England at about the same time. And towards the end of seventeenth century, as was mentioned above, all the Protestant resistance in France was crushed.

This is why in the eighteenth century, when productivity and technology took a great leap in Western Europe (see Fig. 85), the ensuing opposition to the old hierarchy and to the Catholic Church as its paramount manifestation in France, unlike in England, leaned on a secular philosophy, not on a religious protest. It took the shape of a general anti-religious attitude disseminated by eminent philosophical writers such as Voltaire, Rousseau, and Montesquieu.

In spite of the improving economic prospects, and of the inevitable demands for greater freedom and democracy that go with them, hierarchy even deepened to autocracy in France. As a consequence, to put it in a cybernetic language, the curve of actual self-steering dropped somewhat, and landed under the curve of the lowest acceptable self-steering, mainly because of the rise of the latter (see Fig. 90). So the combined effect of an improving economy and persisting social hierarchy produced a genuine revolutionary situation towards the end of the eighteenth century in France.

The revolution was a full-scale one, and went through the four typical phases depicted for the French Revolution in Fig. 91. The unnecessary hierarchy of the *ancien régime* (phase 1) was followed by an outburst of revolutionary frenzy in 1789 (phase 2), which soon took the curve of actual self-steering over its acceptable upper limit and led to the self-destroying anarchy of the *terror régime* in 1792–1794 (phase 3). The starting point was the rising to power of the Jacobins on 10 August 1792, and the bloodbath was over with the execution of their leader, Robespierre, on 28 July 1794. What followed was Napoleon and national restoration accompanied with the inevitable build-up of compensative hierarchy (phase 4).

By way of compensative hierarchy it was necessary to build up, after the Great Revolution, that comprehensive bureaucratic state machinery* that still today is a hallmark – and nuisance – of the French administrative system.

---

*Alexis de Tocqueville, in his *The Ancien Régime and the Revolution* (1856), indicated the role of the Jacobin philosophy in the build-up of compensative bureaucracy.

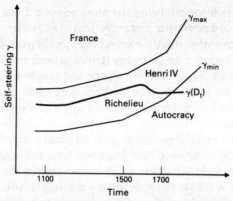

FIG. 90. The emergence in France of a revolutionary situation in the eighteenth century, as a joint effect of improving economy and – because of autocracy – persisting social hierarchy. (Warning: the picture serves only to illustrate the concepts of the qualitative theory of self-steering. The details in the curve of self-steering are just an educated guess, not based on any measurements!–A.A.)

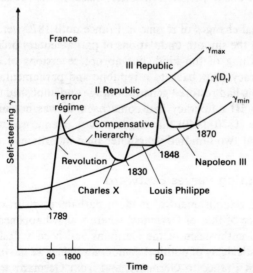

FIG. 91. The abrupt fluctuations in France of self-steering in the wake of the Great Revolution. Note the dilated time scale when compared with Figs 85, 88, 89, and 90 (all of which employed a common scale), which may make the fluctuations seem tamer than they really were – with their repeated approaches to a revolutionary situation in 1830, 1848, and 1870 (the Commune). (Warning: the picture serves only to illustrate the concepts of the qualitative theory of self-steering. The details in the curve of self-steering are just an educated guess, not based on any measurements!–A.A.)

However, French society being the most advanced one on the continent of Europe, the compensative hierarchy that was set up there never quite reached the degree of a strictly hierarchical, totalitarian state as happened in more eastern societies, in Germany (Hitler period) and Russia (Stalinism). The two years of the Jacobin terror régime had taught the quick Frenchmen the same lesson for which the steadier Englishmen had needed the five years of Cromwell's Puritan government: never again a régime of fanatical, self-satisfied ideologues.

But the Great French Revolution is an excellent example of the unsteady, fluctuating course of development launched by a full-scale revolution. The building-up of compensative hierarchy culminated in the truly reactionary reign of Charles X (1824–1830), ended by the July 'revolution' in 1830, not really a proper social revolution, but just an installment of a new king by rather illegitimate means. But Louis Philippe, the new king, was not much better, and the reactionary period was not at an end until after the February 'revolution' in 1848, which established the Second Republic, that of the young socialist Louis Blanc and the poet Alphonse de Lamartine. This was to last only four years, after which a new wave of nationalist enthusiasm helped Napoleon III (1852–1870) to mount the throne. A new reactionary period followed, which was to perish in the bitter defeat in the Franco-German war in 1870–1871.

The periodical changes of régime in France until 1870 were very different by nature from the smooth undulations of parliamentary politics in Britain. They were nothing of the kind, but abrupt reversions of direction, now towards autocracy, now back to a republic and parliamentary power. The Second Republic had restored general suffrage and abolished censorship, but with Napoleon III autocracy had come back. It was not until during the Third Republic (1870–1940) that modern parliamentarism was established in France, about two hundred years later than in Britain.

17.2. Revolution Passes Eastward

Yet the French establishment of modern parliamentarism was seventy-five years in advance of that of Germany, where it first happened in 1945, after that nationalist enthusiasm of the Germans had been at last turned down. The characteristic delay of political democracy, when we move from England to France, from France to Germany, and from Germany to East Europe, where it is still waiting for its turn, is an indication of the more and more hierarchical types of society that we are encountering on such a journey. Social hierarchy and immobility was much stiffer in Prussian feudalism than it ever was in France. That is why the liberal ideas coming from the West after the Great French Revolution, though they did enthuse academic people in the south of Germany, and even in Hamburg (Heine!) or Cologne, never shook the Prussian régime.

United for the first time as a nation state by the war effort in 1870–1871 Germany remained an essentially feudal society, where the general line was asserted by the continuation of Bismark's government (1862–1890) over the reigns of several sovereigns. The first serious threat to German feudalism was posed by the birth of the German workers' movement towards the end of the nineteenth century, after modern manufacture had got started in Germany about in the middle of the century. Defeat in the First World War encouraged communist uprisings in 1919 in Berlin and Munich, both of which stood for a rather small minority opinion and were soon defeated. The mainstream of German politics still ran in the nationalistic channel, and the first attempt at parliamentarism (1919–1933) was doomed to perish. It would be an interesting topic of academic discussion to argue whether Hitler's régime was really absolutism rather than outright totalitarianism – it never had quite the time to pervade all sectors of society. The violent terrorism that followed, after the Second World War, the violent years of nationalism in Germany, Italy, and Japan, indicates that it is not possible to cut off an entire nation from a long hierarchical tradition overnight.

Still better confirmation of the law of historical continuity is offered by the happenings, during the present century and the preceding one, further east. At the end of the nineteenth century, especially during the 1890s, industrialization began in Russia. The resulting boom was vigorous enough to produce rates of growth in the Russian economy which have never since been reached in Russia or in the Soviet Union. It also took the country in 1913 to the fifth place in industrial production in the world, after the U.S.A., Germany, England, and France. But in Russia the tsarist régime, originated in absolutism and verging on outright totalitarianism, was by its hierarchical structure again much more rigid than French or even German feudalism had ever been. So once again, as during the eighteenth century in France, economic growth combined with persisting, immobile social hierarchy created a genuine revolutionary situation (see Figs 88 and 92).

Again the four characteristic phases of a full-scale social revolution are evident. The appearance of unnecessary hierarchy (phase 1), the period of revolutionary enthusiasm (phase 2), including the two revolutions in 1917 and leading to the anarchy and excesses of the first years of the Soviet state (phase 3). When exactly Stalinism began – whether from the defeat of the 'Left Opposition' (Trotsky, Kamenev, Zinoviev) in 1929 or the 'Right Opposition' (Bukharin, Rykov) in 1930 – is irrelevant. At any rate it was the core of the compensative build-up of power in the Soviet Union (phase 4), and at the same time the final climax of totalitarian development in Russia.

The plunge that the curve of actual self-steering (see Fig. 92) took during Stalinism was so deep that most of the post-Stalin period in the Soviet Union is best characterized as a slow, gradual recovery, sometimes quicker – as in Khrushchev's era – sometimes slowing down again. The society is still extremely rigid and hierarchical, but compared with the conditions with

FIG. 92. The Russian revolution and its aftermath, Stalinism, as a continuation of the hierarchial trend illustrated in Fig. 88. Note the dilated time scale. (Of course, the warning given in connection with the earlier historical pictures is again relevant.)

which Stalin started there is one big difference: the country is steadily on a path of a slow but continuous economic advance, and accordingly has reached the period of development. The country, according to this judgement, is no longer developing towards totalitarianism and greater hierarchy, but out of it, towards lesser hierarchy.

Yet the slowly relaxing hierarchy in the Soviet governmental system still preserves many features typical of English society in the early periods of feudalism. The highest legislative body, the Supreme Soviet, very much resembles the king's parliament in England toward the end of the thirteenth century, under Edward I: it assembles for only a few weeks at a time, to accept motions mostly by a unanimous vote. The Central Committee of the Communist Party in turn has an undeniable air of being the House of Peers, as it is manned almost exclusively by men from higher echelons of the Party. Such a system of government leaves practically all power in the hands of a few top Party leaders, with very little change in the leadership over considerable periods of time. A further hierarchical element is built in by the *nomenklatura*,* which fixes the hierarchy within the Party, and determines the economic and other privileges strictly according to one's position in the Party. All these feudal elements of Soviet society can be seen as historically conditioned by the Russian tradition of hierarchy.

The feudal features of the Soviet system are not restricted to the internal structure of society, but appear in the mutual relations of the states within

* Cf. Michael Voslensky, *Nomenklatura: The Soviet Ruling Class* (Doubleday, 1985).

the Soviet bloc as well. What is required by Moscow of the Eastern-European states belonging to the Soviet bloc remembers a vassalage. The *feudal loyalty* obliges the vassal state to obey the foreign policy of its Moscow lord, and to participate in the common defence of the bloc, while leaving each vassal – each Party – in its own country to mind its own internal business. Only when an internal rebellion in some of the vassal states threatens the bond of loyalty is the bloc as a whole entitled to interfere, like the vassals of the same king on each other's behalf.

What seems to be a particularly Marxist ingredient in the Soviet system is the tough persistence of the Gulag and related systems of punishment. Obviously their very existence tells something of the embarrassment of the Marxist leadership in the face of political opposition, which, according to the doctrine, should not be there at all after the last, 'final' revolution. Ideologically, Marxism–Leninism does not have a place for political opposition, except in the 'dustbin of history'.

In view of the law of requisite hierarchy, revolution is never the best solution to a crisis. Instead of full-scale revolutions a gradual dismantling of hierarchy is characteristic of a favourably developing society. In such a society there must also be room for two opposite political forces to function, and to have power as necessary. This is the picture of optimal social development that emerges from the qualitative theory of self-steering given in Part 3.

17.3. The Dominating Developmental Trends in the Contemporary World and the Birth of the European 'Grey Zone'

What is the picture of the real developmental trends in the world of our age as obtained from our qualitative theory of self-steering?

We have seen that during several hundred, and even a thousand, years of historical development in opposite directions, modern Western and Eastern types of societies have separated from each other. While the former entered the orbit of social development in an early stage of this history, the latter have entered it much later, and still exhibit a relative underdevelopment in comparison with the former ones. But modern weaponry, nuclear weapons in particular, has made it possible for the East and the West to appear in the international arena as equal military and political powers. This fact has far-reaching social consequences.

It is for the first time in human history that development and relative underdevelopment can appear as equal influences in the world. In the earlier history of mankind, for the last time in the colonial period, the powers exerting the greatest influence in the world were also among the most developed ones, socially and economically. This is now radically changed.

A visible effect of the new balance of power is the reduced significance in world affairs of Europe, the old centre of social and cultural development, which has suddenly become a no man's land between the two dominating powers. But there is more to it.

While in the earlier multi-centred balance of power some of the most developed countries (especially England) exported their developed social and political institutions to their colonies, the new balance of influence also makes it possible to export some features of underdevelopment, such as a hierarchical, bureaucratic governmental system and, to the Third World especially, even a one-party power.

A complicated case of Eastern influence is the birth of a 'Grey Zone' composed of West Germany, the Scandinavian countries, and some other countries between the East and West in Europe. Although formally belonging to the 'West' the Grey-Zone countries tend towards political and ideological neutrality. But neutrality between a more developed, self-steering type of Western society and the more hierarchical Eastern society is prone to have detrimental effects on the development of Grey-Zone societies themselves, sooner or later: increased bureaucratic structures, reduced individual initiative, shrinking intellectual liberty. In due time it can be expected to show in economic and other social development. Hence the prediction, illustrated in Fig. 93, that the Grey Zone will gradually approach the slower tempo of development characteristic of the hierarchical Soviet society.

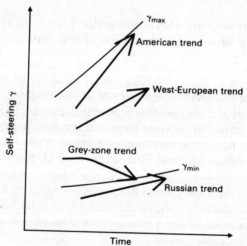

FIG. 93. The contemporary trends of self-steering in Europe and North America:

the U.S. trend leading the development along the utmost limit of self-steering,
the Western-European trend of slower development,
the Russian trend of sluggishly getting rid of unnecessary hierarchy, and
the Grey-Zone trend of gradual recession toward the Russian one.

(Note: this is an educated guess on trends in self-steering, which is related but *not* identical with economic growth!)

But the Grey-Zone effect is likely to be felt, to a lesser extent, elsewhere in Western Europe as well. Hence the prediction (cf. Fig. 93) that Western Europe too will be somewhat thwarted in development when compared with the highly self-steering North America (and possibly the rising societies of the Pacific, where the Grey-Zone effect is not experienced).

As to the world outside Europe and North America, it is first to be noted that most Third-World societies of our time are not, in the sense of the law of requisite hierarchy, underdeveloped to the same extent as were the ancient archaic societies. Unlike the archaic communities they live in a modern world, and are necessarily influenced by the technical and social progress made by the developed countries. But in the Third World, because of the support they get from their own hierarchical traditions, the bureaucratic structures and one-party system of the Marxist state have a greater chance to block development than they have in Europe. Furthermore, many Third-World countries are facing the political choice between greater self-steering and greater hierarchy in the circmstances, where the ordinary citizens of these countries have very little say. It follows that a hierarchical order, bound to foster prolonged underdevelopment, can be forced upon them.

The so-called Third World, however, is not a homogeneous group of societies. It includes a huge amount of various cultural and social traditions, whose comparative analysis in view of the theory of Part 3 is more difficult than that of European societies because of the lack of available common standards, such as the statistics of yield ratios in Europe. Hence the illustrations of the theory given here have been confined to European societies.

Are we in for a Deceleration or even Stagnation?

The above picture of developmental trends in technologically advanced countries focused on the expected relative trends of these countries in proportion to each other, in the near future. An impressive overall view of global economic development in the foreseeable future was recently suggested by one of the pioneers of the dynamic input–output models, the Hungarian mathematical economist Andrew Bródy. In his remarkable book entitled *Slowdown: Global Economic Maladies* (Bródy, 1985) he develops a pessimistic argument, far more convincing than the usual catastrophic sensationalism that is represented today by so many futurologists and end-of-the-world ideologues. Bródy argues that we are facing a deceleration, or even stagnation, of global economy because of the neglect of science and the steady lowering of the social position of scientists that has been going on for some time.

Statistically the weakening of the position of scientists can be observed, say, from the decrease of their relative salaries, and in some cases even absolute real salaries, since the 1930s and even before that. The pace of technological innovations has decelerated, and the resources of technical innovation created by science in the earlier periods, when the work of scientists was better appreciated and remunerated, has now been used up. As a consequence, the growth of the global economy has been decelerating simultaneously with decreasing productivity since about the beginning of the 1970s. It is the gloomy but credible prediction of Bródy that, because of the long period of gestation of the products of science and technology, the global slowdown could not be turned to vigorous growth until some time beyond the turn of millenium, even if the improvement of the position of science were to start now (of which, I am afraid, there is little visible sign in the science policies of the advanced countries).

There is reason to return to Bródy's interesting argument at the end of Part 5 of this book, after having discussed the formation of knowledge, and some major obstacles to it, in that part.

Part 5

The Problem of the Origin: The Self-steering and Steered-from-outside Layers of Consciousness?

(A Thought Experiment)

CHAPTER 18

The Self-steering and Steering-from-outside of Man

We shall now continue the study of human self-steering from where it was left in Part 3, and ask:

Where in human consciousness is the source of self-steering? What kind of dynamical properties of consciousness can account for human self-steering?

Obviously, we are here moving on much more speculative ground than Part 3, where we could lean on mathematical dynamics and on cybernetic entropy laws. Yet the questions are so interesting and important that even educated guesses may be welcome by way of working hypotheses. Indeed, without any suggestions for answering the above questions, a big black hole would remain in the middle of the theory of self-steering: we must be able to offer some, however preliminary, ideas for filling that gap. Still better, if they can be given support by some well-confirmed scientific theory.

The properties we are looking for have to do with dynamical processes of human consciousness. There are not many well-established theories to cope with such processes. A recent attempt at a theory of (among other things) human consciousness is *autopoiesis* (see Varela, 1979; Zeleny, 1981a). However, it is only seemingly a dynamical theory. Actually the 'processes' are there introduced *ad hoc*, as ready-made external elements, not derived from any genuinely dynamical theory.

A dynamical theory has to be mathematical. For a mathematician it is fairly clear that the best-confirmed mathematical theory of human consciousness today is the *theory of subjective probability*. The Filter Theorem of that theory will be the foundation of our investigation. But before we can proceed to discussing it we must have some acquaintance with some elementary concepts of the theory of subjective probability.

18.1. The Mathematical Concepts of *a priori* and *a posteriori* Subjective Probability

(1) The measure of probability

We can think of the amount 'one' of 'probability mass' spread, possibly in an uneven way but smoothly, on the set X of the thinkable world-states. Let

$P(X) = 1$ be the amount of the total mass, and $P(A)$ the part of it spread over a subset $A \subset X$. The subset A, on the other hand, stands for a certain state of affairs (see Section 9.2). We shall call $P(A)$ the probability of this state of affairs. The remaining mass $1 - P(A)$ is spread over the complement set $\mathscr{C}(A)$. Thus we have:

$$0 < P(A) < 1, \quad P(\mathscr{C}(A)) = 1 - P(A). \tag{18.1}$$

Obviously, for two subsets A and B having no common elements we have

$$P(A + B) = P(A) + P(B). \tag{18.2}$$

Here $A + B$ is the set-theoretical sum of the sets A and B. In fact what we have said above is nothing but the definition of the measure P of probability, in an axiomatic form given as follows.

Definition. A *measure of probability* P is a function from 2^X to the closed interval $[0,1]$ of real numbers, such that

(I) $P(X) = 1$;
(II) $0 \leqslant P(A) \leqslant 1$ for all $A \subset X$, i.e. for all $A \in 2^X$; and
(III) $P(A + B) = P(A) + P(B)$.

(2) The acts of observation

Let us now consider an actor facing the world whose thinkable states are the elements of X. Let us single out one of these elements, $\hat{x}$, as representing the true state of the world at a certain moment, and let E be a state of affairs that is true on that moment, so that (see Section 9.2)

$$\hat{x} \in E. \tag{18.3}$$

We say that the actor 'observes' the 'fact' E in an *act of observation*

$$f_z : X \to 2^X, \tag{18.4}$$

such that

$$f_z(\hat{x}) = E. \tag{18.5}$$

Thus, an act of observation does not change the world, but only picks up the fact E, or *makes the observation E*.

Each act of observation f_z is associated with a mechanism of observation, mathematically represented by an *observable z*, which is function from the set X of world states to the consciousness Z of the actor,

$$z : X \to Z, \tag{18.6}$$

such that

$$E = \{x; z(x) = z_E\} \Leftrightarrow z_E \quad (1\text{--}1). \tag{18.7}$$

Here z_E is the *image* standing for the fact E in the actor's consciousness, and there is, as expressed by the formula (18.7), a one-to-one correspondence between the facts E about the world and their images z_E in the actor's consciousness. The mechanism of observation, or the observable z, accordingly conveys the image z_E of the fact E into the consciousness of the actor.

Different actors observing the same world have different consciousnesses, say Z, Z', Z'', etc. and thus different images z_E, z_E', z_E'', etc. of the same fact E. This is precisely why the function z is needed: while the true states of affairs E belong to the objective world, their images z_E belong to a subjective world, viz. the consciousness of the actor in question, and the function z conveys the messages from the former to the latter.

(3) The effect of observation of a fact on the subjective probabilities of the actor

Let us then assume that we are given a measure of probability P that stands for the *cognitive beliefs* of a conscious actor. In other words, $P(A)$ for any $A \subset X$ gives the probability of the truth of the state of affairs A, as judged by the actor. It is called the *subjective probability* of A for the actor in question.

It is the basic idea of the theory of subjective probability that the knowledge of each fact E changes the cognitive beliefs of that actor in a lawful way. To find out the law, two axioms are set forth:

(i)   The actor already has, prior to any observed facts being in his possession, a pattern of *a priori beliefs* represented by a measure of probability P standing for his *a priori subjective probabilities*.

(ii)   The observation of a fact E changes his cognitive beliefs in a way, say $P(A) \rightarrow P(A/E)$, that makes the state of affairs E 'a posteriori' true, i.e.

$$P(E/E) = 1. \tag{18.8}$$

The changed cognitive beliefs of the actor are called his *a posteriori beliefs*, and they are mathematically represented by his *a posteriori subjective probabilities* $P(A/E)$.

The simplest law of transformation that on the condition (i) makes the equation (18.8) true is the Bayesian formula

$$P(A/E) = P(A \cap E)/P(E) \quad \text{for all } A \subset X. \tag{18.9}$$

It can also be considered as a generalization of the *a priori* probability $P(A)$, which obeys

$$P(A) = P(A \cap X)/P(X) = P(A/X). \tag{18.10}$$

Given a piece of 'empirical knowledge', i.e. a sequence of facts $E_1, E_2, \ldots, E_n$ in possession of the actor, how have the cognitive beliefs of the actor changed in this case? By a direct application of the Bayesian formula, we get first

$$P(A/E_1) = P(A \cap E_1)/P(E_1), \tag{18.11}$$

and then

$$P(A/E_1 E_2) = P(A \cap E_1 \cap E_2)/P(E_1 \cap E_2) \quad \text{for } E_1 \cap E_2 \neq \varnothing. \tag{18.12}$$

But if

$$E_1 \cap E_2 = \varnothing, \tag{18.13}$$

we obviously must conclude that between the observations E_1 and E_2 the true state of the world has changed, i.e. that we have

$$\hat{x}_1 \rightarrow \hat{x}_2, \quad \hat{x}_1 \in E_1 \text{ and } \hat{x}_2 \in E_2, \quad \text{with } \hat{x}_1 \neq \hat{x}_2. \tag{18.14}$$

Here $\hat{x}_1$ is the true world-state at the moment of the observation of E_1, and $\hat{x}_2$ is that at the moment of the observation of E_2. Hence, we go on as if E_2 were the first observation made, by writing

$$P(A/E_1 E_2) = P(A/E_2) \quad \text{for } E_1 \cap E_2 = \varnothing. \tag{18.15}$$

By continuing in this way, using either rule (18.12) or rule (18.15), we can construct the *a posteriori* subjective probabilities

$$P(A/E_1 E_2 \ldots E_n) \quad \text{for any } A \subset X, \tag{18.16}$$

and accordingly the *a posteriori* cognitive beliefs of the actor who is in possession of the empirical knowledge $E_1, E_2, \ldots, E_n$.

Numerical Example. Suppose that we are given the following distribution of probability mass: $X = a + b + c + d$ and

$$P(a) = 1/2, \quad P(b) = 1/4, \quad P(c) = P(d) = 1/8. \tag{18.17}$$

If we are interested in the cognitive beliefs, of an actor having this distribution of *a priori* probability mass, in the state of affairs

$$A = a + b \quad \text{and} \quad B = a + c, \tag{18.18}$$

we first calculate, using axiom (III):

$$P(A) = P(a) + P(b) = 3/4, \quad P(B) = P(a) + P(c) = 5/8. \tag{18.19}$$

Thus, the actor has the *a priori* belief that the state of affairs A is slightly more probable than B. Suppose then that the actor has made the successive observations

$$E_1 = b + c \quad \text{and} \quad E_2 = c, \tag{18.20}$$

How have his cognitive beliefs changed? By applying the Bayesian formula we first get:

$$P(A/E_1) = P(A \cap E_1)/P(E_1) = P(b)/P(b + c) = (1/4)/(3/8) = 2/3,$$

(18.21)

$$P(B/E_2) = P(B \cap E_1)/P(E_1) = P(c)/P(b + c) = (1/8)/(3/8) = 1/3.$$

(18.22)

Thus, we have

$$P(A/E_1) > P(B/E_1),$$

(18.23)

and the first observation E_1 accordingly has confirmed the actor's *a priori* belief in A as the more probably true state of affairs of the two. But then we calculate, following the judgement of the author after the observation E_2:

$$P(A/E_1 E_2) = P(A \cap E_1 \cap E_2)/P(E_1 \cap E_2) = 0/(1/8) = 0, \quad (18.24)$$

and

$$P(B/E_1 E_2) = P(B \cap E_1 \cap E_2)/P(E_1 \cap E_2) = P(c)/P(c) = 1. \quad (18.25)$$

Thus, the second observation showed to the actor that he had been wrong: the state of affairs B was after all true, while A was untrue.

Note 1: What are the 'Observations'? It is obvious from what has been said above that in the theory of subjective probability it is irrelevant how the actor has come to the possession of an empirical fact E, whether by using his own sense organs or by being credibly told of the observation made by another actor. It is implied that the Bayesian mechanism of formation of cognitive beliefs is working in both of these cases. Therefore, it is also possible that the actor may be afterwards told of the falsity of some of his supposed 'facts', in which case his cognitive beliefs are changed accordingly. What is essential is that an observation E is always a *singular* state of affairs, i.e. one that is related to a given singular point of time and space. We shall return to this issue later.

Note 2: Whence Come the 'a priori beliefs'? If the *changes* of cognitive beliefs are due to an accumulating empirical knowledge, and accounted for by the theory of subjective probability, how is the *existence* of *a priori* beliefs to be explained in the first place? We have seen that the existence of these beliefs is one of the axioms on which the theory is based. Thus, the existence of *a priori* beliefs is just assumed, not explained by the theory of subjective probability. How to account for the existence and formation of *a priori* beliefs is another problem, and we shall take it up later on, after having learned what the theory of subjective probability has to teach.

18.2. The Filter Theorem

The elementary concepts of the theory of subjective probability were introduced in the preceding section in an elementary way, with a didactic purpose, since it is helpful for even a non-mathematical reader to have an idea of this important part of modern mathematical conceptual apparatus. The proof of the theorem to be introduced now cannot be given on the same elementary level of exposition. This is why it is suggested that a non-mathematical reader should bypass it, and move directly to the discussion of its significance in the later sections (beginning with Section 18.3).

On which conditions and to what extent is the human actor himself able to correct his possibly erroneous *a priori* beliefs on the score of his accumulating empirical knowledge? A general answer to this question is given by the fundamental theorem of the theory of subjective probability, which will be called 'Filter Theorem', in the following.

We assume that A is an objectively true state of affairs, and that the actor is in possession of the empirical knowledge consisting of the observations $E_1, E_2, \ldots, E_n$ obtained by the successive acts of observation $f_1, f_2, \ldots, f_n$, respectively. Each of the acts f_i is associated with the same real-number-valued observable z, whose different values in a unique way correspond to the different observations made:

$$z_i \Leftrightarrow E_i = \{x; z(x) = z_i\} \quad (1\text{--}1) \quad \text{for } i = 1, 2, \ldots, n. \tag{18.26}$$

The observable z is a Borel-measurable function in X, i.e. that

$$\{x; z(x) < r\} \subset \mathscr{B} \subset 2^X \quad \text{for all } r \in R. \tag{18.27}$$

Here $\mathscr{B}$ is a Boolean algebra in 2^X, and R is the set of real numbers.

It is also assumed that the observations $E_1, E_2, \ldots, E_n$ are inductively independent of each other, i.e. that

$$P(E_1 E_2 \ldots E_n/A) = P(E_1/A)P(E_2/A) \ldots P(E_n/A) \quad \text{for all } A \subset X. \tag{18.28}$$

(Note: this does not make the proof easier but the undertaking more ambitious.)

Now the theorem can be derived in the following three major steps (cf. Savage, 1971; Aulin-Ahmavaara, 1977).

(1) The application of the Bayesian formula

From

$$P(A/E_1 E_2 \ldots E_n) = P(A \cap E_1 \cap E_2 \cap \ldots \cap E_n)/P(E_1 \cap E_2 \cap \ldots \cap E_n)$$

and

$$P(E_1 E_2 \ldots E_n/A) = P(A \cap E_1 \cap E_2 \cap \ldots \cap E_n)/P(A) \tag{18.29}$$

we get

$$P(A \cap E_1 \cap E_2 \cap \ldots \cap E_n) = P(A)P(E_1 E_2 \ldots E_n/A)$$
$$= P(A/E_1 E_2 \ldots E_n)P(E_1 \cap E_2 \cap \ldots \cap E_n), \tag{18.30}$$

so that

$$P(A/E_1 E_2 \ldots E_n) = P(A) \cdot \frac{P(E_1 E_2 \ldots E_n/A)}{P(E_1 \cap E_2 \cap \ldots \cap E_n)}. \tag{18.31}$$

In a similar way we get, for the opposite state of affairs, that which we shall here denote by $\bar{A}$:

$$P(\bar{A}/E_1 E_2 \ldots E_n) = P(\bar{A}) \cdot \frac{P(E_1 E_2 \ldots E_n/\bar{A})}{P(E_1 \cap E_2 \cap \ldots \cap E_n)}. \tag{18.32}$$

Thus, the ratio of the *a posteriori* probabilities of A and $\bar{A}$ are, according to (18.31) and (18.32):

$$\frac{P(A/E_1 E_2 \ldots E_n)}{P(\bar{A}/E_1 E_2 \ldots E_n)} = \frac{P(A)}{P(\bar{A})} \cdot \frac{P(E_1 E_2 \ldots E_n/A)}{P(E_1 E_2 \ldots E_n/\bar{A})}. \tag{18.33}$$

By using the assumption of inductive independence of the observations, (18.28), this is transformed into

$$\frac{P(A/E_1 E_1 \ldots E_n)}{P(\bar{A}/E_1 E_2 \ldots E_n)} = \frac{P(A)}{P(\bar{A})} \cdot \frac{P(E_1/A)P(E_2/A) \ldots P(E_n/A)}{P(E_1/\bar{A})P(E_2/\bar{A}) \ldots P(E_n/\bar{A})}. \tag{18.34}$$

(2) The replacement of observations by the respective values of the observable

By applying the one-to-one correspondence (18.26) we can write (18.34) in the form

$$\frac{P(A/z_1 z_2 \ldots z_n)}{P(\bar{A}/z_1 z_2 \ldots z_n)} = \frac{P(A)}{P(\bar{A})} \cdot R(z), \tag{18.35}$$

where the likelihood ratio

$$R(z) = \frac{P(z_1/A)P(z_2/A) \ldots P(z_n/A)}{P(z_1/\bar{A})P(z_2/\bar{A}) \ldots P(z_n/\bar{A})} \tag{18.36}$$

appears.

But now we can consider $z_1, z_2, \ldots, z_n$ as n stochastic variables having the same distribution of probability (as usual in sampling theory). It follows that

$$\log R(z) = \sum_{i=1}^{n} \log \frac{P(z_i/A)}{P(z_i/\bar{A})} \tag{18.37}$$

is a sum of n stochastic variables having the same distribution of probability.

(3) The application of the law of large numbers

The law of large numbers, showing that the observed average log $R(z)/n$ approaches the theoretical expectation value

$$J = \int_{-\infty}^{+\infty} P(s/A)\log\frac{P(s/A)}{P(s/\overline{A})}\,ds \geqslant 0 \qquad (18.38)$$

however close when $n \to \infty$, can accordingly be applied to (18.37). This gives

$$\lim_{n\to\infty} P(\log R(z)/n < J - \varepsilon|A) = 1 \quad \text{for any } \varepsilon > 0. \qquad (18.39)$$

In other words,

$$\lim_{n\to\infty} P((R(z) > e^{n(J-\varepsilon)}|A) = 1. \qquad (18.40)$$

It follows that

$$R(z) \to \infty \quad \text{with } n \to \infty, \quad \text{if } J > 0. \qquad (18.41)$$

But we have

$$J > 0 \Leftrightarrow P(z/A) \neq P(z/\overline{A}) \quad \text{at least for some } z \in R. \qquad (18.42)$$

Under the assumption $J > 0$ the results (18.35) and (18.41) thus give

$$P(A/E_1 E_2 \dots E_n) \to 1 \quad \text{for } n \to \infty \Leftrightarrow P(A) > 0, \qquad (18.43)$$

$$P(A/E_1 E_2 \dots E_n) = 0 \quad \text{for all } n \Leftrightarrow P(A) = 0. \qquad (18.44)$$

Combining the results (18.42)–(18.44) we have the following theorem.

Theorem 18.2.1 ('Filter Theorem'). The *a posteriori* subjective probability $P(A/E_1 E_2 \dots E_n)$ of a true state of affairs A grows with a sufficient accumulation of the empirical knowledge $E_1 E_2 \dots E_n$ however close to sureness, provided that

   (1) $P(z/A) \neq P(z/\overline{A})$ at least for some $z \in R$, and
   (2) $P(A) > 0$.

If the *a priori* probability $P(A)$ is zero, the *a posteriori* probability too remains zero, however large the accumulated empirical knowledge contrary to such a judgement.

18.3. A Discussion of the First Condition of the Filter Theorem: Empirical and Inductive Knowledge

Condition (1) of the Filter Theorem states that the distribution of subjective probability over the values of the observable z in the case that A is true must be different from that in the case that A is untrue. This is to say: the mechanism of observation has to be such that it distinguishes between the state of affairs A and its opposite $\bar{A}$. This in itself trivial condition has a content that depends on whether A is a singular state of affairs or a generalization.

(1) Singular states of affairs and the empirical knowledge concerning them

A singular state of affairs A can be defined as one whose verification and refutation are possible with only one observation, made by an 'eyewitness' of the 'event' A. The definition implies that such a state of affairs is always associated with a certain fixed point of time and space, i.e. it is an 'event' occurring in some definite place at some definite moment. An eyewitness is a person who was present when the event in question is claimed to have happened, and saw it either taking place or not taking place. In the former case he made an observation E that implies A, i.e. $E \subset A$, and in this case

$$P(A/E) = 1. \tag{18.45}$$

In the latter case he observed a fact E' that implies $\bar{A}$, so that

$$P(A/E') = 0. \tag{18.46}$$

Thus, a singular state of affairs, if true, is itself a piece of empirical knowledge. In the case (18.45) the actor, on hearing the testimony of an eyewitness (he himself of course may be one), comes into possession of the empirical fact A. If E is included in the empirical knowledge $E_1 E_2 \ldots E_n$ of the actor, the *a posteriori* subjective probability of A on the evidence $E_1 E_2 \ldots E_n$ is exactly one, A being now part of the empirical knowledge he has. Condition (1) makes it sure that with increasing n such a fact E is, if not reached, at least approached enough for the *a posteriori* probability $P(A/E_1 E_2 \ldots E_n)$ to approach one. This is the rather trivial content of the Filter Theorem for a singular state of affairs A.

On the other hand, it is quite possible that condition (1) is not fulfilled for a given singular state of affairs A. This is the case in which no eyewitnesses of the event A are available to our actor and, accordingly, his mechanism of observation cannot distinguish between the truth and untruth of A.

(2) Inductive knowledge

A *generalization* can be defined as a state of affairs that no finite amount of observations is sufficient to confirm true, but which even a single observation may refute. The opposite $\bar{A}$ of a generalization A is a statement of existence. The observations E of the former type obey

$$E \cap A \neq \varnothing, \tag{18.47}$$

while an observation E' that refutes a generalization A satisfies

$$E' \cap A = \varnothing, \quad E' \cap \bar{A} \neq \varnothing, \quad E' \subset \bar{A}. \tag{18.48}$$

When applied to a generalization A, the Filter Theorem has a highly interesting non-trivial content. It indicates that while the *a posteriori* subjective probability $P(A/E_1E_2...E_n)$ of a generalization A remains smaller than one, i.e. short of certainty, it approaches it asymptotically with $n \to \infty$, provided that conditions (1) and (2) are fulfilled. That condition (1) is met means only that the mechanism of observation z is capable of distinguishing between the generalization A and its opposite statement $\bar{A}$. We shall soon take up the significance of the other condition. Assuming that both conditions are satisfied we get, on the basis of the Filter Theorem, the following picture of the accumulation of *inductive knowledge* (whose very existence is disputed by philosophers but is contested by no practical man in his everyday life).

At first the attention of the actor is focused on an object (for the meaning of 'object' in this connection see Section 9.1). His imagination makes up a hypothesis, i.e. a generalization A, concerning this object. He then checks hypothesis A by means of the empirical knowledge $E_1E_2...E_n$ he has. As to the outcome of checking there are three possibilities:

(i)   hypothesis A is given support by the empirical evidence so that the *a posteriori* subjective probability $P(A/E_1E_2...E_n)$ grows all the time with increasing evidence $E_1E_2...E_n$; or

(ii)   hypothesis A is refuted by the inclusion in the evidence $E_1E_2...E_n$ of an observation E' that counters it giving $P(A/E') = 0$; or

(iii)   the *a posteriori* subjective probability of hypothesis A does not substantially grow, but neither does it become zero: this is the case, very frequent in everyday life, in which the issue remains unsettled.

In case (i) above we can speak of the accumulation of inductive knowledge and, in view of the Filter Theorem, we can say that with increasing evidence $E_1E_2...E_n$ the *degree of verification* of the generalization A increases ever closer to full certainty. This is what happens in our everyday life when, for instance, we see the same pieces of furniture and other material objects around us time and again, and can even touch them to verify their existence. Nevertheless, our knowledge of them remains inductive, i.e. our conceptions of those objects are just generalizations supported by our sensory observations, and thus hypothetical (which makes some philosophers, called subjective idealists, deny their very existence elsewhere than in our imagination). This despite the fact that their *a posteriori* subjective probability for most of us surely is rather close to certainty.

The great message of the Filter Theorem, however, is that human actors are capable of correcting their erroneous *a priori* beliefs, provided that they have accurate enough observations of the world (condition 1) and that the *a priori* subjective probability is not zero (condition 2). It remains to consider the significance of the latter, very interesting, condition.

18.4. A Discussion of the Second Condition of the Filter Theorem: The Filter in Human Consciousness

In view of the theory of subjective probability it is useful to adopt the following terminology:

Definition. The actor has an *a priori subjective truth* concerning a state of affairs A, if his *a priori* subjective probability $P(A)$ is either zero or one. In the former case he is *a priori* convinced that A is not true; in the latter case that it is true.

The latter part of the Filter Theorem says that however accurate are the observations that the actor makes, he is unable to correct his erroneous *a priori* belief by means of accumulated empirical knowledge, if this belief is an *a priori* subjective truth that denies the truth of a true state of affairs A, i.e. if $P(A) = 0$ despite the fact that $\hat{x} \in A$.

But the *a priori* subjective truth $P(A) = 1$ is also unaffected by empirical knowledge, whether A be true or not. This is trivial, and can be seen from the formula (18.33); for $P(A) = 1$ we have $P(\bar{A}) = 0$ so that the ratio $P(A)/P(\bar{A})$, and thus $P(A/E_1 E_2 \ldots E_n)/P(\bar{A}/E_1 E_2 \ldots E_n)$ are infinite, which shows that the *a posteriori* subjective probability $P(\bar{A}/E_1 E_2 \ldots E_n)$ remains zero and $P(A/E_1 E_2 \ldots E_n)$ one for all integers n.

Thus, the latter part of the Filter Theorem is tantamount to the statement that

if the actor has an *a priori* subjective truth concerning a state of affairs A, he has lost the ability to learn about A from empirical knowledge and is unable to correct his possibly erroneous *a priori* belief about A.

Such a loss of ability to learn from experience is, of course, a fatal blow to the intellectual capacities of whoever it may concern. But an even more interesting message of the Filter Theorem is what follows from the statement given above concerning the structure of human consciousness:

It follows that human consciousness works as if there were two layers and a filter between them. The upper layer (see Fig. 94) is composed of *a posteriori knowledge* and stands for the rational formation of knowledge where

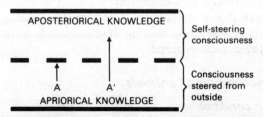

FIG. 94. The two layers of human consciousness and the functioning of the filter between them, according to the Filter Theorem. The suggested cybernetic interpretations are given in the right margin.

knowledge is being checked by means of observations. Under this layer there is the layer of *a priori knowledge*, and between them a *filter* that blocks *a priori* subjective truths but is permeable to uncertain *a priori* knowledge, i.e. to *a priori* beliefs represented by *a priori* subjective probabilities between one and zero: only the latter lead to the formation of *a posteriori* knowledge.

But we can also give a dynamical interpretation, in terms of self-steering and steering from outside, to the two layers of consciousness separated by the filter (cf. Fig. 94):

(i)   The layer of *a posteriori* knowledge is the *self-steering consciousness*, because in the formation of this kind of knowledge the actor himself chooses the objects of his thinking, makes generalizations concerning them, and checks the results by acquiring empirical knowledge: the actor is himself in control of his knowledge, as far as *a posteriori* knowledge is in question.

(ii) The layer of *a priori* knowledge is the *consciousness steered from outside*: the actor himself does not control the formation of his *a priori* knowledge. *A priori* knowledge, by its very definition, is already there, in the consciousness of the actor, when he starts consciously to check his beliefs. Whatever the mechanism of formation of *a priori* knowledge, it is something that the actor cannot control by his conscious acts. The controller of *a priori* knowledge, whatever it might be (the genetic information in his cells? some outer actor?), is outside the consciousness of the actor. Hence the term 'consciousness steered from outside'.

18.5. On the Mechanisms of the Formation of *a priori* Knowledge

It seems a likely assumption that the genetic information that shapes the structure and modes of functioning of our nervous system, also sets a general invariable frame within which all the formation of knowledge in the brain takes place. But instead of the invariant limitations imposed by the genetic information, we shall now concentrate on the possible variable factors that may control the formation of our *a priori* knowledge. More precisely, we ask: are there any known mechanisms with the help of which an outside actor is able to feed information into the human or animal brain, which acts in the process only as a passive receiver?

Oh yes, there is the mechanism of conditioning. But as the available evidence differs very much in moving from animals to man, it is advisable to deal with these cases separately.

(I.) The conditioning of animals

(1) Pavlovian conditioning

We know the general course of events: a dog is given food, and just a little (a second or a few seconds) before it a bell is rung, or a door is knocked, a light is flashed, or some other clear *signal* is given. When the sequence of the signal and food is repeated a few times, while the dog has been hungry enough in each experiment, a *conditioned reflex* is born: the dog has learnt to secrete saliva on hearing or seeing the signal. Hereafter the reaction takes place every time the signal is given while the dog is hungry. In other words, a permanent entirely mechanical association between the signal and food has been established in the brain of the dog.

A conditioned reflex can be dismantled only by *contraconditioning*, i.e. by not giving the food, or whatever has been used as the reward, following the signal. If this has happened a few times the reflex is weakened, and if the process is continued, by giving the signal without reward, the reflex can be totally extinguished.

Anything wanted by the animal can be used as a reward. Instead of a reward one can even set up conditioned reflexes between a signal and a punishment, i.e. something that the animal would rather avoid, such as an electric shock, or a blow by a cane, or touching cold, etc. In fact punishments are just as good as tools of conditioning as are rewards.

As to the test animal, almost any animal will do, not just dogs, even man performs well in these experiments.

(2) Instrumental conditioning

The general set-up is not much different from that of Pavlovian conditioning. However, this time the use of an instrument instead of the meaning of a signal is taught to the animal by means of conditioning. A typical experiment runs as follows. A rat is given food, or a sexual partner, or some other reward, provided that it is able to open the door behind which the reward is waiting, or run through the maze at the end of which the reward has been placed. When the experiment has been repeated a sufficient number of times the animal again has learned to perform the requisite act, viz. to open the door, or run the labyrinth, etc., provided that it is hungry for the reward.

Again, instead of rewards punishments can be used as well. To avoid an electric shock a rat learns to quickly turn a lever, etc. Thus, by using rewards and punishments an animal can be taught not only to recognize signals but also to use different tools.

In both kinds of animal conditioning the test animal learns only what has been wanted by the experimenter: the brain of the animal has been fed information from outside. Is this possible with man?

II. The conditioning of man

The human evidence is different for different kinds of conditioning. We have to distinguish between three cases as follows.

(1) Conditioned reflexes can be established with ease in man as well as in animals, by using simple rewards such as food, or simple punishments such as an electric shock.

(2) Instrumental conditioning of man meets with considerable difficulties, since it is difficult to instal in an experimental situation such rewards or punishments that would motivate human beings to learn a new instrument. You are motivated to learn the use of a new computer, if the reward is a more or less permanent job, and you happen to need such a job. But you are less likely to take the trouble just for the sake of an experiment, where the profit goes to the experimenter only. Or someone hopes to get good vocational training, or a certain girl as a wife, or to become a famous author or maybe the president of a country – such objectives cannot be realized in an experimental situation.

To put it in a general way: the fears and hopes of people are more complex than those of animals, and often pertain to social life, and thus cannot be realized as rewards or punishments in experimental situations. This is why instrumental conditioning of man is in most cases unrealistic as an experiment. But this fact does not exclude even in the case of man what could be called

The principle of conditioning: the use of fears and hopes for steering the consciousness from outside by means of various 'rewards' or 'punishments'.

It is only that instead of experimental evidence we have to be content, in the case of man, to resort to *historical evidence* in favour of the principle of conditioning. It is to be frankly admitted that historical evidence is less binding than experimental evidence. Historical evidence refers to situations in ordinary life, where the external conditions are never the same, and the observed behaviour can accordingly be interpreted by referring to numerous different factors.

In the realm of historical evidence no doubt we come closest to the rigour of experimental conditions in careful anthropological observation. This may be because of the relatively simple and very invariant patterns of behaviour in primitive communities, and also since the observer is there able to adopt the role of an uncommitted outsider better than is otherwise the case.* Historical evidence, obtained especially from anthropological observation of primitive communities, seems to indicate that a major means of conditioning human beings is a third type of conditioning that makes full use of man as a social animal:

(3) Collective conditioning

This means feeding information into human consciousness from outside in social situations, which are prone strongly to excite the emotions of the

---

* This is no more the case. Anthropology is today, like many other 'soft' sciences, to a great extent politicized. Thus, to find uncommitted observers we have to resort to the great anthropologists, such as Bronislaw Malinowski or Mircea Eliade, of the pre-politicized era. We have also to note that genuine primitive communities do not exist any more. This brings us to the anthropological observations coming from earlier periods, e.g. from the beginning of the century.

participants and are of utmost importance for the individuals taking part in them.

Excellent examples of such social situations are the initiation ceremonies in primitive communities, the harvest or the New Year festivals in the archaic temple states, for instance in ancient Egypt and Babylonia, and the collective demonstrations, marches, and other forms of emotional 'mass communication' exercised publicly before a great mass of people convened expressly for the occasion to confess their common belief in some religious or political cause.

We are now in a position to state a hypothesis on the role of conditioning in the formation of the specific content of our *a priori* beliefs:

> *The conditioning hypothesis.* *A priori* subjective truths are transplanted in the human individual by means of collective conditioning from the human collectives to which he belongs during his lifetime.

Our *a priori* truths, according to the hypothesis, reflect the beliefs of our social environments, and are not our 'own' knowledge but have been fed into us by collective conditioning, beyond our conscious control. Only *a posteriori* knowledge is self-acquired and based on our own individual experience. *A priori* truth is collective by its nature, and is implanted in the minds of people by collective conditioning.

Let it be emphasized that, according to the conditioning hypothesis, even our *a priori* truths are *variable*: they vary all the time with our social environments. A new social environment, such as school, or a new job, or a new circle of friends, is conditioning, contraconditioning, and reconditioning our *a priori* beliefs all the time. Thus, these beliefs are by no means invariant. On the other hand, they can be changed only by collective contra- and reconditioning, not by rational argument. Reason and observation matter only in the formation of *a posteriori* knowledge.

As a matter of course, the conditioning hypothesis cannot be tested by experimental means. You cannot in experimental situations produce the social conditions that are necessary for genuine collective conditioning to take place. It follows from the conditioning hypothesis itself that, if you try to verify it experimentally, you are bound to have an entirely misleading idea of the real strength and significance of collective conditioning in human life.

The only means of testing the conditioning hypothesis is historical evidence, which comes closest to experimental verification in careful anthropological observation of primitive communities. That, accordingly, is the major source we have to use for an empirical check of the conditioning hypothesis, and also in order to test the two-layer theory of consciousness involved in the theory of subjective probability.

18.6. The Anthropological Evidence for the Two-layer Theory and the Conditioning Hypothesis: A Survey of the Main Points

The issue will be taken up in Chapter 19 in more detail. Here are only the main points for a hurried reader:

(1) The two layers of the primitive mind

A founder of social anthropology, Bronislaw Malinowski, wrote in his *Magic, Science, and Religion* (New York, 1948, p. 17):*

> In every primitive community, studied by trustworthy and competent observers, there have been found two clearly distinguishable domains, the Sacred and the Profane; in other words, the domain of Magic and Religion and that of Science.

What is by social anthropologists called 'the Profane' is exactly the same thing that in the theory of subjective probability is termed *a posteriori* knowledge. It includes the use of reason based on common sense and observation as applied to everyday chores in a primitive community, such as sailing, construction, gardening, etc. In this respect primitive man in no way differs from modern man: he has a rational 'science' of his own, which is perfectly logical, and well checked by the empirical knowledge he has.

But in a pursuit where his science lets him down, and the undertaking thus becomes risky or even dangerous, primitive man takes recourse to *magic*. Fishing on the high seas is dangerous and risky, thus much magic is associated with it. Fishing in a lagoon by poisoning is secure, thus no magic is involved in it. The weather cannot be governed by science, thus there is much magic connected with it. The reasons for diseases were unknown, thus magic had to be resorted to. Love and hate are no rational things, and magic accordingly was invoked. A war is an insecure undertaking and much magic is again needed.

Magic is based on *myths*, but magic is only that part of the mythology of a given community that has to do with earthly affairs. The most important myths of every primitive community are those that formulate the beliefs of that community in matters transcendental, that is, the *religion* of that community. But all myths, the religious as well as magical ones, are collectively conditioned in the primitive mind. The evidence for this will be discussed below. Provided that the evidence is accepted, the myths accordingly are the *a priori* truths of primitive man. But they are exactly what is meant by the domain of 'the Sacred'. Thus, the anthropological division between the Profane and the Sacred seems to be identical with the distinction made in the theory of subjective probability between *a posteriori* knowledge and *a priori* truths: the filter predicted by the Filter Theorem is also present in the primitive mind.

---

* This work will be referred to as M I.

(2) Myths as a priori truths

Not all of the religious ideas expressed in a primitive community have the position of sacred truths, i.e. myths. Malinowski, one of the sharpest anthropological observers, makes a clear distinction even in the domain of religion between what belongs to the Sacred and what belongs to the Profane in the primitive mind. He summarized his findings as follows (MI, pp. 252–253):

1. Social ideas or dogmas. – Beliefs embodied in institutions, in customs, in magic–religious formulae and rites and in myths. Essentially connected with and characterized by emotional elements, expressed in behaviour. [This is the Sacred part of primitive religion.]
2. Theology or interpretations of the dogmas:
   (a) Orthodox explanations, consisting of opinions of specialists.
   (b) Popular, general views, formulated by the majority of the members of a community.
   (c) Individual speculations. [Theology is the Profane part of primitive religion.]

Thus, myths are the verbally formulated sacred truths, which Malinowski called 'social ideas', i.e. ideas held by all the members of community. He adds: 'Such social ideas can be treated as the "invariants" of native belief' (M I, p. 245). In addition to their verbal formulation in myths, sacred truths are also 'embodied' in institutions, customs, and rites of the community. We shall return to this later.

By calling the social ideas expressed in myths 'dogmas' Malinowski referred to the fact that no member of the community was permitted to dissent from these truths, and indeed they did not. There is no doubt that the content of myths in the primitive mind is what in the theory of subjective probability is called *a priori* truth. The anthropological observations confirm the above formulations by Malinowski.

But how were these *a priori* truths established in the primitive mind? Let us now take up this essential point.

(3) The collective conditioning of myths

The explanation of the *a priori* truth of myths in terms of collective conditioning rests essentially on five arguments.

(I) The strict connection of religious myths with collective ceremonies

In primitive communities religious myths are permitted to be read only in connection with solemn collective ceremonies, such as the initiation of young men, the harvest or New Year festivals, the funerals, and the like. Even the magic myths, though often read by the magician alone or by a group of magicians together, are backed by the collective authority of the community enjoyed by the magicians, and the most important magic too is observed collectively by the community as a whole.

(II) *The strict parallelism of myth and ritual*

The reading of a myth is *always* accompanied by the performance of a rite. No exceptions to this strict rule are permitted in a primitive community, What links the verbal performance, myth, and the acts connected with it, i.e. the accompanying ritual, is dictated by rigorous and detailed rules as to *how* each phase of the combined reading of the myth and the movements related to it are to be made. The ritual has to be repeated in exactly the same way down to the minutest detail every time. Now it is a well-known fact about conditioning, from both animal and human experiments, that the experiment has to be repeated in the same way as before, even in its seemingly inessential details, in order for conditioning to be successful.

Thus, the Conditioning Hypothesis not only explains why myths are connected with collective ceremonies, but also what has been inexplicable, viz. the appearance of scrupulously performed rites *always* in connection with myths in primitive communities, while any rituals are entirely lacking in connection with profane reasoning (as noted by many). Rituals are necessary in connection with the reading of myths for the conditioning of myths as *a priori* truths to succeed.

But the crucial test of the Conditioning Hypothesis is the following fact:

(III) *The collective-specific nature of myths*

The content of myths is characteristic of a primitive community. The myths of each primitive community are unique: the next village may already have entirely different myths, which, however, are believed in with the same absolute trust, allowing no dissent among the members of that community. The geographical borders of belief in a certain mythology are the same as those of the collective that takes part in the common religious ceremonies and rituals. This fact seems to exclude any attempt at an explanation of myths as *a priori* truths other than the one given in terms of the collective conditioning of myths. How else could the collective-specific nature of myths be accounted for? How else could one explain why the absolute belief in a certain mythology in one community is in the next community replaced by an equal belief in an entirely different mythology?

Thus, the remaining two arguments are only augmentations supporting the crucial test:

(IV) *The role of taboos*

The taboos known from primitive communities are necessary, in view of the Conditioning Hypothesis, in order to prevent the contra-conditioning of myths. If the things or ideas connected with a collectively conditioned myth were allowed to associate freely to whatever situations may appear in everyday life, contra-conditioning is likely to happen sooner or later, and the belief in the myth would have gone. Therefore, such things and ideas must be declared sacred, and strictly tabooed, to exclude any contact with profane situations. In view of the central position of myths in the primitive mind it is not surprising that the punishments for breaking taboos were among the strictest employed in real primitive communities, and used to include the penalty of death and exile from the community, which in those ancient times we are here referring to must have been equal to death.

(V) *The connection of myths with power*

Even though other anthropologists do not disagree, we can again resort to Malinowski, who often gave the sharpest formulation of a generally accepted fact (was it due to his university education as a physicist?). In his introduction to the concept of myth, he stated the connection of myths with social power with a special emphasis (M I, p. 84):

Myth, it may be added at once, can attach itself…to any form of social power or social claim. It is used always to account for extraordinary privileges or duties, for great social inequalities, for severe burdens of rank, whether this be very high or low.

[Therefore] 'myth, taken as a whole, cannot be sober dispassionate history, since it is always made ad hoc to fulfill a certain sociological function, to glorify a certain group, or to justify an anomalous status (M I, p. 125).

Here Malinowski is speaking of the role of myths not only in primitive communities, but in societies generally. In view of the Conditioning Hypothesis he is here stating the obvious: collective conditioning and power are deeply connected. Real power can be acquired in any society only by adhering to what passes for collective truths in that society. And the more collective conditioning the power-holder is able to exercise in a society, the more concentrated a system of power he is able to set up. Hence, the more hierarchical is society, the more collective conditioning of *a priori* truths there is.

The close connection between hierarchy and the amount and intensity of collective conditioning can be used to explain an important feature of underdevelopment: how cultural and religious or ideological tradition, when maintained by means of collective conditioning, in the course of time becomes the strongest factor prolonging underdevelopment.

CHAPTER 19

The Primitive Mind: Anthropological Evidence for the Two-layer Theory and the Conditioning Hypothesis Scrutinized*

To the connoisseur it may be immediately clear that the two-layer theory of consciousness implied by the theory of subjective probability (Section 18.4), when completed by the Conditioning Hypothesis (Section 18.5), in what concerns the primitive mind is strikingly well confirmed by anthropological and ethnological findings. However, because of the utmost importance of the matter of the identification of the source of, and obstacles to, self-steering in human consciousness, we shall study the anthropological evidence in more detail in the following.

Meanwhile, let me note to the initiated reader that the 'basement' of human consciousness, consisting of *a priori* knowledge and steered from outside by means of conditioning, is not to be identified with the 'ghost of Collective Soul' so heartily rejected by Bronislaw Malinowski. Neither does it mean to imply an identification of the 'social' with the 'religious' as advocated by Emile Durkheim. The lower layer of consciousness in the two-layer theory is in itself nothing collective, though it is imposed on the individual by the collective: it remains a part of the mind of the individual. Durkheim's identification could be true only of an extreme type of totalitarian society.

As far as Durkheim's 'collective ideas' go, I cannot but undersign Malinowski's remark (M I, pp. 273–274) to the effect that 'there does not seem to be anywhere a clear candid statement of what they [i.e. Durkheim's school] mean by "collective idea", nothing approaching a definition.' But we shall see that Malinowski's 'social ideas of dogmas' exactly correspond to what is called *a priori* truths in the theory of subjective probability.

19.1. The Two Layers of the Primitive Mind: The Sacred and the Profane

When Bronislaw Malinowski in May 1915 set out for the Trobriand Islands situated in what was then British New Guinea, he had in front of him a

---

* See the footnote on p. 274.

pioneering piece of anthropological fieldwork which was to alter many conceptions held by eminent scholars. One of them had been advocated by the 'brilliant French sociologist', to use Malinowski's own words, Professor Lévy-Bruhl who held 'that primitive man has no sober moods at all, that he is hopelessly and completely immersed in a mystical frame of mind' (M I, p. 25). Malinowski's opposite view, gained on the spot, was to be 'that every primitive community is in possession of a considerable store of knowledge, based on experience and fashioned by reason' (M I, p. 26.)

He found the Trobrianders to be expert fishermen, industrious manufacturers and traders, and skilled gardeners, their main means of subsistence.

The success in their agriculture depends... upon their extensive knowledge of the classes of soil, of the various cultivated plants, of the mutual adaptation of these two factors, and, last not least, upon their knowledge of the importance of accurate and hard work. They have to select the soil and the seedlings, they have appropriately to fix the times for clearing and burning the scrub, for planting and weeding, for training the vines of the yam plants. In all this they are guided by a clear knowledge of weather and seasons, plants and pests, soil and tubers, and by a conviction that this knowledge is true and reliable, that it can be counted upon and must be scrupulously obeyed (M I, pp. 27–28).

A similar picture of rational knowledge as a basis of work is obtained of their canoe-building, so essential for seafaring and trade. But an interesting and even crucial test, as noted by Malinowski, is provided by fishing in the Trobriand Islands.

While in the villages on the inner lagoon fishing is done in an easy and absolutely reliable manner by the method of poisoning, yielding abundant results without danger and uncertainty, there are on the shore of the open sea dangerous modes of fishing and also certain types in which the yield greatly varies according to whether shoals of fish appear beforehand or not. It is most significant that in the lagoon fishing where man can rely completely upon his knowledge and skill, magic does not exist, while in the open-sea fishing, full of danger and uncertainty, there is extensive magical ritual to secure safety and good results (M I, pp. 30–31).

The conclusion is confirmed by the appearance of magic in gardening, trade and seafaring as well. Even the best work of gardening may be entirely spoilt by unfavourable weather conditions, by unseasonable droughts or rains, for instance, and the yield may also be destroyed by bush-pigs, locusts, and other natural enemies. It is here, in the realm of hope and fear, that magic steps in to replace the missing intellectual means of combating these enemies. Likewise sudden gales, wrong directions of wind, and even a hostile reception by expected trading partners may offset the efforts undertaken in trading expeditions. All this is the domain of magic, for the simple reason that normal intellect does not help – most seamen are still today notably superstitious. Health care and healing of illnesses was a domain where primitive man had to rely almost exclusively upon sorcery and other magic, as the necessary medical knowledge did not exist, or was rudimentary.

This is how Malinowski arrived at his famous conclusions now rather universally accepted among experts:

In every primitive community, studied by trustworthy and competent observers, there have been found two clearly distinguishable domains, the Sacred and the Profane; in other words, the domain of Magic and Religion and that of Science. (M I, p. 17.)

The identification of the upper, self-steering layer of consciousness, that is, of rational, *a posteriori* knowledge with the domain of the Profane, also called that of Science, is rather obvious. For what Malinowski understands by the latter is

that primitive man can observe and think, and that he possesses, embodied in his language, systems of methodical though rudimentary knowledge, [and even] words which serve to express general ideas such as existence, substance, and attribute, cause and effect, the fundamental and the secondary; words and expressions used in complicated pursuits like sailing, construction and checking, numerals and quantitative descriptions, correct and detailed classifications of natural phenomena, plants and animals (M I, p. 33).

Thus, primitive man is able to observe and make generalizations tested by experience, in short, he is able to gain *a posteriori* knowledge.

The same thing is, by the way, attested by the points where primitive knowledge is defective. Malinowski perceived, for instance, that the Trobrianders were ignorant of the fact of procreation. Really, there was for them no observable link between sexual intercourse and pregnancy. The girls started regular sexual intercourse, exerted as a rule every night, at the age of seven to nine, while they normally bore a child only two or three times during their entire life. The logical contradiction was duly indicated to the ethnographer when he tried to point out the connection to the natives. They politely listened to the 'custom' of the West but remained firm in their belief, which in this case was that it is the soul of a dead, baloma, that sometimes enters the womb of a woman, to be so reincarnated at birth. They had observed that pregnancy was impossible without preceding sexual intercourse, but obviously the latter was only needed to open the way for the baloma to come. No doubt this accounts for the facts as far as facts were known to primitive man.*

19.2. The Nature of Magic and the Conditioning of Myths

But we still have to show that it is the *a priori* beliefs established by collective conditioning that correspond, in the case of the primitive mind, to what Malinowski called the domain of the Sacred, or that of magic and religion. Let us first take magic for a closer examination.

Let the Trobrianders, as depicted by Malinowski, again usher the way:

All important economic activities are fringed with magic, especially such as imply pronounced elements of chance, venture, or danger. Gardening is completely wrapped up in magic; the little hunting that is done has its array of spells; fishing, especially when it is connected with risk

---

* The discovery concerning the natives' ignorance of the connection of sexual intercourse to pregnancy comes from E. S. Hartland, *The Legend of Perseus* (1894–1896).

and when the results depend upon luck and are not absolutely certain, is equipped with elaborate magical systems (M I, pp. 190–191).

The same holds for canoe-building, but only in the case of the larger sea-going canoes made for venturous undertakings.

Weather – rain, sun, and wind – have to obey a great number of spells, and they are especially amenable to the call of some eminent experts, or, rather, families of experts, who practice the art in hereditary succession. In times of war – when fighting still existed, before the white man's rule – the Kiriwinians availed themselves of the art of certain families of professionals, who had inherited war magic from their ancestors. And, of course, the welfare of the body – health – can be destroyed or restored by the magical craft of sorcerers, who are always healers at the same time (M I, p. 191).

Love too has its magic.

Magic accordingly pertains to those precarious situations of everyday life where hopes and fears reign. However:

One thing is certain: magic is not born of an abstract conception of universal power, subsequently applied to concrete cases. It has undoubtedly arisen independently in a number of actual situations (M I, p. 78).

But such situations where man feels let down by his rational knowledge must abound in primitive life.

Magic is so widespread that, living among the natives, I used to come across magical performances, very often quite unexpectedly, apart from the cases where I arranged to be present at a ceremony (M I, p. 191).

To understand the 'native's' situation it is of paramount importance to note that the rational knowledge of primitive man is mostly, perhaps quite exclusively, what philosophers call the *maker's knowledge*. He has learned to make canoes, huts, utensils, just by doing it. And he is able to master the use of his tools in gardening, fishing, hunting, seafaring, and other pursuits of life just because he has himself originated these tools. Ergo: if he only could know how weather, wind and rain, drought and sickness originated, that is, made at the beginning of times, they too would be at his command. This is how the maker's knowledge, to which primitive man is accustomed, explains why the spell, the essential part of magic, must tell the *origin of the thing* to be mastered by magic.

At this point the self-steering intellect of primitive man, his playing fancy, readily supplies him with stories about the Makers that created, at the marvellous beginning times, all those interesting things to which man attaches his best hopes or worst fears. But they are not yet myths. To put the magic into effect primitive man has to repeat the deeds of the old Makers, thus re-creating the things in question and so becoming their master. First the act of re-creation, that is, the *ritual*, makes a *myth* of the story that tells how the thing was made at the beginning of time, and how, accordingly, it must be re-created to give the magician the power over it. For only the repeated ritual can give the myth a position as an absolute belief, as a dogma, as a piece of conditioned belief.

Here is the essence of our theory of myth generally, and also of the *myths of origin* and the *myths of the perfection of the beginnings* that constitute the main contents in magical spells. Among the myths of origin the cosmogony myth, telling how the universe itself was originated, takes a special place, and a spell often begins with a cosmogonic opening. The variegated set of different cosmogony myths and other myths of origin, obtaining in different tribes of natives living in different places, tells us that the creative imagination of man has been at work, producing various results in various surroundings. But it takes conditioning by means of a repeated, strictly invariant ritual to make people believe in a certain myth in a certain tribe as an absolute truth.

Indeed, there is no exception to the rule that *myths and rituals always go together*. There is, and cannot be, any myth in any community without the corresponding ritual that establishes the myth as a piece of sacred, absolute truth that no-one within the community can doubt with impunity. In this way all the other myths bearing on the same thing but believed in by other communities are strictly excluded and tabooed, so that, with all the variety of different myths over any larger area with a primitive population, *there remains for any single community only one myth for each thing*.

But not everything can become the subject of a myth. Absolute beliefs, such as myths, can only be established by conditioning for those things to which people attach strong feelings of desire, hope, or fear, and of which they are uncertain. This, precisely, is the stuff of which myths are made. Speaking the language of conditioning, desires and hopes are associated with some promises of reward, and fears with threats of punishment. Wherever hopes and fears are entertained rewards and punishments, and thus conditioning, are involved in human action. In the case of myths, quite generally, the conditioning is collective: the myth in question is solemnly recited, as the core of the ritual, every time that the ritual confirming it is performed. Thus, the rewards or punishments connected with the consequences of the ritual are directly linked to the myth, reinforcing its position as an *a priori* truth.

The details of reinforcement vary from case to case. A necessary condition of any conditioning of truth is that the actual performer of the rite, if not the community as a whole, stands high in the social hierarchy and thus is generally held to represent the collective. A magician is always a highly esteemed expert, often the head of the village concerned or of some other village, or clan, or sub-clan traditionally specializing in the magic in question. In proper religious rituals, related to death and resurrection, and often even in magical rituals, the whole community participates. But whether the engagement of the community be immediate or not – by a person or group representing, by the principle of *pars pro toto*, the community – the collective nature of rituals is a strong conditioning factor.

The conditioning of man is at its most effective when it is collective, because the implied punishments and rewards then gain further force from the social pressure imposed by the collective upon its individual members, and because only in this way can contra-conditioning, e.g. the violation of taboos, be prevented. Another strongly conditioning factor in all magico-religious ceremonies undertaken in primitive communities is the common opinion, generally held by all the participants, that in order to be effective the ritual has to follow the traditional model to the minutest detail. Such a rule excludes the negative cases so that only positive conditioning remains. If the expected result fails to appear, the poorly performed rite is to blame. This in fact turns the negative case to positive conditioning: be more careful the next time you perform the ritual. If, on the other hand, the desired reward appears, this of course reinforces the belief in the power of the ritual and in the truthfulness of the myth in which the ritual was based, according to the rules of conditioning.

The psychological fact of conditioning was not known to Bronislaw Malinowski, who perhaps more than anyone else contributed to the establishment of social anthropology as a well-founded science. But he knew the problem, which he put, in connection with religious myths, as follows:

Why ritual?...Why has man to express such simple affirmations as the belief in the immortality of the soul, in the reality of a spiritual world, by antics, dramatized performances, by dancing, music, incense, by an elaboration, richness, and an extensiveness of collective action which often consumes an enormous amount of tribal or national energy and substance.*

It may sound rude even to suggest a definite, exact answer to such a profound question, but I cannot help feeling that Malinowski is best answered by referring to our first factor of conditioning: it is the conditioning power of collective action that necessitates collective rituals and ceremonies, particularly when proper religious myths have to be reinforced. In magical rites, where the expected reward has to appear in this world, the second factor attains a special significance. As a matter of fact Malinowski, in his own explanation of why magic is believed in, comes rather close to both factors of conditioning mentioned above:

The answer to this is that, first, it is a well-known fact that in human memory the testimony of a positive case always overshadows the negative one. One gain easily outweighs several losses.... But there are other facts which endorse by a real or apparent testimony the claims of magic.... It is an empirical fact that in all savage societies magic and outstanding personality go hand in hand. Thus magic also coincides with personal success, skill, courage, and mental power. No wonder that it is considered a source of success (M I, pp. 82–83).

It remains to be noted that such an outstanding man, the magician, embodies the collective values and beliefs of the community, and thus acts as a conditioning factor through his very personality as the performer of magic.

Further support for the Conditioning Hypothesis comes from the fact that *no rituals or ceremonies are met, in primitive communities, in connection with the transmission of ordinary, profane knowledge.* Speaking of the aborigines of Central Australia, Malinowski emphasizes:

*Bronislaw Malinowski, *The Foundations of Faith and Morals*, first printed by Oxford University Press in 1936; second printing by Folcroft Library Editions in 1974; pp. 4–5. This book will be henceforth referred to as M II.

There is a body of rules, handed from one generation to another, which refers to the manner in which people live in their little shelters, make their fire by friction, collect their food and cook it, make love to each other, and quarrel. This secular tradition consists partly of customary or legal rules, determining the manner in which social life is conducted. But it also embodies rules of technique and behaviour in regard to environment. It is this aspect of secular tradition which corresponds to knowledge or science. The rules which we find here are completely independent of magic, of supernatural sanctions, and they are never accompanied by any ceremonial or ritual elements (M II, p. 32).

The same obtains in every primitive community known to ethnographers.

We have to conclude that myths, really, tell the *a priori* truths of primitive man, clearly separated from his rational, profane wisdom. The absolute truth of myths is established by collective conditioning in connection with the rituals of which they are essential parts. The myth is always recited when the corresponding ritual is performed, and thus becomes connected with the same rewards or punishments as the ritual as a whole. What is new in our interpretation of myths is that they, according to this conception that derives from the two-layer theory and conditioning, owe their truth entirely to rituals, and not vice-versa (as has usually been presumed, on some vaguely felt grounds): *the strength of belief in a given myth is exactly the same as the conditioning power of the repeated ritual in question.*

Once a myth has been established by collective conditioning, it cannot be refuted except by contra-conditioning. Thus, the existence of myths (i.e. *a priori* truths) limits the domain where primitive man can trust his reason. Myths, accordingly, stand as effective obstacles to any attempt to widen the sphere of profane knowledge by new inventions and experiments. Or, stated in terms of the two-layer theory, the existence of *a priori* truths restricts and thwarts the formation of *a posteriori* knowledge. Or, in cybernetic terms, myths prevent human self-steering, and indeed are tools for steering man from outside. The steering actor, of course, is the conditioner; in this case the collective of which the individual is a member.

It follows from our theory of myths as collectively conditioned sacred truths, that the doings of the primordial Makers told in a myth assume a supernatural role in the imagination of primitive man. These Makers, whether fancied as ancestors, fairies, animals, or proper gods, all grow in the process of conditioning into supernatural beings. Thus, we can accept Mircea Eliade's descriptive definition of myth, according to which

myth tells how, through the deeds of Supernatural Beings, a reality came into existence, be it the whole of reality, the Cosmos, or only a fragment of reality – an island, a species of plant, a particular kind of human behaviour, and institution. Myth, then, is always an account of 'creation'; it relates how something was produced, began to be... The actors in myths are Supernatural Beings. They are known primarily by what they did in the transcendent times of the 'beginnings'.*

---

*Mircea Eliade, *Myth and Reality*. Allen & Unwin, London (1964), pp. 5–6. We shall refer to this book as E I.

But we cannot endorse the sentence added by Eliade in explanation of the significance of myths: 'Because myth relates the gesta of Supernatural Beings and manifestation of their sacred powers, it becomes the exemplary model of all significant human activities' (E I, p. 6). We have to oppose this seemingly innocent statement, since according to our conditioning theory of myth it is not the *content* of the myth that gives it its power and significance, and to the related supernatural beings their sacred quality, but the collective conditioning of the myth and the accompanying ritual as a whole. Starting off with the two-layer theory and the Conditioning Hypothesis we have emphasized that the sacredness and inviolability of myths, as *a priori* truths, cannot be established just by telling so: only collective conditioning can make a myth.

A mathematical reader of anthropological literature cannot avoid the impression that the explanatory power of the theories offered is mostly rather weak, and that an obvious *circulus vitiosis* is often involved. No doubt anthropological science as it stands is at its best when it focuses on empirical observations and on immediate inductive generalizations from that knowledge. An example of the latter is the insight concerning the *parallelism of myth and ritual*. Referring to the pioneering works of James Frazer, *Malinowski wrote:

In these works, as in so many of his other writings, Sir James Frazer has established the intimate relation between the word and the deed in primitive faith; he has shown that the words of the story and of the spell, and the acts of the ritual and ceremony are the two aspects of primitive beliefs (M I, p. 147).

What our present theory of myth has to add to this fundamental insight is the bond of conditioning that inseparably links together the myth and the corresponding ritual.

After this excursion into the concept of myth let us come back to magical myths, i.e. to those *a priori* truths to which primitive man resorts in order to overcome the hazards of earthly life. If primitive man, faithful to his conception of all knowledge as the Maker's knowledge, cannot master rain or storm, drought or fertility, beasts or diseases except by repeating the creative acts of the original Makers, how was such a thing possible at all for mortal man? After all, primitive man was a perfectly normal human being, in possession of rational intellect like ours, and he very well grasped that only a *symbolic* repetition of those acts was possible.

Words are symbols of real things, and material things can be made into symbols of other material things, and deeds of other deeds. The recitals of spells contained in magical rites were already symbolic repetitions of the primordial deeds told. But the ritual conditioning gave a sacred meaning to all the acts and material things involved in the ceremony, thus making them symbols of primordial acts and things. In this way little things came to

* James George Frazer, *The Golden Bough*, 12 volumes, London (1907–1915).

signify places or deeds, or the whole stages of primordial acts, which is known as the principle of *pars pro toto* in *primitive symbolism*. Or, as told by Eliade:

Take the commonest of stones; it will be raised to the rank of 'precious', that is, impregnated with a magical or religious power by virtue of its symbolic shape or its origin: thunderstone, held to have fallen from the sky; pearl because it comes from the depths of the sea. Other stones will be sacred because they are the dwelling place of the souls of ancestors, or because they were once the scene of teophany, or because a sacrifice or an oath has consecrated them.*

Add to this that the causal bond here referred to with the word 'because' is accomplished by collective conditioning, and the description given is also impeccable from our present point of view.

How is magic actually made? The magician usually prepares himself for the ceremony in seclusion from other people, but also from the object of his magic. For instance, often the animals that the magic is to concern are tabooed to him altogether. He has to possess a good memory, for

a magical formula is an inviolable, integral item of tradition. It must be known thoroughly and repeated exactly as it was learned. A spell or magical practice, if tampered with in any detail, would entirely lose its efficancy (M I, pp. 213–214).

This indeed is an essential part of its conditioning effect, as was noted previously. On the other hand, this makes the magician an expert, and indeed, as a rule, he has inherited the magic of his specialty from a kinsman who was a magician. In addition to being associated with a certain clan or subclan, magic is often also linked with a definite place, so that certain villages 'own' the magic of a definite kind. This is the case for instance with the totemic system of Central Australian tribes.

After all the solemn preparations, the performance of the magic rite itself has often struck ethnographers as a rather commonplace act, almost a routine affair.

In Timor, for instance, when a rice field sprouts, someone who knows the mythical traditions concerning rice goes to the spot [Eliade tells (E I, p. 15)]. 'He spends the night there in the plantation hut, reciting the legends that explain how man came to possess rice... Those who do this are not priests.†

Or, coming back to the Central Australians:

Australian totemic myths usually consist in a rather monotonous narrative of peregrinations by mythical ancestors or totemic animals. They tell how, in the 'Dream Time'–that is, in mythical time–these Supernatural Beings made their appearance on earth and set out on long journeys, stopping now and again to change the landscape or to produce certain animals and plants, and finally vanished underground. But knowledge of these myths is essential for the life of the Australians. The myths teach them how to repeat the creative acts of the Supernatural Beings, and hence how to ensure the multiplication of such-and-such an animal or plant. These myths are told to the neophytes during their initiation. Or, rather, they are 'performed', that is, re-enacted (E I, p. 14).

---

* Mircea Eliade, *The Myth of the Eternal Return or, Cosmos and History*, 2nd Princeton/Bollingen paperback printing (1974), p. 4. Henceforth referred to as E II.

† A. C. Kruyt, quoted by Lucien Lévy-Bruhl, *La mythologie primitive*, Paris (1935), p. 119.

Most of our knowledge of magic comes from studies of the Central Australians, the Melanesians, and the North-American Indian tribes. Of these, the Central Australians are most primitive and represent the Stone Age people, while the Melanesians stand highest on the steps of development. Considering the state of the art in anthropology in 1934, Bronislaw Malinowski came rather close to the conditioning theory of myths in saying:

It is only because man is in need of magic wherever his forces fail him that he must believe in his magical power, which is vouched for by myth. On the other hand, the primeval myth of the magical miracle is confirmed and repeated in every act of subsequent magic (M II, p. 30).

19.3. The Crucial Test of the Collective Conditioning of *a priori* Truths

To obtain the crucial test of the Conditioning Hypothesis we have to discuss the birth and establishment of the central religious myths of primitive communities in more detail. In a deservedly famous passage, Malinowski summed up his view of the origin of religion as the magic of death as follows:

And here we can see how the substance of religious belief is not arbitrary. It grows out of the necessities of life. What is the root of all the beliefs connected with the human soul, with survival after death, with the spiritual elements in the Universe? I think that all the phenomena generally described by such terms as animism, ancestor-worship, or belief in spirits and ghosts, have their root in man's integral attitude towards death. It is not mere philosophical reflection on the phenomena of death, nor yet mere curiosity, nor observations on dreams, apparitions, or trances, which really matter. Death as the extinction of one's own personality, or the disappearance of those who are near, who are loved, who have been friends and partners of life, is a fact which will always baffle human understanding and fundamentally upset the emotional constitution of man. It is a fact about which science and rational philosophy can tell nothing. It cuts across all human calculations. It thwarts all practical and rational efforts of man. And here religious revelation steps in and affirms life after death, the immortality of the spirit, the possibilities of communion between living and dead. This revelation gives sense to life, and solves the contradictions and conflicts connected with the transience of human existence on earth (M II, pp. 27–28).

Yet, it is conceivable that merely overcoming death is not sufficient, that man could have set his hopes higher from the very beginning, and seek liberation from all the pain and suffering of life, not just from death. But the history of religions assures us that the idea of total salvation from *all* earthly miseries was rather late to come. It did not appear until the Judaeo-Christian eschatological visions much later. The origin of religion is in the magic of death, whose purpose it is to ensure to man his immortality, in a spiritual form, after death.

Primitive man, who knew knowledge only in the form of the Maker's knowledge, was bound to ask: who invented death? His creative fancy, once stirred to movement, filled his mind with fascinating tales of those more or less plausible ancient happenings since which death falls to the lot of man. A number of personages and animals were necessarily entwined with these events, and the next question was to be: who made them? And, finally, who made the world in which all these deeds occurred? So we are back in the *cosmogonic myth* with which the greatest part of all powerful magic has to start. The creation of the world, the creation of man and of other protagonists in the drama of mythical times, and how man lost immortality – this is the succession of events that to primitive man answers the questions evoked by his terror of death.

Fear and hope are two facets of the same sentiment. Where there is fear, there too hope mingles. It is the medley of hope and fear that, under the compulsiveness of a natural law, made primitive man pose the problem: how death originated? He won't see the possibility that death could have been there from the outset, that being mortal had ever been the fate of man, as such a fate would not have left him any hope of winning against death. Furthermore, putting the question that way was fully in line with his thinking generally: if there is something that I cannot do, some hurdle perhaps that I cannot overcome, there is always someone who is able to do that, viz. he who made that hurdle at the beginning of time. This is the common idea behind all magic, and the magic of death is no exception to the rule.

The world of primitive man consisted of the immediate surroundings of his life. The cosmogonic myths vary, according to the environment.

On the shores of the Pacific and in its many islands [Malinowski writes], we are told how the world was fished out of the sea or moulded out of slime; or again, from other continents, we have stories relating how out of chaos the various parts or elements of the Universe have been shaped in succession, or how the earth was hurled from space, or out of darkness, by a divine maker. The wide range of beliefs which are usually labelled 'nature worship' again have a rich mythology of totemic ancestors, of the early appearance and miraculous, though not always moral, behaviour of Nature Gods (M II, p. 9).

What strikes one especially in these cosmogonic myths is the frequent appearance of *deus otiosus*, the Creator, who, after having created the world and man, withdrew into the Sky or some remote part of the universe, without caring much for humanity ever since. Sometimes he even left part of the Creation to other divine beings, his son or his representatives. Is the myth of an otiose god a manifestation of hopelessness, of a feeling of being deserted, in those often extremely primitive hunting and gathering tribes, in which the myth has been recorded in our age by ethnographers? Or does it express genuine feelings of archaic communities?

That a weariness of life can take possession of a whole primitive tribe was demonstrated by the desperate wanderings of the Guarani Indians in search of the 'land without Sin', wanderings that began in the nineteenth century and carried on until 1912. They covered long distances under the guidance of their shamans whose ideas were written down by Curt Nimuendaju* in the following moving words:

Not only the Guarani but all the Nature is old and weary of life. How often the medicine-men,

* Curt Nimuendaju, 'Die Sagen von der Erschaffung und Vernichtung der Welt als Grundlagen der Religion der Apapocuva-Guarani', *Zeitschrift für Ethnologie*, Vol. 46 (1914), p. 335 (translated by Eliade, E I, pp. 57–58).

when they went to meet, in dream, Nanderuvuvu, have heard the Earth imploring him: 'I have already devoured too many corpses, I am filled from it, and I am exhausted. Do make an end of it, my Father! The water also beseeches the Creator to let it rest, disturbed no longer, and so the trees…and also the rest of Nature.'

Deus otiosus is met not only in the very primitive Pygmies but in the Bantus as well, and also in the Yorubas of the former Slave Coast. Likewise, many American Indian tribes have the idea of the Creator who deserted them. One can well imagine that all these tribes could have reason to feel abandoned, after all those thousands of years during which they had invoked their God in vain. But the myth of an otiose Creator was familiar to many historical civilizations too, for instance, in Mesopotamia. And the Greeks' Ouranus was both otiose and impotent! Is it possible that people were by then already tired of life? We shall get a sort of answer, or at least an idea of the supposed general feeling of life in the first great urban societies of the Near East, when we later investigate their conditions of living and their mythologies generally.

But as far as the ancient, truly primitive communities are concerned, we can confidently go by the assumption that weariness of life was rather uncommon. It seems that primitive man did not even recognize any natural cause of death, as people in the Trobriand Islands still did not when Malinowski visited them in 1915, but used to ascribe any death to some evil magic by a sorcerer in the pay of an enemy. And there is the testimony of all central religious myths, telling in one form or another of a genuine longing for immortality. In these myths how man lost his original immortality is a recurring theme. Their general message is this:

Man is what he is today because a series of events took place *ab origine*. The myths tell him about these events and, in so doing, explain to him how and why he was constituted in this particular way. For *homo religiosus* real, authentic existence begins at the moment when this primordial history is communicated to him and he accepts its consequences…. For example: man is mortal because a mythical Ancestor stupidly lost immortality, or because a Supernatural Being decided to deprive him of it, or because a certain mythical event left him endowed at once with sexuality and mortality, and so on. Some myths explain the origin of death by an accident or an oversight: God's messenger, some animal, forgets the message, or lingers idly on the way, arrives too late, and so on. This is a plastic way of expressing the absurdity of death (E I, p. 92).

Take any of those stories, called by Malinowski (M II, p. 16) the 'myth of myths', which are quite essential to primitive faith, and you will find in most cases quite a trivial tale displaying familiar human sentiments in the ancient protagonists of the drama. In a typical myth telling how human beings arrived on earth and lost their immortality, the Trobrianders attributed this paramount loss to the passion of an insulted old woman. Originally mankind was living, leading much the same kind of existence as nowadays but underground. They ascended on to earth by crawling out of certain 'holes', also called 'homes', which were still known to the natives. At that time men were able to rejuvenate just by sloughing the skin like the snakes still do. But once, after an old woman from the village of Bwadela had thrown her old skin in the waters of a creek her granddaughter no longer recognized her, and drove her away. Hurt and angry, the old woman took back her old skin, and put a curse on her daughter and granddaughter. From that moment human beings were mortal. However, their ghosts went on 'living' after death in the villages, until another disaster occurred. One of the invisible ghosts was scalded with hot broth by her daughter who did not recognize her. To the spirit's protest the daughter replied: 'Oh, I thought you were away, I thought you were only returning after harvest.' Since that day the souls of the dead do return to their original villages in the Trobriand Islands only after harvest, living the rest of the time in the island of Tuma, not far away.

Accordingly, it is the soul of man, called *baloma* in the Trobriand Islands, that in the primitive faith already stands for the continuance of human life after death. And all material things do have their own baloma, so that already in the primitive faith we have the dual world of all religions, where behind the spurious world of our sensory observations there is another, real and sacred world of which only religious revelations can tell. The sacred wisdom of the second, real world is solemnly entrusted to each new generation in the ceremonies of *initiation*.

The rituals of initiation provide a lesson in the craft of collective conditioning. They are preceded by a period of preparation during which the youngsters to be initiated live in seclusion from other people. In the initiation proper the novices have to suffer a series of ordeals culminating in an act of bodily mutilation.

At the mildest [Malinowski writes, (M I, p. 38), it consists of] a slight incision or the knocking out of a tooth; or, more severe, circumcision; or, really cruel and dangerous, an operation such as the subincision practiced in some Australian tribes. The ordeal is usually associated with the idea of the death and rebirth of the initiated one, which is sometimes enacted in a mimetic performance.

And all the time, besides this well-known inquisitorial technique of intensifying, by means of torture, one's consciousness of the delivered message, the systematic instruction of the youth in sacred, religious myths is going on. The ceremonies, including all the ordeals that accompany them, were, according to myths, instituted by ancestors or superhuman Superior Beings in mythical times, and it is a further purpose of initiation to make the youth familiar with the stories bearing on them. So, as remarked by Malinowski:

the youth is taught...the sacred traditions under most impressive conditions of preparation and ordeal and under the sanction of Supernatural Beings – the light of tribal revelation bursts upon him from out of the shadows of fear, privation, and bodily pain (M I, p. 39).

It is hard to imagine any setting which could be more efficacious than that described above, if one had to imprint, by means of collective conditioning, the absolute truth of some ideas on someone's mind. It is beyond any doubt that primitive man, in what concerns the efficacy of brainwashing has already produced, in the form of initiating rituals, something that neither the Holy Office of the Middle Ages nor Stalin's Secret Police in the thirties could surpass. It adds to the solemnity of the sacred wisdom that it is usually not told to women or children, that is, to the uninitiated. Here the difference between profane tales and sacred myths is conspicuous. Whereas, as was noted by Eliade, 'false stories' can be related anywhere and at any time, myths must not be recited except during a period of sacred time, usually in autumn or winter, and at night.

This custom [he adds] has survived even among peoples who have passed beyond the archaic stage of culture. Among the Turco-Mongols and the Tibetans the epic songs of the Cesar cycle can be recited only at night and in winter (E I, p. 10).

The period of sacred time during which myths are normally recited centres around the harvest or New Year's festival, to which we shall return later. This is also the preferred time of initiation ceremonies. But there are some occasions, like *birth or death*, in which the recital of central religious myths usually occurs even outside the sacred period. From North-American mythology comes the report* telling of a custom of the Osage Indians. Whenever a child is born, 'a man who had talked with the gods' is called to recite the cosmogonic myth followed by the myth of the origin of animals to the newborn infant. This recital must happen before the child is given breast. And before it is given water to drink, the holy man is fetched again, this time to recite the myth of the Creation now followed by the origin of water. And when the child wants solid food, the origin of grains and other foods, preceded by the cosmogony myth, is recited. These kinds of religious rituals are, obviously, more prone to resuscitate faith in the parents and other relatives rather than in the child, who hardly grasps their meaning. They are also rare examples of religious ceremonies that are private and not collective, though no doubt they are performed in turn before every family of the tribe.

The funerals of eminent persons, of course, are collective happenings where the conditioning impact of collective action must be very intense. 'Mortuary ritual is the enactment of the truth of immortality', Malinowski says (M II, p. 28), and one cannot but agree. The importance of *mortuary ritual* in primitive religion is sometimes appraised to be as great as that of the ceremonies of initiation, as is indicated by the custom, encountered in the Santals of India,† to let the cosmogonic myth be recited by the guru to the individual only on two occasions during his life: first, when a Santal is granted full social rights, and for the second time during the funeral service. But by far the most impressive shows of conditioning religious rituals, including those of initiation, used to be enacted collectively in connection with the annual festivals celebrating the new harvest and the New Year.

When speaking of the *New Year* as a part of primitive religion it must be remarked that it was not a feast occurring on the same fixed date everywhere.

* Alice C. Fletcher and F. La Flesche, *The Omaha Tribe*, Bureau of American Ethnology, 27th Annual Report, Washington (1911), p. 116.
† P. O. Bodding, 'Les Santals', *Journal Asiatique* (1932), pp. 58 ff.

The length of a year was not connected with precise astronomical observations concerning the solar year until much later in Egypt. In primitive societies the New Year was, and still is, the occasion on which the taboo on the new harvest is lifted, so that people can safely start eating it without having to fear the punishments of ancestors or other transcendental dangers. That is why the beginning of a new year varied greatly, not only from one region to another but also in different periods. It could happen in March–April, or on 19 July (in Egypt), or in August–September as in the Trobriand Islands with their milamala festivals that were made so famous by Bronislaw Malinowski; or it could take part even later in autumn, or in December–January as in the Christian world today. Sometimes in primitive communities, where various species of grains or fruits ripen at different seasons, one can find several New Year festivals.* But all over the primitive world, both the ancient and the still-existing ones, the period around the combined festival of harvest and the New Year was, and is, a sacred time during which the central religious myths – that of cosmogony, of the creation of man, and of the spiritual immortality of man – are solemnly recited and enacted in numerous collective rituals.

The liveliest description of this festive period comes as usual from Malinowski, who attended the milamala festival in the Trobriand Islands twice, and reported his experiences in one of his classical essays.† The milamala is held after the yam crops have been harvested – this timing of the annual feast is common in all protoagricultural societies, that is, among the cultivators of tubers. It is opened by certain ceremonies connected with dancing and with the first beating of drums, both of which are activities that are tabooed outside the milamala month. The milamala indeed takes a whole moon having the same name, but dancing can sometimes be extended for another moon, or even for two.

All dancing had an extrahuman origin and was thus in primitive societies considered sacred:

Choreographic rhythms have their model outside of the profane life of man; whether they reproduce the movements of the totemic or emblematic animal, or the motions of the stars; whether they themselves constitute rituals (labyrinthine steps, leaps, gestures performed with ceremonial instruments) – a dance always imitates an archetypal gesture or commemorates a mythical moment. In a word, it is a repetition, and consequently a reactualization, of illud tempus, 'those days' (E II, pp. 28–29).

Dancing, repeated day after day and night after night, interspersed with ceremonial visits to neighbouring villages, is for some time the only visible mark of the milamala. But from the beginning of the festive period the souls of the dead, the baloma, have been present too. They arrived simultaneously with the inauguration of the dancing, to enjoy the repast given to them at

* Martin P. Nilsson, *Primitive Time Reckoning*, Acta Societatis Humaniorum Litterarum Lundensis, I, Lund (1920), p. 270.

† Bronislaw Malinowski, 'Baloma; the Spirits of the Dead in the Trobriand Islands', reprinted in M I, pp. 149–274.

the katukuala, the opening feast of the milamala. There are platforms built for them in the villages, even 5 to 7 metres high, and the baloma are expected to observe the festivities from these platforms. Also displays of valuables and food are arranged, meant to please both the living and the dead.

Throughout the course of the milamala moon, several kinds of meals are offered to the baloma. They are supposed to eat the baloma of the food exposed, whereafter – but only after the ghosts have taken their share – men can eat the material part of the food. While the baloma are in the village, a number of phenomena loaded with a sacred religious significance occur. Not only can the spirits, if they are not content with the rituals and with the behaviour of people generally, spoil the feast by sending thunder, rain, and bad weather, but they are also able to show their discontent by spoiling next year's crops. This indeed is the reason why often several bad years follow each other, since a bad year makes it difficult for people to arrange a good milamala, which in turn angers the dead. It becomes the religious character of the festival that the souls, especially of those relatives who are recently deceased, appear to men in dreams during this sacred period. Other unusual things also happen: coconuts fall down more frequently than is usual, as the baloma are eating them. All this goes on for two to four weeks: the milamala has a fixed end, that is, the second day after the full moon, but it may have began any time between the previous full moon and the new moon. The presence of the souls ends at the second night after the full moon when, about one hour before sunrise, the morning star appears and the dancing ceases, while the drums intone a peculiar beat calling the baloma to leave the village and go back to their Tuma-island. As a matter of fact the souls of the dead are rather unceremoniously, even irreverently, driven out of the village.

Malinowski wrote: 'At Milamala we see how the ritual of give and take, the sacrifices to the spirits and their response, are the expression of the truth contained in mythology' (M II, p. 25). And we can also see how this give and take is connected with the collective conditioning of the central religious myths. In the major annual festival, the very survival of the tribe is connected with the existence and content of the immortal souls of the dead: if they are angry, whole villages will suffer. And again, a good harvest is there to testify to the goodwill of the dead. There are other, more immediate, rewards too: the milamala, like similar feasts in other primitive communities, is for the younger people in the village the blessed period of sexual licence, the proper time for amorous advances. All in all, both the punishments and rewards that are associated with the presence of the dead at the great annual festival are concrete enough to reinforce, by way of collective conditioning, the belief in the existence of man's immortal soul.

The combined harvest and New Year festivities, such as the milamala, are the major collective ceremonies of primitive religion. Malinowski, with his unerring gift of putting the right questions, was led to ask:

Why are most religious acts in primitive societies performed collectively and in public? What is the part of society in the establishment of the rules of moral conduct? Why are not only morality but also belief, mythology and all sacred tradition compulsory to all the members of a primitive tribe? In other words, why is there only one body of religious belief in each tribe, and why is no difference of opinion ever tolerated? (M I, p. 60).

With his questions Malinowski, unaware of it himself, has unequivocally defined the general circumstances that make possible the conditioning of *a priori* truths in man. In accordance with the Conditioning Hypothesis their establishment presupposes collective, public confessions of the common belief and a common display of taboos on forbidden ideas, performed in unison. This is why the central ceremonies, not only of archaic religions but also those of modern totalitarianism, must include the *whole collective*. This is equally true of the milamala, as it is of the collective happenings of the Nazis, Fascists, or Communists.

The role of society is decisive in the establishment of *a priori* truths in its members' minds. As far as this sacred absolute knowledge is concerned, dissident opinions must be excluded in order to prevent contra-conditioning of such *absolute beliefs*. This is why we have here the crucial test of the explanation of myths as *a priori* truths established and reinforced by collective conditioning:

The Crucial Test of the Conditioning Theory of Myths:
The collective conditioning of *a priori* truths is possible only, if any dissident opinion is strictly forbidden under the threat of severe punishment. This is why there cannot be but one body of religious beliefs and myths in each primitive community, just as there can be only one orthodoxy, demanded of all members of society, in a totalitarian state.

The fact that this is true of primitive communities, and explains the collective-specific nature of myths, is not to say that primitive communities were totalitarian societies. They were not, because they were lacking an important element of totalitarianism, viz. the concentration of absolute power in the hands of a privileged, religious or ideological elite. What primitive communities and totalitarian societies do have in common is the mechanism of collective brainwashing by collective conditioning.

Though Malinowski so accurately formulated the essential problem of primitive religion, viz. 'the identification of the whole social unit with its religion; that is, the absence of any religious sectarianism dissension, or heterodoxy in primitive creed', and grasped that 'religion is a tribal affair' (M I, p. 55), his solutions to it are far from being satisfactory. They share the common tendency of most contemplations about the origins of religion to disappear into a mist of rhetoric, or alternatively to be entangled in a vicious circle. First, Malinowski seems to suggest that even primitive faith is not quite as collective as it appears: 'The saving belief in spiritual continuity after death is already contained in the individual mind; it is not created by society' (M II, p. 62). Here, faithful to the Conditioning Theory, I must bluntly oppose. What is contained in the individual mind is the *hope* of immortality, and the *fear* of its non-existence as well, but positive *belief* is created through collective conditioning by means of rituals.

Next, Malinowski refers to a case where the sheer motif of survival seems to postulate a collective faith:

Totemism, the religion of the clan, which affirms the common descent from or affinity with the totemic animal, and claims the clan's collective power to control its supply and impresses upon all the clan members a joint totemic taboo and a reverential attitude towards the totemic species, must obviously culminate in public ceremonies and have a distinctly social character (M I, p. 65).

True as this is, does it really *explain* why dissident opinions are not tolerated? After all, the motif of survival is shared by most developed societies as well, in which dissident opinions are not punished. The same methodological objection applies to what Malinowski considered to be 'the essentially sound methodological principle', viz. 'that worship always happens in common because it touches common concerns of the community' (M II, pp. 6–7). After all, there are 'common concerns' that do not require common worship, and indeed may require no worship at all. The mere fact that people have common interests does not explain why, in a primitive community, they have common absolute beliefs, excluding all dissidence, and why, in a neighbouring community, they have other absolute beliefs, again compulsory to all the members of the community.

In the midst of what seems to be the general vagueness of anthropological explanations of why the central myths of primitive religion are believed, a sudden flash may appear, promising to solve the problem in one stroke. So Eliade simple assures: 'The cosmogonic myth is [felt] true because the existence of the World is there to prove it; the myth of the origin of death is equally true because man's mortality proves it, and so on' (E I, p. 6). After the first shock one has to demur. If the faculty of normal, profane logical reasoning in primitive man has to be taken seriously, we must endow him with the capability of logical distinction between a sufficient reason and a necessary reason. If some incident that occurred in mythical times effected the curse of death upon mankind, that incident can be regarded a sufficient reason of mortality, but never a necessary one as Eliade would have primitive man believe. Indeed some *other* myth of the origin of death would have done the job as well. You cannot infer from the mortality of man the truth of any *particular* myth of the origin of death, and you cannot conclude from the existence of the world the truthfulness of any *particular* cosmogonic myth, and we have many reasons to suppose that primitive man did understand such logical distinctions. So the firm belief in the particular myths that make up the religion of a particular primitive community must be accounted for in some other way, without violating the faculty of profane intellect in primitive man.

A good candidate is the explanation given above in terms of Conditioning

Theory. It holds that the central myths of primitive religion are established as absolute, sacred truths through conditioning by means of collective rituals, the core of which is the recital of myths, in the presence of the community in a body. It follows that other primitive communities will have other absolute beliefs, expressed in other myths, and equally compulsory to all members of those communities. This is because the public expression of any dissident idea would amount to contra-conditioning, which indeed is the only way to dismantle dogmatic belief.

If we accept the Conditioning Theory of myths, this makes it possible for us to explain not only why different primitive communities have different beliefs and myths, but also why these beliefs are felt by primitive man to be so obvious that the idea of questioning them does not easily come to him. That myths are so experienced is an empirical fact that was recorded by Malinowski in his pioneering journey.

An example given by Malinowski concerns the female gender of certain spirits in the Trobriand Islands: 'But it was clear that to all my informants the fact of women being tolipoula (the spirits who are the masters of fishing) was so natural that it had never occurred to them to question it previously' (M I, p. 208). Indeed, conditioned ideas have been fed into the mind past the conscious control of man. Often the conditioning is so strong that powerful negative emotions are associated with taboos: 'The forbidden totem animal, incestuous or forbidden intercourse, the tabooed action or food, are directly abhorrent to him' (M I, p. 57). In terms of the Conditioning Hypothesis this is easily explained, as such beliefs have been established in the consciousness by making them become associated with rewards or punishments, i.e. with positive or negative emotions. Well-conditioned *a priori* truths can always be expected to be loaded with strong emotions.

Primitive religion and its obvious connection with collective conditioning having been discussed sufficiently for the present purpose, it only remains to make some completing notes.

Note 1 (Malinowski Near the Conditioning Theory). In his more consistent theoretical contemplations Bronislaw Malinowski, mainly known for his perceptive and accurate anthropological observation, came fairly close to the conditioning theory of myths. Take, for instance, the following three excerpts from his Riddell Lectures that jointly, in fact, sum up some essential features of this theory:

Thus during harvest and after, at the season of Milamala when the spirits come for a few weeks and settle again in the villages, perched upon the trees or sneaking about the houses, sitting on high platforms specially erected for them, watching the dancing and partaking of the spiritual substance of the food and wealth displayed for them, the knowledge of the whole dogmatic system concerning spirits is necessary, and it is then imparted by the elder to the younger. Every one as yet uninformed is told that after death the spirit has to go to Tuma, the nether world associated with the small island of that name.... In order to keep in touch with the supernatural realities and happenings of the Milamala, it is necessary for every one to be instructed in the ways of spirits and on their behaviour (M II, pp. 14–15).

And a little later

Dogma, ritual, and ethics are therefore inseparable. Take the facts presented above about the mythological cycle of death, immortality, and the spirit world. We started from stories, but we were immediately led into a discussion of action directed towards the subject-matter of these stories, that is, a discussion of ritual (M II, p. 25).

And, as a final confirmation:

The ritual of sacrifices and devotion to ancestors is essential. Ritual, in turn, is necessarily permeated with ethics: the ancestors in spirit form, as the living parents, punish for bad behaviour and disrespect, and reward for good conduct and dutiful services (M I, p. 27).

Note 2 (The Profane Elements of Religion). Profane elements connected with the normal use of reason are involved even in primitive religion, whose central parts belong to the Sacred, to the realm of myths. In accordance with the two-layer theory, in all questions that are not tabooed by one's *a priori* truths, man is able to apply his rational intellectual abilities. Even primitive man, as we have seen in Section 19.1, has his profane reason as well, but he can use it only in issues that remain uncontaminated by his mythical truths. Malinowski has pointed out there is a division inside the primitive mind that clearly separates what is held to be sacred dogmatic knowledge from what is left for free individual imagination in religious ideas too. In addition to 'social ideas' or 'dogmas' told in myths, which are believed in by every member of the community, there are public opinions and individual speculations about religious matters that do not belong to the collectively conditioned truths (for their systematics as given by Malinowski see Section 18.6, passage 6).

For instance, you have to believe in the existence of the baloma, one's immortal soul, and in its visiting the home village during the milamala festivities, as is told in myths. But myths do not tell all the details of the baloma or their visits, such as where exactly the spirits dwelt during the milamala, or how they were transported from and to the Tuma Island. All this is left for everybody's free fancy.

Some of this profane theological imagery was, according to Malinowski, widespread among the members of the community, some other ideas were speculations of some intelligent individual, who had a longing for further details. Characterizing the difference between the Sacred and the Profane in matters of religion, Malinowski wrote:

In the domain of the purely intellectual aspect of belief, there is room for the greatest range of variations. Belief, of course, does not obey the laws of logic, and the contradictions, divergencies, and all the general chaos pertaining to belief must be acknowledged as a fundamental fact (M I, pp. 247–248).

Note 3 (The Positive Role of Conditioning). We have in this book, because of the special topic of the book, discussed only one aspect of conditioning, viz. the conditioning of *a priori* truths as an obstacle to human self-steering. There are other, positive aspects as well. One of the greatest psychologists of our century, H. J. Eysenck, has emphasized the positive role of conditioning in learning, or rather in training, and especially in the formation of social conscience that enables men to live social life in the first place (Eysenck, 1977). It is obvious that conditioning set in this larger context is an important challenge to scientific research.

Note 4 (Modern Means of Collective Conditioning).   One can ask what are the modern means of collective conditioning applied to the formation of collective, *a priori* truths in an industrial or post-industrial society. This topic has not been discussed in detail in the present book, but some obvious hints may be given. Collective marches, demonstrations, and mass meetings are, of course, just the kinds of occasions where collective conditioning of *a priori* (religious or political) truths can be established. However, in a Western, non-totalitarian society there are rather few of such occasions today. In the West, no doubt the most effective way of collective conditioning of common social ideas is mass communication: TV, radio, and the press – TV especially.

Note 5 (The Long Cycles of Civilization?).   It can be argued that the self-steering and steerable-from-outside layers of human consciousness, discussed in this part of the book, may be favoured, each in turn, at the cost of the other layer. In fact, it was suggested above that mythological thinking, and hence the steerable-from-outside layer of consciousness, dominated over the self-steering layer in archaic societies. But we can well think of the formation of cycles that now indicate a growth of mythological, now a domination of scientific–rational layers of consciousness even in the later history of human societies. The mythical, steerable-from-outside layer being assumed to be closely connected with the (feeling of) lack of security in human community, such cycles could be expected to vary together with the dominance of either the arguments related to security or the arguments associated with intellectual freedom and the search for truth on the political stage.

Could it be that such security versus truth cycles explains some of the long-term variations in human history? Have not the arguments related to security often been used to suppress individual freedom and the search for truth? There have surely been some periods of history, for instance the eighteenth and nineteenth centuries in Europe, when intellectual and other individual freedoms together with science were particularly flourishing. Such long cycles would essentially be cycles of civilization, and here we come back to Andrew Bródy's ideas about the contemporary deceleration of global economic growth due to the decreased appreciation of science and scientists in advanced societies. Are we now living on a downhill slope of a long cycle of civilization, whose ascending period coincided with the beginning and early blooming of Western exact sciences in the eighteenth century or even before that?

Scientific References

(Historical and anthropological references in Parts 4 and 5 are indicated in the text)

Abraham, R. H. and Shaw, C. D. (1983), *Dynamics: The Geometry of Behavior I–III*. Aerial Press, Santa Cruz.

Akerlof, G. A. and Stiglitz, J. E. (1969), Capital, wages and structural unemployment, *Economic Journal*, **79**, 269–281.

Andronov, A. A., Vit, A. A., and Khaikin, C. E. (1966), *Theory of Oscillations*. Pergamon Press, Oxford.

Ashby, W. R. (1970), *An Introduction to Cybernetics*. Chapman & Hall, London (first published in 1956).

Ashby, W. R. (1972), *Design for a Brain*. Chapman & Hall, London (first published in 1952).

Aubin, J. P., Bensoussan, A., and Ekeland, I. (eds) (1983), *Advances in Hamiltonian Systems*, Birkhäuser, Basel.

Aulin, A. Y. (1982), *The Cybernetic Laws of Social Progress*. Pergamon Press, Oxford.

Aulin, A. Y. (1985), Cybernetic causality: a unitary theory of natural and social systems, *Mathematical Social Sciences*, **10**, 103–130.

Aulin, A. Y. (1986a), Notes on the concept of self-steering. In Geyer, F. and van der Zouwen, J. (eds).

Aulin, A. Y. (1986b), Cybernetic causality II: Causal recursion in goal-directed systems, with applications to evolution dynamics and economics, *Mathematical Social Sciences*, **12**, 227–264.

Aulin, A. Y. (1987a), Cybernetic causality III: The qualitative theory of self-steering and social development, *Mathematical Social Sciences*, **13**, 101–140.

Aulin, A. Y. (1987b), The method of causal recursion in mathematical dynamics: the interruptions of Feigenbaum bifurcations in Verhulstian ecosystems and other applications, *International Journal of General Systems*, **13**, 229–255.

Aulin, A. Y. (1987c), Methodological criticism, *Systems Research*, **4**, 71–82.

Aulin-Ahmavaara, A. Y. (1977), A general theory of acts, with application to the distinction between rational and irrational social cognition, *Zeitschrift für allgemeine Wissenschaftstheorie*, **8**(2), 195–220.

Aulin-Ahmavaara, A. Y. (1979), The law of requisite hierarchy, *Kybernetes*, **8**, 256–266.

Aulin-Ahmavaara, P. (1987), *A Dynamic Input–Output Model with Non-homogeneous Labour for Evaluation of Technical Change.* Annales Academiae Scientiarum Fennicae, Ser. B, No. 242, Helsinki.

Ayres, F. Jr. (1974), *Theory and Problems of Matrices.* McGraw-Hill, New York.

Bhatia, N. P. and Szegö, G. P. (1967), *Dynamical Systems: Stability Theory and Applications.* Springer-Verlag, Berlin.

Bohr, H. (1924, 1925, 1926), Zur Theorie der fastperiodischen Funktionen I, II, III, *Acta Mathematica*, **45**, 29–127; **46**, 101–214; **47**, 237–281.

Brock, W. A. and Scheinkman, J. A. (1976), Global asymptotic stability of optimal control systems with applications to the theory of economic growth, *Journal of Economic Theory*, **12**, 164–190.

Bródy, A. (1970), *Proportions, Prices and Planning.* North-Holland, Amsterdam.

Bródy, A. (1985), *Slowdown: Global Economic Maladies.* Sage, London.

Bródy, A. (1988), a review of P. Aulin-Ahmavaara, *op. cit.*, in *Finnish Economic Journal*, **82**, No. 1.

Cass, D. and Shell, K. (1976a), Introduction to Hamiltonian dynamics in economics, *Journal of Economic Theory*, **12**, 1–10.

Cass, D, and Shell, K. (1976b), The structure and stability of competitive dynamical systems, *Journal of Economic Theory*, **12**, 31–70.

Eysenck, H. J. (1977), *Crime and Personality* (revised edition). Routledge & Kegan Paul, London (1st edition published by Houghton Mifflin, Boston, 1964).

Feigenbaum, M. J. (1978), Quantitative universality for a class of nonlinear transformations, *Journal of Statistical Physics*, **19**, 25–52.

Frommer, M. (1928), Die Integralkurven einer gewöhnlichen Differentialgleichung erster Ordnung in der Umgebung eines singulären Punktes, *Mathematische Annalen*, **99**, 222–272.

Geyer, F. and van der Zouwen, J. (eds) (1986), *Sociocybernetic Paradoxes: Observation, Control and Evolution in Self-steering Systems.* Sage, London.

Goodwin, R. M. (1967), A growth cycle. In C. H. Feinstein (ed.), *Capitalism and Economic Growth.* Cambridge University Press, Cambridge.

Higashi, M. and Klir, G. J. (1982), Measures of uncertainty and information based on possibility distributions, *International Journal of General Systems*, **9**, 43–58.

Hirsch, M. W. and Smale, S. (1974), *Differential Equations, Dynamical Systems, and Linear Algebra.* Academic Press, New York.

Iooss, G. (1979), *Bifurcation of Maps and Applications.* North-Holland, Amsterdam.

Jantsch, E. (1980), *The Self-organizing Universe.* Pergamon Press, Oxford.

Jorgenson, D. W. (1960), On stability in the sense of Harrod, *Economica*, August.

Kondratieff, N. D. (1935), The long waves in economic life, *Review of*

Economic Statistics, **17**, No. 6.

Korpinen, P. (1987), A monetary model of long cycles. In Tibor Vasko (ed.), *The Long-wave Debate*. Springer-Verlag, New York.

Kurz, M. (1968), The general instability of a class of competitive growth processes, *Review of Economic Studies*, **35**, 155–174.

Lanford, O-E. (1977), Turbulence seminar. In Bernard, P. and Rativ. T. (eds), *Lecture Notes in Mathematics*, **615**. Springer-Verlag, Berlin.

Lange, O. (1965), *Wholes and Parts: A General Theory of System Behaviour*. Pergamon Press, Oxford.

Leontief, W. W. (1953), General numerical solution of the simple dynamic input–output system, *Report on Research for 1953, Harvard Economic Research Project 5–15*, pp. 160–161.

Le Roy Ladurie, E. (1973a), Un concept: l'unification microbienne du monde, *Revue suisse d'histoire*, tome XXIII, fasc. 4.

Le Roy Ladurie, E. (1973b), *Le Territoire de l'historien*. Gallimard, Paris.

Le Roy Ladurie, E. (1974), L'histoire immobile, *Annales* (E.S.C.).

Lorenz, E. N. (1963), Deterministic nonperiodic flow, *Journal of Atmospheric Science*, **20**, 130.

Lucas, R. E., Jr. (1988), On the mechanics of economic development, *Journal of Monetary Economics*, **22**, 3–42.

Nemytskii, V. V. and Stepanov, V. V. (1972), *Qualitative Theory of Differential Equations*. Princeton University Press, Princeton (first published in 1960).

Nicolis, G. and Prigogine, I. (1977), *Self-organization in Nonequilibrium Systems*. Wiley-Interscience, New York.

Peitgen, H. O. and Richter, P. H. (1984), *Harmonie in Chaos und Kosmos* and *Morphologie komplexer Grenzen: Bilder aus der Theorie dynamischer Systeme*, obtainable from: Forschungsschwerpunkt Dynamische Systeme, Universität Bremen, D-2800, Bremen.

Pohjola, M. (1981), Stable and chaotic growth: the dynamics of a discrete version of Goodwin's growth cycle model, *Nationalökonomie*, **47**, 27–38.

Popper, K. R. (1972), *Objective Knowledge*. Oxford University Press, Oxford.

Ruelle, D. (1980), Strange attractors, *Mathematical Intelligencer*, **2**, 126.

Savage, L. (1971). *The Foundations of Statistics*. Wiley, New York (first published in 1954).

Schuster, H. G. (1984), *Deterministic Chaos*. Physik-Verlag, Weinheim.

Smale, S. (1980), *The Mathematics of Time: Essays on Dynamical Systems, Economic Processes, and Related Topics*. Springer, New York.

Steinmann, G. and Komlos, J. (1988), Population growth and economic development in the very long run: A simulation model of three revolutions, *Mathematical Social Sciences*.

Stove, D. (1982), *Popper and After: Four Modern Irrationalists*. Pergamon Press, Oxford.

Tsukui, J. and Murakami, Y. (1980), *Turnpike Optimality in Input–Output Systems*. North-Holland, Amsterdam (first edition 1979).

Varela, F. J. (1979), *Principle of Biological Autonomy*. North-Holland, New York.

Verhulst, P. F. (1838), Notice sur la loi que la population suit dans son accroissement, *Correspondance Mathématique et Physique 10*, 113–121.

Verhulst, P. F. (1845), Reserches mathématiques sur la loi d'accroissement de la population, *Mémoires de l'Académie Royale Bruxelles*, **18**.

Verhulst, P. F. (1847), [the same title] *Mémoires de l'Académie Royale Bruxelles*, **20**.

Volterra, V. (1936), *Lecons sur la Théorie Mathématique de la Lutte pour la Vie*. Gauthier-Villars, Paris.

Wan, H. Y. Jr (1971), *Economic Growth*. Harcourt Brace Jovanovich, New York.

Webb, G. F. (1985), *Theory of Nonlinear Age-dependent Population Dynamics*. Dekker, New York.

Woods, J. E. (1978), *Mathematical Economics*. Longman, Harlow (Essex).

Wright, G. H. von (1971), *Explanation and Understanding*. Cornell University Press, Ithaca.

APPENDIX

Why Mathematical Foundations Are Important in Science*

1. Introduction

For some decades, certain misconceptions concerning the role of mathematics in the fundamental theories of exact science have been successfully spread in the market-place of popular philosophy. Admittedly these notions have not notably affected the professional philosophy of science, and they have indeed very little touched, say, the American tradition of analytic philosophy. But they have attained an enthusiastic reception not only among laymen but among thousands of scientists, especially in the biological, behavioral and social sciences, and in part of the systems-research movement.

The misconceptions of science which I am referring to are those of *extreme empiricism*, such as has been most spectacularly represented by Sir Karl Popper and Imre Lakatos in England, and by Thomas S. Kuhn and Paul K. Feyerabend in the U.S.A. Since the mistakes involved are rather obvious, even trivial, and since they have been rarely spelled out in written text despite the considerable damage they have inflicted on the mentioned sciences, these mistakes, and their historical roots and consequences, will be discussed in some detail in the following.

The crucial point is the accumulation of exact scientific knowledge, and the role played by mathematical fundamental concepts in this connection. We can approach the problem indirectly, by starting with a phenomenon that is familiar to many scientists working in applied fields today.

2. The Inflation of Various Models and Paradigms

The influence of ultra-empiristic ideas today could be traced in any field of applied science where a solid foundation is lacking. We could choose our preliminary examples from works of much lower scientific standard to be met in current social research, or even in some areas of popular biology. However, to avoid a too specialized choice, we can take for an introductory example the well-known books published on the initiative of the Club of Rome, one of the most venerable international organizations to encourage

* The essay here reproduced was first published in *Systems Research*, 4, 1987, pp. 71–82.

the inquiry into the future prospects of mankind.

The choice of the Club of Rome is all the more better for the present purpose, as the reports to the Club are known to have preserved a high scientific standard in each special field involved. This cannot – and will not – be questioned. Many of the writers of these reports deservedly have a high status in scientific community – to challenge their ability and integrity as scientists is out of question. The motivation of the Club is most respectable. Still, even these in so many ways excellent reports to the Club of Rome show something of the ultra-empiristic bias that is characteristic of our age.

Only top specialists in each particular field of study involved are enlisted to make the research reported to the Club, and afterwards publicized world-wide. And yet an authoritative and balanced critic, such as the late Raymond Aron, could write [1, pp. 661–662, 664] in an utterly critical tone of what he called 'the pseudo-scientific prophecies of the Club of Rome'. Among other things he wrote: 'The first report to the Club of Rome was to me a glaring example of pseudo-science, of catastrophic millenarism in disguise of numbers. The Great Fear of the year 2000 followed in these years 1974–75 the Great Hope of the 1960s'.

I appeal to the reader not to take Aron's criticism for just a political or ideological attack, and rather to consider the methodological point involved. Aron hardly had in mind any doubts about the accuracy of the details reported. But Aron was a philosopher of the rationalist school, and I think one can easily see what was the target of his suspicions. While the details may stick you come to think whence did the writers of those reports get the whole, viz. the overall picture of social development. Aron expressly refers to the fact that the first report reflected the popular ideas of the day, where zero-growth ideology played a central part. Later on, when the political momentum of zero-growth subsided, the idea no more figured prominently in the reports to the Club of Rome either.

Aron may have asked himself: if all that science can produce regarding an overall view of society is a repetition of the prevalent public opinion, then why all these calculations with so many different models?

The present writer will by no means be implicated in sharing any political or ideological attack against the Club of Rome, whose aims I support, or against the writers of its reports, whom I respect. But I have to insist that Aron's criticism should be taken seriously. A rationalistic philosopher of history is well equipped to see the lack of any rationally based view of society and social development as a whole in what could be called the *model-construction strategy* of scientific research in most reports to the Club of Rome.

What is exactly meant by the term 'model-construction strategy' will be explained later on, in Section 3. Let me here just anticipate a result of later analysis, and state that unless a solid scientific foundation exists giving the whole, into which to place the pieces of knowledge collected by separate models, the model-construction strategy can have only a modest success. It is good in applications of sciences that already have a firmly established conceptual foundation, such as exact natural sciences. But in the social science or in futurology, such a foundation does not exist.

If Raymond Aron did not feel persuaded by the reports to the Club of Rome, this may have been because he sensed that in the applied model-construction strategy an important rational element was lacking, viz. a rationally based general theory of society and social development as a whole. Hence, the lacking rational foundation of interpretation had to be compensated for by a political interpretation in the terms of the day.

Such a situation, of course, is not the fault of the Club of Rome: it is common in the social sciences. Even in the most part of mathematical economics, a solid general foundation is lacking. Thus an enormous number of various models are being published, based on rather arbitrary assumptions, and often without any conceivable connection to any general foundation or to each other. Add to this the immense variety of widely different 'paradigms' used to justify the mostly mutually incompatible sociological modes of approach, and you have a truthful picture of the social sciences today.

It is thinkable that these methodologically undeveloped sciences, of which futurology and social science at large were used as examples above (and parts of biology too conform to this picture), will never have anything like a general theory or solid fundamental concepts. This would mean that they never would have any foundation upon which to build consistent interpretations of single results related to different sectors of research. In that case all our knowledge of social development as a whole, or societies as wholes, for instance, would be merely reflections of the varying political pursuits, inherently subjective, with no invariant element in it, and really not worthy of being called 'knowledge' in any objective sense of the word.

This is in fact what the ultra-empiristic philosophy, in its most rabid forms, says that all science and all knowledge is (see e.g. Feyerabend [7] or Chalmers [4]). To stress the point some representatives of ultra-empiricism are actually ready to put the words 'science' and 'knowledge' in quotation marks in any connections. They maintain that science – whatever science – and any model or paradigm is just an opinion among others (Feyerabend: 'Everything goes!'), and that there is no such thing as the accumulation of scientific knowledge.

To an ordinary scientist, accustomed in his daily work to distinguish between facts and hypotheses, and between well-grounded theories and wishful speculations, such a stand is far too extreme. To do him justice, let us briefly discuss some characteristic features of the accumulation of scientific knowledge and its theoretical foundations, such as they have been best discernible in the development of exact science.

3. The Testimony of the Accumulation of Fundamental Scientific Knowledge: Reality is Mathematical

One of the most striking features connected with the fundamental theories of mathematical sciences, both in pure mathematics and in exact natural science, is that they have not been acquired by a model-construction strategy but by a profoundly different approach that will be here called the *strategy of generalizing proofs*. To see the difference, let us first consider the concept of model.

Let any set (or aggregate or class or collection or domain) of objects, among which certain relationships hold good, be called a system S of objects. Then a 'model' of this system, in the broadest sense of the word, can be called any system of some other objects intended to represent, more or less accurately, the same relationships as are valid in the modelled system S. In pure mathematics we may have complete models, i.e. mathematical theories that give isomorphic representations of some other mathematical theory, in terms of some other objects. (For instance, in mathematical logic such models of axiomatic theories have been used to prove the consistency of the modelled theories.) In empirical sciences every model, whether mathematical or not, is always only an approximative representation, in terms of a theoretical construct, of a system of real-world objects, because we cannot claim to have a complete knowledge of the latter.

In applied science the situation is often met where a practical need urges one to construct a model of some real-world system. The usefulness of the model for the practical purpose it is intended for is mostly greater the more accurately the model reflects the properties of the modelled system. Hence it is important in model construction to get at as detailed a representation as possible of this particular piece of the real world. This accordingly is the general idea of the model-construction strategy.

Good model-construction is always more or less governed by a clear-cut practical end. But its success essentially depends on whether a general scientific foundation, i.e. a general fundamental theory, exists in the field of study or not. Such a general theory, if it exists, functions as the basis of interpretation of the results obtained with models, and often also as the source of the principal variables included in the models. This is the case with the technological models that can draw from the treasury of fundamental physical and chemical theory. If a well-grounded fundamental theory does not exist, as is mostly the case with models applied in social science, models become arbitrary and may often be misleading.

The fundamental theories have never been made to order, nor to respond to any practical challenge. Instead they have been created by following the inherent logic of the available theoretical–mathematical tools, asking: What general theorems and laws pertaining to the whole field of research in question can possibly be derived from the still more general assumptions underlying the existing tools of theoretical thought? This is the general idea of what was called above the strategy of generalizing proofs – the name will be justified later. The resulting fundamental theories do not give a full description of any particular system S of real-world objects, but they give common properties of a very large set of such systems, either accurate common properties (e.g. the conservation of energy) or approximative ones (e.g. harmonic oscillations), sometimes of all systems S.

It is by way of such an approach that the fundamental theories of exact natural sciences, such as the Hamiltonian theory of classical physics, or the theory of relativity, or quantum theory were created. And, of course, it is this strategy of generalization with which pure mathematics – and even mathematical philosophy – works.

We can also say that an imitation of this strategy, in a less rigorous form, has been pursued by every non-mathematical philosopher too who ever attempted to create a philosophical system of his own. And even the classics of sociology seem to have had something like this in mind. With the sole exception, perhaps, of Emile Durkheim, who leaned on statistical models, the rest of them from Auguste Comte and Vilfredo Pareto to Karl Marx and Max Weber seem to have been trying to apply something like a generalistic, proof strategy to their subject.

At this stage should be mentioned the attempt by Comte to show that the history of mankind could be divided in two (three if the 'transitional stage' is included) great periods that were profoundly different from each other. Let us call it the Comte hypothesis. But a proof of the hypothesis was out of reach with the mathematical–theoretical tools then available. It is interesting to notice that with modern tools, something like the Comte hypothesis can indeed be derived from cybernetic entropy laws. But such general issues can be to some extent mathematically discussed only by starting with the recently developed topological concepts of mathematical dynamics [2].

A rigorous strategy of generalizing proofs has to be based on mathematical tools. This is because proofs, i.e. deductions of general theorems and laws from general assumptions, play such a central role in this strategy, and only mathematical proofs are rigorous. Indeed only a mathematical proof is:

(1) Intersubjective in itself, i.e. entirely transparent and immediately acceptable to every connoisseur – such a consensus is strikingly universal among competent mathematicians, apart from extreme subtleties concerning the philosophical interpretation, not the practical performance of mathematics.

Furthermore, a mathematical proof is:

(2) 'Refutable' only by further generalization of the underlying general assumptions, but not in any other sense. Of course mathematicians are

human beings, and errors of proof are possible, but not errors that a competent mathematician would not recognize when they are pointed out to him.

An allowance must be made for the fact that the concept of mathematical rigor has itself changed, viz. improved, during the development of mathematics. Thus, what was acceptable rigor before Weierstrass and Dedekind, was no more so after them. And what was a rigorous formulation in the 'classical mathematics' of Weierstrass, Dedekind and Cantor, was no more so for the 'intuitionists' Brouwer and Hermann Weyl, but was partly justified again, albeit in a different form, by Hilbert, Bernays and Ackermann. And if you compare the proofs given by the Bourbakists to those given earlier of the same propositions, their form is again new. But what has changed is just the form of expression and the required rigor of proof, not the content of theorems.

Compared, say, with the history of sociology, where every new sociological thinker seems to slap on the ear every previous one, and sets up an entirely new system of ideas, the development of mathematics shows a steady growth of mathematical knowledge. The characteristic feature of this growth of knowledge is the generalization of old theories to more comprehensive ones, where the old theories remain true as special cases.

This accumulation of mathematical knowledge is made possible by the unique property of mathematical proof as a proof that is not refutable but only generalizable, i.e. by the property (2) mentioned above. The strategy followed by fundamental mathematical study is to prove that a given theory can be obtained as a special case of a more general theory. Hence the term generalizing proofs that was used above to characterize this strategy.

Everybody knows that pure mathematics advances by means of generalization of theories toward ever more abstract theories: from Euclidean geometry to analytic geometry to topology to algebraic topology, etc. Or from linear algebra to Hilbert spaces to operator algebras, etc.

But it is precisely the same strategy of research that is used in the foundational theoretical study of exact natural science. In this case, of course, not all theories are permitted, but only those ones that permit an explanation of some new phenomena in addition to the phenomena explained by the old theories. The latter must remain true in special cases, i.e. for the phenomena that they explain.

For instance, as every mathematical physicist knows, the classical theory of physics has never been refuted, nor abolished, by the theory of relativity, nor by quantum theory. Newton was never 'refuted by Einstein', against a fashionable claim of contemporary popular philosophy and folklore. Nor was the validity of classical physics in its own domain in any way 'refuted' by quantum theory. In both of these greatest revolutions of science, as they are often misleadingly called, the fundamental physical theory was only generalized so as to cope with new observed phenomena. The old theory, now called classical, remained valid in a special case.

The special situations, in which classical theory still today holds true, are easy to pinpoint in each case. In the theory of relativity, classical Newtonian or Galilean theory of motion is valid for velocities that are small in comparison with the velocity of light. Thus in the realm of our everyday macroscopic observations on this planet the classical theory of physics remains as valid as ever. According to quantum theory, classical physical theory is perfectly valid in the limit where the intervention of research instruments ceases to affect essentially the properties of the investigated object, i.e. in macrophysics. (This is the message of Heisenberg's principle of indeterminacy.)

Example 1 (*Newtonian vs Einsteinian theory*). The observed bending of light in passing of the sun is a typical Einstein effect, derivable only from the theory of relativity, but not from the classical, Newtonian theory of celestial mechanics. This is because the phenomenon does not belong to the domain of validity of classical theory: the velocity involved is that of light itself. You can roughly estimate the magnitude of the difference between the results given by classical and relativistic theory by calculating the deviation from 1 of the expression

$$1/(1 - v^2/c^2)^{1/2},$$

where c is the velocity of light and v is that of the movement we are investigating. For a velocity $v = 3\,\text{km/sec}$, which is about ten times as large as the velocity of propagation of sound, the deviation of the predictions of the two theories would be already as small as of the order of 0.000000005%. It goes without saying that in the mechanics concerning our everyday macrophysical observations the difference is negligible, and the Newtonian physics is valid. On the other hand, for $v = c$ the difference becomes infinite: indeed even the 'classical' Maxwell equations of electromagnetic radiation, which propagates with the velocity of light, are Lorentz-invariant, i.e. satisfy the requirements of the special theory of relativity. But the Maxwell equations do not give the Einstein effect related with the bending of light, because the latter is an effect of the general theory of relativity: to get it predicted by the theory you have to take into account the curvature of space–time in an inhomogeneous gravitational field.

Example 2 (*Classical vs quantum theory*). In the Hamiltonian formalism, which is the fundamental general theory of classical physics, a complete state-description of dynamical system is given by canonical co-ordinates and momenta, which together form the 'total state' x of the system. This means that every property of the system is a function of x. A simple example of such a Hamiltonian system is a set of moving material bodies, whose positions and velocities at a given moment t define the total state $x(t)$ of this system at the moment t. In macrophysical circumstances a simultaneous measurement of positions and velocities is always possible, and may take only some technical inventiveness. Furthermore, the equations of motion define a causal recursion, by means of which one can compute the total state $x(t)$ of the system from any of its past or future states $x(t')$. Hence the French mathematician Pierre Simon de Laplace (1749–1827) could state that the Newtonian theory enabled you to predict the future. This is still true, as far as the movements of material bodies go, and provided relativistic effects are duly taken into account. But in the microworld of atoms and elementary particles, your instrument (e.g. a ray of light, i.e. of photons) required for the measurement of the position of a particle pushes it, thus making impossible a simultaneous accurate measurement of its velocity, and vice versa: if you measure the velocity, you cannot ascertain simultaneously its position. Heisenberg's relation of indeterminacy gives a mathematical formulation of the way in which a simultaneous measurement of two canonically conjugated variables, such as position and velocity, disturb each other's accuracy. What can be measured is only the probability distribution of positions and velocities of the investigated particles. Hence, if the complete state-description of dynamical system is now connected with such a probability distribution, as it really is in quantum theory, we regain the possibility of predicting the future even in the case of quantum particles. Indeed the causal recursion of the total state $x(t)$ to any past or future total state $x(t')$ is again there. Only the states $x(t)$ and $x(t')$ now determine distributions of probability: this is now all that there is to be predicted, not for lack of causality, but because more than that cannot be observed of quantum particles. In the limit case, where the variances of these distributions become zero, we are back in classical mechanics, with its non-probabilistic predictions. Classical theory thus appears as a special case of quantum theory.

To resume the general discussion we can state:
The progress of fundamental mathematical theory both in pure mathematics and in exact natural science happens through generalization, not by refutation of earlier theory. It is based on the strategy of generalizing proofs which, because of the property (2) of mathematical proofs, leads to an accumulation of theoretical knowledge.

But the accumulation of theoretical knowledge is the distinctive characteristic of all profound scientific knowledge. Hence the success of the strategy of generalizing proofs, and of it alone, strongly suggests the idea inherent in that strategy: *Reality is mathematical.*

4. Where and how Ultra-empiricism got it Wrong

The maxim 'reality is mathematical' underlying the strategy of generalizing proofs has come under a philosophical attack twice in this century, from two different quarters. The first of them, anti-positivism, was politically motivated, and actually an inheritance from the 19th century (from Hegel and Marx mainly). It held the notion of mathematical reality an abhorred 'reductionism' to be avoided by all means. But the appeal of anti-positivism, along with the associated political ideologies of Nationalism and Marxism, has much declined in the West, and we can overlook it here.

The second attack came from the school of ultra-empiricism, whose basic tenets concerning exact science were formulated by a disciple, Ludwig von Bertalanffy, as follows:

One of the important aspects of the modern changes in scientific thought is that there is no unique and all-embracing 'world system'. All scientific constructs are models (sic) representing certain aspects or perspectives of reality. This applies even to theoretical physics (Von Bertalanffy [22], as quoted by Klir [10]).

Even ultra-empiricism has an air of ideological conviction about its tenets. 'The very title of this book speaks volumes', David Stove, an Australian philosopher, writes [20] of A. F. Chalmers' book *What Is This Thing Called Science?* [4]. Stove goes on to say 'It has the true Feyerabendian ring – a combination of levity and menace'.

(a) Popper

To the most well-doing effects of Popper the political philosopher (*The Open Society and Its Enemies*) there corresponds Popper the philosopher of science, whose influence, unfortunately, can hardly be described otherwise but harmful. To his admirers Popper seems to stand for a balanced mixture of empiricism and rationalism (the latter because of his suggested hypothetico-deductive method). But the core of his philosophy of science, viz. the principle of falsifiability, became the starting point of the modern school of ultra-empiricism and has had disastrous consequences all over the 'softer' end of the spectrum of sciences.

The mistaken notion that 'all scientific constructs are models' (see von Bertalanffy's text quoted above) has its roots in the colossal misunderstanding of the mutual relation between modern and classical physics, viz. in the erroneous belief that modern physics has 'refuted' classical physical theory. It seems to have originated in the young Popper's mind, in Vienna in the 1920s, where not only 'was the Austro-Hungarian empire...to crack up at this time. For the mind of the young Popper, the fall of another and far more soundly-based empire was no less formative: I mean the Newtonian empire of physics', writes Stove [20].

Another historian of philosophy, Provost [18], agrees, saying this:
'Where Kant derived his inspiration from Newtonian absolutism, Popper derived his from Einstein's refutation of Newton. It must have struck Popper that if the perfect science, that is, Newtonianism, which had stood unchallenged for over two hundred years, had been refuted, then it must be that any theory that can be called "scientific" must be refutable'.

According to Provost (*ibid*.):

This led to the logical conclusion that the demarcation between science and metaphysics must be falsification. In particular Popper needed some clear and simple standard that would separate such acceptable science as Einsteinian physics from theories he considered pseudo-scientific, like Marx's theory of history or Freud's psychoanalysis. His criterion was that, in order to qualify as science, a theory had to be stated in such a way that it was possible to refute it. Part of every Popperian theory had to be a set of potential falsifiers – empirical facts, which if found true, or crucial tests if failed, would decide the theory unequivocally false.

One can very well accept the idea of scientific hypotheses as something to be tested, either mathematically or empirically or both ways. But genuine mathematical proofs can never be falsified, only generalized. Hence, Popper's principle of falsifiability excludes from science not only pseudosciences but the foundation of exact sciences, viz. their fundamental mathematical theories, including pure mathematics.

From this basic error of Popperian philosophy (e.g. Popper [17]), it follows that no genuine accumulation of theorectical knowledge is possible at all. This is an inevitable fact, and makes of Popper the founder of the 20th-century ultra-empiricism (a predecessor in the 18th century of course was Hume). Power made this conclusion in his claim that – as formulated by Stove (*ibid.*) – 'no scientific conclusion can ever be probable, that no theory ever becomes even more probable, when evidence in its favour is discovered, than it was beforehand; indeed that every scientific theory not only begins by being infinitely improbable, but must always remain so'.

Such an extreme view, well documented as being Popper's own [19], can hardly be characterized otherwise than ultra-empiristic. We have seen that underlying Popper's principle of falsifiability, and this his ultra-empiricism, probably was his (and many others') mistake concerning the mutual relation of Newtonian and Einsteinian physics. In reality 'Newton' was never 'refuted by Einstein', as we have already pointed out. In believing erroneously that such a refutation had taken place Karl Popper, and many other ultra-empiricists after him, seem to have been led into their extreme philosophical position.

As far as pure mathematics was concerned, Popper first admitted that his principle of falsifiability was not valid. But he seems to have accepted, not without some enthusiasm when his pupil and successor in the Chair of Philosophy, Imre Lakatos, made an attempt to extent the realm of Popperian refutationism to mathematics. The attempt [12] was doomed to a failure [cf. 19, 20].

(b) Kuhn

Lakatos' philosophy of empirical science, also published posthumously [13], was already eclipsed by the world-wide enthusiasm evoked by the American philosopher Thomas S. Kuhn, whose book on scientific revolutions [11] came out at the same time as Lakatos was developing his 'scientific research programmes' [13]. The name of Kuhn's well-known book could as well have been The Structure of Scientific Refutations. Indeed it could be read as an application of Popperian refutationism to the development of science, although Kuhn of course – like other ambitious ultra-empiricists – strictly rejects all Popperian influence.

Science, according to Kuhn, starts time and again from a zero-point, with a new 'paradigm' that refutes everything that had been stated in terms of the previous paradigm. After a successful revolutionary campaign of the forces of the new paradigm against the defenders of the 'normal science' a new normal science is created, to give way on time to the next scientific revolution, another start from the zero-point. What is really extreme in Kuhn's view is that our knowledge is not improved in these Kuhnian revolutions: again, as in Popper's philosophy, there is no place for the accumulation of scientific knowledge. Science is just an intellectual power game, and every new paradigm is as improbable as was its predecessor.

Or, as put by Stove in his critical appraisal of the ultra-empiricists:

Now you could, of course, take all this just as an account of the history of science, and find more or less of value in it, according to as you consider it more or less accurate history. But that is not at all how it is intended to be taken, or how Kuhn himself takes it. He takes it as a sufficient reason to accept a certain philosophy of science, and a philosophy of the most uncompromisingly relativist kind. He will not talk himself, or let you talk if he can help it, of truth in science,... As for 'knowledge', 'discovery', 'progress': why, all that, of course, is no more than the language which the partisans of any paradigm will apply to their own activities... all such talk is 'paradigm-relative'. There is nothing rational about paradigm-shift in science, according to Kuhn [20, p. 68].

But in reality there is much rationality in a change of scientific paradigm, if by a new paradigm we mean a new fundamental theory in exact science. Surely one of the greatest changes of paradigm in science was the replacement of macrophysical explanations by molecular ones, i.e. the change from the macrophysical complete state-description of classical physics to the molecular state-description of statistical thermodynamics, which occurred in the 19th century. For the first time in the history of science, the then still unobservable molecular movements were taken as causal explanations of observable phenomena (Dalton's atom theory in 1808 has been important for chemistry but first Ludwig Boltzmann's thermodynamics created a new fundamental theory of physics). Yet the molecular state-description was no start from a zero-point, but was in a perfectly rational relation (illustrated in Fig. 1 on p.9) to the macrophysical one.

The mutual relation between molecular and macrophysical paradigms can be mathematically expressed by the formula

$$X^* = \cup \{A_x; x \in X\}.$$

Here X is the set of world-states in a complete macrophysical state-description, X^* is the corresponding set in a complete molecular state-description. In words: to every macrostate x there corresponds a set A_x of molecular states, which may be exchanged with one another within the set A_x without any change in the macrostate x.

For instance, the same macrostate determined by a given temperature and pressure of an ideal gas closed in a certain volume can be produced by different states of molecular movements in the gas, so that each of the latter states gives the same observed temperature and pressure. It follows that all the information that can be expressed in terms of the macrostate-description can be carried over to the molecular state-description, but not vice versa. Hence, instead of being a start from a zero-point, the new paradigm even increased the total amount of scientific information. Thus we have a case of accumulation of theoretical knowledge in an important change of paradigm, ignored by Kuhn.

Surely the greatest of all paradigm-shifts has been the creation of quantum theory, which among other things replaced the two-valued logic of the macro-physical and molecular paradigms by a three-valued logic. Quantum theory brought with itself another extension of complete state-description, according to the formula

$$X^{**} = \cup \{B_y; y \in X^*\},$$

where y is a molecular state and B_y the corresponding set of quantum states. As soon as we go inside the molecules we enter the realm of the quantum-theoretical state-description. Each molecule and atom has an infinite number of internal quantum states. In quantum theory even statistical dynamics gets a new content: the molecular Boltzmann statistics is replaced either by a Bose or Fermi statistics, depending on whether the quantum particles you are dealing with are 'bosons' or 'fermions'.

But again, as indicated by the above formulae, all the information contained in the ealier complete state-descriptions can be carried over, in principle at least, to the quantum-theoretical state-description, but not vice versa. The capacity of information of the quantum-theoretical paradigm X^{**} is larger than that of the molecular paradigm X^*, which again is larger than that of the macrophysical paradigm X, as is illustrated in Fig. 1.

Thus theoretical knowledge is accumulated from X to X^* to X^{**}, in a striking contradiction to what Kuhn said about paradigm-shifts in connection with 'scientific revolutions'.

Example 3 (*Classical vs relativistic theory revisited*). The reader may wonder, where in the scheme of Fig. 1 belongs home another generalization of physical theory, which we already know, viz. that from the Newtonian to Einsteinian theory. Both of them are macrophysical theories, and thus belong to the level of macrophysical state-description in Fig. 1 (although their impact of course extends, in accordance with the picture, onto the other two levels). Thus both the classical, Newtonian and the Einsteinian, relativistic fundamental theories of physics can be formulated in terms of the macrophysical state-description. Only the form of the equations of motion are different in the different theories. The equations of, say, the special theory of relativity have to be invariant with respect to the Lorentz group of transformations of space-time. The Galilei group of transformations, with respect to which the equations of Newtonian mechanics are invariant, is a part of the Lorentz group. This is just another way of saying that the Einsteinian theory is a generalization of the Newtonian one, the latter being valid, according to the theory of relativity, in a special case. Among other things it follows directly from the wider invariance group of relativistic mechanics that what in the Newtonian mechanics appears as mass becomes in the relativistic mechanics a combined mass-energy variable: the law of conservation of matter becomes that of conservation of energy, whereby matter may change to energy, and vice versa.

Example 4 (*The three-valued logic of quantum theory*). When speaking of the three-valued logic of quantum theory it is not meant, of course, that the mathematical formalism of quantum theory was that of a three-valued propositional calculus (Lukasiewicz, Kleene, or other), or that of a three-valued predicate calculus (e.g. of Rosser and Turquette). The formalism is one of an operator algebra defined on a Hilbert space, but the meaning of three-valued logic in this case can be explained simply as follows. Each total state of the investigated dynamical system is in quantum theory mathematically represented by a vector x starting from the origin of the (Hilbert) space of states, and being of unit length. If E is the state of affairs stating that an observable O has a definite value c, then only the probability

$$P_t(c) = |(x_t, x_c)|^2$$

of the appearance of c at a given moment t can be derived from the theory. Here x_t is the state of the system at this moment and x_c is the eigenstate of the operator O associated with the eigenvalue c. The parentheses indicate the hermitean inner product. The probability P_t is not a subjective probability but an objective one: it gives the theoretical value beyond which the truth or untruth of the state of affairs E cannot be determined, according to quantum theory. Hence, not only truth and untruth but also a third truth-value, called indeterminacy, must now be taken into account, according to the following three mutually excluding cases:

Case 1: $x_t = x_c$, thus $P_t(c) = 1$, thus E is true;
Case 2: x_t is orthogonal to x_c, thus $P_t(c) = 0$, thus E is false;
Case 3: x_t is neither equal nor orthogonal to x_c, thus $0 < P_t(c) < 1$, thus E is indeterminate.

An alternative terminology would be to speak of an infinite-valued logic of quantum theory, giving to each magnitude of the probability P_t its own truth-value. But most often, when a term has been used, it has been three-valued logic. In the special case, where quantum theory reduces to classical theory, i.e. when the variances of the probability distributions related to the quantrum states become zero, the three-valued logic reduces to the two-valued logic, of course. Then every state of affairs will again be at any moment either true or false.

To return to the general discussion: Kuhn's interpretation of scientific revolutions and of the changes of paradigms is deeply misleading, not in a trifling historical detail, but in its very core. The fundamental theories of exact science are never refuted but remain true in special cases even in the new theories. The essential new theoretical inventions in fundamental exact science do not refute earlier knowledge but add to it, thus making scientific theoretical knowledge to accumulate.

(c) Feyerabend

Kuhn in his turn has been outflanked by...P. K. Feyerabend...Feyerabend calls himself a 'Dadaist' and his philosophy 'epistemological anarchism'. He maintains that science knows, and should know, of no rules of method, no logic – inductive, deductive, or whatever. And for his slogan he actually chose Cole Porter's old title Anything Goes! Feyerabend's main manifesto is a book called Against Method (1975). Of all the productions of the human mind in any age, this must rank as one of the most curious. It is impossible to convey briefly the unique absurdity of the book, but one or two of the author's foibles may be mentioned. Feyerabend says – not as a joke or schoolboyish pun on the word 'law', but in all seriousness – that scientific laws ought to be decided in the way that other laws are decided in an open society; by a democratic vote. He is also a sturdy partisan of the claims of witchcraft, voodoo, and astrology (Stove [20, pp. 68 – 69]).

How true is the above description can be easily ascertained by reading the above mentioned book of Feyerabend [7]. There is no doubt that the mistaken ideas of ultra-empiricism have in Feyerabend's philosophy reached their extreme logical end. What was somewhat covered in earlier representations (by Popper or by Kuhn) is there spelled out with an admirable clarity:

1. If mathematical reality is discarded (the principle of falsifiability by Popper and Lakatos), and thus the accumulation of fundamental theoretical knowledge denied in science (Kuhn's paradigm-shifts), we are bound to end up with intellectual anarchy: Anything Goes!

2. In such a situation, all theories are models (von Bertalanffy), and all models are equal (Feyerabend). Science is reduced to a power game (Kuhn), and the only non-violent way of reaching solutions in science would be by a vote (Feyerabend).

5. Repercussions of Ultra-empiricism in Philosophy and in the 'Soft' Sciences

We have to distinguish clearly between the repercussions of ultra-empiricism in the philosophy of science, on the one hand, and the 'soft' sciences on the other.

(a) No greater damage inflicted on the professional philosophy of science

In the ranks of the professional philosophers of science the 20th-century school of ultra-empiricism (Popper, Lakatos, Kuhn, Feyerabend) never enjoyed any great appreciation. Soon after David Stove had published his biting criticism of extreme empiricism indicating Sir Karl Popper as the (reluctant) founding father [20], he got a favourable reception from one of the grand old men of the modern philosophy of science, Ernest Nagel, saying this:

In my judgement Stove's criticism is sound, it in no way distorts Popper's view of science, and by and large leaves much of Popper's main thesis pretty much in ruins. It is high time someone called a spade a spade... Until Stove's essay appeared I had the impression that Popper was a sacred cow about whom nothing but agreement could be expressed. I am glad that my impression was mistaken [14].

Nagel's reply no doubt reflects the majority opinion among the professionals in the field. It has been counted, for instance, that only three of the professors of philosophy in the U.S. universities are 'Popperians'. The influence of ultra-empiricism has been mainly confined to the market-place of popular philosophy, where it has been very successful indeed, having been followed sometimes moderately (neofinalism, e.g. G. H. von Wright [23]), sometimes in a rather wild form (Derrida's deconstructionism!).

(b) The damage is greater in the 'soft' sciences

The strongest repercussions of ultra-empiricism have been recorded in the softer end of the spectrum of the sciences, from sociology to biology to non-mathematical systems research. We may quote Stove, writing on the bad influence of Kuhn's book [11]:

This book has sold well over half a million copies in the English edition alone, and has been translated into innumerable languages. Even among philosophers, who are a notably resistant strain of people, not a few have succumbed, or 'gone Kuhn'. In the intellectual slums, where resistance of any kind is weak – among sociologists, eductionalists, anthropologists, and the like – the execution done by this book has been simply terrific. What is more important, the book has been and still is being read, with the effects to be expected, by scientists in their thousands [20].

On Popper's damaging influence in biology Stove writes:

The case of biologists is especially important, and grimly amusing as well. For some decades, thousands of biologists, although they are people otherwise sensible enough, have been worrying themselves sick over the news that their theory of evolution is unfalsifiable... Popper has said that falsifiability is the mark of science, and therefore it must be so, in spite of sense and reason. And the amusing part is, that it is precisely this faith which has now exposed biologists to the tender mercies of their revitalised Christian tormentors. For Creationism is (as its adherents claim) no less scientific than Evolutionism, *if falsifiability is made the test of science.* Could there be a more cruel irony than this? Still, we should not waste sympathy on self-inflicted wounds: the biologists asked for trouble, and they got it [21].

As to the systems-research movement, an ultra-empiristic bias is no less discernible there in some contemporary methodological writing. If we compare, say, the recent articles by Glanville [8], Bråten [3], or de Zeeuw [5, 6] with the papers read by von Neumann and McCulloch at the Hixon Symposium in 1948 (reprinted in [9]), we see the change from foundational mathematical concepts, characteristic of von Neumann and McCulloch, to a blatant refusal of any 'fundamentals' (Glanville) in the later writers. Glanville's comments on elementary particles physics [8] reveal a layman ignorant of exact science. (Compare this with von Neumann, the mathematician–cybernetician, who wrote a book on the Hilbert-space formalism of quantum theory!)

(c) The problematic 'second-order cybernetics': The intellectual domain reduced to the political one?

The 'second-order cybernetics' advocated by the mentioned recent writers contains valuable elements, such as the emphasis laid on the autonomous, self-steering nature of man. But at the same time the second-order cybernetics seems to be hopelessly corrupted by the anti-mathematical, anti-intellectual climate created by ultra-empiristic popular philosophy.

Thus Bråten's paper [3] is entirely dominated by the typical ultra-empiristic idea that all scientific theories are models. The writer also seems to share Feyerabend's conviction of the ultimate equality of all models: he is on the warpath against what he calls the 'model monopoly' of any particular model over any other one.

De Zeeuw's problematics [5, 6] likewise arises from the everything-is-models view of science, but he concentrates on the so emerging further problem of science as a means of power, emphasized by Kuhn and Feyerabend. De Zeeuw's presentation of this theme is in fact more sophisticated than that of Kuhn or Feyerabend, and deserves a closer inspection.

The ethical starting point of de Zeeuw is most sympathetic. He regards human actors as autonomous or, as I would prefer to say, self-steering, creative beings. Every individual is justified in his pursuit to extend his self-steering, creative faculties, provided that he does not violate the corresponding right of others, but rather helps them – and the collectives of which he is a member – to do the same. The old and evergreen Pareto problem of how to let the individuals flourish to the common good of all gets in de Zeeuw's paper the pointed form of how to create 'an actor that is making other actors strong without making any one of them stronger than others' (De Zeeuw [6, p. 139]).

Obviously, to that end the control of any actor, individual or collective, by any other actors should be the minimum possible still keeping society and its economy going. This is because, probably, 'actors must be able to change unpredictably, to be unpredictable from the point of view of the activities of other actors. Otherwise no initiative for change is possible' (De Zeeuw [6, p. 136]).

Here the serpent enters the garden of Eden in the shape of science. De Zeeuw, quite rightly, emphasizes that every model construction implies a choice of the 'group of phenomena' that is isolated from the rest of the world for being analyzed in that particular model. Furthermore, he states quite rightly that the choice reflects the values and imperfect knowledge of the researcher, i.e. it is subjective. It follows that the application of the model becomes, in social science especially, a means of power, i.e. a means of control of other actors, a means of subordinating other actors to the pursuits and will of the model-maker and model-user. All this is perfectly true of all models.

But to de Zeeuw there is nothing but models in science. Then, how to adjust scientific model construction to the egalitarian end declared above? Within the framework of ultra-empiricism, where nothing but models exist in science, the solution in fact is impossible.

No wonder that de Zeeuw's suggested way out of the dilemma is problematic. He proposes the scientific truths to be decided upon in a 'multi-actor process' of all the interested researchers and other actors (how exactly?), since 'collective competence cannot be reached without the implementation of social systems' (De Zeeuw [6, p. 141]).

Does not such a multi-actor process make the scientific truths a matter of political power struggle between the researchers themselves or other 'interested' parties? Yes, indeed, that is the idea, for it appears that there is a 'general multi-actor process, the political process, of which specific multi-actor processes like the research process form a part' (De Zeeuw [6, p. 138]).

Like Kuhn and Feyerabend, de Zeeuw thus ends up with the interpretation of scientific research as a part of the political power-game. Like these philosophers he reduces the scientific truths to a powergame (possibly between the researchers themselves), because model construction without a fundamental general theory is a power-game, and since to the ultra-empiricists there is nothing in science but models. It follows that the decisions about scientific truths could only be made either by a democratic vote, as suggested by Feyerabend, or by some non-democratic kind of political free-style wrestle: through a power-game.

(d) The intellectual content and the power-game element in what is considered true in different sciences

Fortunately, we needn't make that choice. We needn't subordinate the scientific truth to social power. What is basically wrong with the second-order cybernetics, of de Zeeuw and others, is the ultra-empiristic dogma that all scientific theories are models. If, but only if, this dogma is accepted, science indeed is reduced to politics.

There is, however, deeply planted in human mind, a central domain of thought, where scientific truths are not affected by political power: mathematics. No political power-holder has ever had any real influence on what has been considered true in mathematics, not even in Galileo's time (despite forced public confessions). Even the Soviet Communist Party has been powerless in front of the exact sciences, although it has willingly imposed its authority on the soft sciences, such as biology or social science. Mathematics is the sphere of *pure intellect*, whose truths are not contaminated by social power of any kind, neither that of politicians nor that of the researchers themselves upon each other.

Reality is mathematical. Therefore reality can be investigated by the mathematical strategy of generalizing proofs. In this way the domain of pure intellect can be extended from pure mathematics to the fundamental mathematical theories underlying the special sciences. There probably are questions, such as the purpose of the existence of the universe and mankind, which can never be answered by science. But as far as the sciences go, there are no predestined absolute limits to the expansion of pure intellect, in the form of fundamental mathematical theories, into the sciences.

On the other hand, the fundamental mathematical theory underlying any particular science is the only domain of pure intellect in that science. At any given point of time the limits of the domain of pure intellect, indicating how far this domain extends at that moment in each science, are finite. Therefore the intellectual content of any applied science, i.e. the fundamental mathematical theory underlying the basic concepts of that science, has to be complete-mented by model construction in order that one could respond to the practical challenges posed by everyday life. This brings in a political element: the choice of a model is always partly a political choice.

The more advanced is the mathematical theory underlying a given science, the more there is of the intellectual content and the less of the power-game involved in what is considered true in that science, and vice versa. I think no one can deny that, for instance, sociological models and 'paradigms' still today are indeed largely expressions of the political views of their authors, and involve not much of the intellectual element.

Thus what passes for the sociological truth in each country and within each school of sociological thought indeed still is, to a great extent, a political verdict, either resulting from a power game between the sociologists themselves (in the West), or being dictated by the political authorities (in the East). Here the emergence of scientific truths from a political process in some 'social systems', as stated by de Zeeuw, applies. But the situation of the least advanced sciences, such as sociology, cannot be generalized to other sciences.

Econometric models already contain an intellectual, mathematical component, in addition to a notable political one. The same applies to, say, the models of the experimental psychology at their best. Closest to the ideal of pure intellect, as far as models go, are the technological models based on exact natural sciences. But even in the construction of technological models too a political element is involved. (This is to be distinguished from the politics involved in decisions concerning the implementation or non-implementation of technological or other models in society: such decisions of course are always purely political choices.)

Ultra-empiricism is bound to reduce the intellectual content of science exhaustively to the social power game, as is well illustrated by the philosophies of Kuhn and Feyerabend, and by the methodology of de Zeeuw. This reduction, this *total* reduction of science to politics, which is inherent in ultra-empiricism, in the idea that all science is models, amounts simply to a denial of science. Compared with the infinite reality, science (and mathematics as shown by Gödel) is always to be something finite and incomplete. However, even though reality can never be exhausted by science, science is our only way toward an intellectual understanding of reality, even better and better ...

Unfortunately, the anti-intellectualism of the ultra-empiristic type is now widespread in the 'soft' sciences, including the soft, non-mathematical formulations and applications of that loose compound of ideas called 'general system theory'. Worse still, the ultra-empiristic trend seems to have a large following on both sides of the Atlantic, within the systems research movement generally.

(e) A further problem of contemporary cybernetics: computer-definitions of concepts?

Even in the case of systems reseach there is a certain irony associated with the development. The maxim 'reality is mathematical', inherent in fundamanetal mathematical theories, was respected by the pioneers of the systems movement, such as Turing, von Neumann, McCulloch, Wiener, or Ashby. But later on, in strange ways even the ease with which modern computers can be handled and played with has come to encourage the anti-mathematical, anti-intellectual theoretical attitude.

As a matter of fact, you can feed almost anything, any sequence of formal symbols, obeying some syntactic rules of operation, into a computer. But the fact that you can do that, and play with the rules you have invented, is no guarantee at all of any meaningful scientific concepts being involved in your formal language.

Therefore the habit, ever more often applied in systems research of trying to *define* scientific concepts by means of their 'simulations' in a computer (e.g. Pask [15, 16], Zeleny [24]) is disastrous from a methodological point of view. It deprives systems methodology of the very demanding semantic control that is involved in a set-theoretical, mathematical construction of fundamental scientific concepts.

(f) Stove

The final irony is that ultra-empiricism has left one of its traces even on its sharpest critic: such is the dominance of this popular philosophy in mass communication and other publicity in our age. Also David Stove seems to believe that Einstein really 'refuted Newton'. Obviously no textbook of theoretical physics ever passed his hands. Understandable as it is, he therefore lost some of the most efficient rational arguments against ultra-empiricism. (His own philosophical position is that of inductive probabilism, the next-best rational choice.)

The most valuable contribution of Stove's brilliant book [19] is, I think, a detailed and careful construction of the connection between David Hume, the common ancestor of all ultra-empiricists, and Popper. Whatever elements of rationalism there may be in Popper's philosophy, he seems to have obtained directly from Hume all the basic ultra-empiristic elements, including the non-accumulation of scientific knowledge and the irrelevance in science of general mathematical theory. (But Hume was a historian!)

6. Conclusions

(To whom it may concern:) Give up the vain attempts of justifying models without any fundamanetal mathematical theory in your special field of science, and take a leave from model construction. Improve instead your mathematics. But do not mistake logic or metamathematics for mathematics! Logical theory is good for linguistic analyses only. What you probably need most is the modern science of mathematical dynamics. Find out what it is, and try what the most advanced theoretical tools of thought of our age can give you in your special field of research, whatever it is.

Literature Referred to in the Appendix

1. R. Aron, *Mémoires*. Julliard. Paris (1983).
2. A. Aulin. Cybernetic causality III. *Math. Soc. Sci.* **14** (1987), 71–82.
3. S. Braten, The third position. Beyond artificial and auto-poietic reduction. In F. Geyer and J. van der Zouwen (eds), *Sociocybernetic Paradoxes*, pp. 193–205. Sage, London (1986).
4. A. F. Chalmers, *What Is This Thing Called Science?* Open University Press, London (1985).
5. G. de Zeeuw, Can social change be supported by inquiry? *Kybernetes* **13** (1984), 165–171.
6. G. de Zeeuw, Social change and the design of enquiry. In F. Geyer and J. van der Zouwen (eds), *Sociocybernetic Paradoxes*, pp. 131–144. Sage, London (1986).
7. P. K. Feyerabend, *Against Method*. New Left Books, London (1975).
8. R. Glanville, The same is different. In M. Zeleny (ed.), *Autopoiesis: A Theory of Living Organization*, pp 252–262. North-Holland, New York (1981).
9. L. A. Jeffress (ed.), *Cerebral Mechanisms of Behavior: The Hixon Symposium*. Hafner, New York (1967).
10. G. J. Klir, *An Approach to General Systems Theory*. Van Nostrand Reinhold, New York (1969).

11. T. S. Kuhn, *The Structure of Scientific Revolutions.* Chigaco University Press, Chicago (1962).
12. I. Lakatos, *Proofs and Refutations.* Cambridge University Press, Cambridge (1976).
13. I. Lakatos, *The Methodology of Scientific Research Programmes.* Cambridge University Press, Cambridge (1978).
14. E. Nagel, A letter to the editor. *Encounter,* December (1985), 78.
15. G. Pask, *Conversation, Cognition and Learning.* Elsevier, Amsterdam (1975).
16. G. Pask, Organizational closure of potentially conscious systems. In M. Zeleny (ed.), *Autopoiesis: A Theory of Living Organization.* North-Holland, New York (1981).
17. K. R. Popper, *Conjectures and Refutations.* Routledge, London (1963).
18. W. H. Provost, Science as paradigmatic complexity. *Int. J. Gen. Syst.* **10** (1985), 257–278.
19. D. C. Stove, *Popper and After: Four Modern Irrationalists.* Pergamon Press, Oxford (1982).
20. D. C. Stove, Karl Popper and the jazz age. *Encounter* June (1985), 65–74.
21. D. C. Stove, A reply to J. S. Liebersohn. *Encounter* March (1986), 76.
22. L. von Bertalanffy, General system theory. *Gen. Syst.* **7** (1962), 1–20.
23. G. H. von Wright, *Explanation and Understanding.* Cornell University Press, Ithaca (1971).
24. M. Zeleny, What is autopoiesis? In M. Zeleny (ed.), *Autopoiesis: A Theory of Living Organization,* pp. 4–17. North-Holland, New York (1981).

Index
(mathematical concepts: the pages of definition)

327